ELSEVIER GEO-ENGINEERING BOOK SERIES
VOLUME 1

# Stability Analysis and Modelling of Underground Excavations in Fractured Rocks

ELSEVIER GEO-ENGINEERING BOOK SERIES
VOLUME 1

# Stability Analysis and Modelling of Underground Excavations in Fractured Rocks


**Weishen Zhu**
*Shandong University, Jinan, China*

**Jian Zhao**
*Nanyang Technological University, Singapore*

Geo-Engineering Book Series Editor

**John A. Hudson**
*Imperial College of Science, Technology and Medicine,*
*University of London, UK*



2004

ELSEVIER

Amsterdam – Boston – Heidelberg – London – New York – Oxford
Paris – San Diego – San Francisco – Singapore – Sydney – Tokyo

ELSEVIER Ltd
The Boulevard, Langford Lane
Kidlington, Oxford OX5 1GB, UK



First edition 2004
Library of Congress Cataloging in Publication Data
A catalog record from the Library of Congress has been applied for.

British Library Cataloguing in Publication Data
A catalogue record from the British Library has been applied for.

ISBN: 0-08-043012-0
ISSN: 1571-9960 (Series)

♾ The paper used in this publication meets the requirements of ANSI/NISO Z39.48-1992 (Permanence of Paper).
Printed in The Netherlands.

# Series Preface

I am delighted to introduce the new Elsevier Geo-Engineering Book Series. Our objective is to publish high quality books on subjects within the broad Geo-Engineering subject area — e.g. on engineering geology, soil mechanics, rock mechanics, civil/mining/environmental/petroleum engineering, etc. The topics potentially include theory, ground characterization, modelling, laboratory testing, engineering design, construction and case studies.

The principles of physics form a common basis for all the subjects, but the way in which this physics is manifested across the wide spectrum of geo-engineering applications provides a rich variety of potential book themes. Accordingly, we anticipate that an exciting series of books will be developed in the years ahead.

We welcome proposals for new books. Please send these to me at the email address below.

Professor John A Hudson *FREng*
Series Editor
jah@rockeng.co.uk

# Preface

It is well recognised that rock masses are very different from common man-made engineering materials. The nature-created geological medium is a discontinuous, inhomogeneous, anisotropic and non-linearly elastic medium. The characteristics of rock masses vary with time, location and direction. Due to the nature of most of underground rock excavations, those excavated structures are usually of a large size and require long-term serviceability. Therefore, the way rock mass characteristics vary with time, location and direction will be reflected in the surrounding rock masses of the excavations. Such variation should, ideally, be captured in the modelling and analysis of rock masses.

This book attempts to tackle the problems of modelling and analysis of excavations in rock masses, by taking into account discontinuity influence, time dependence behaviour and construction method.

The content of this book is largely based on many years of research work by the senior author and his research group in China, supplemented by the contribution by the junior author and his research groups in Singapore and China. Much of the content of this book has been published in the form of technical papers, while some content is presented for the first time. The physical and numerical modelling work described in this book is conducted with several research grants awarded to the senior author, including the China National Natural Science Foundation (NSF) program "Interaction between geomaterials and hydraulic structures" sub-project "Mechanical characteristics of jointed rock masses and construction mechanics of rock structures", other NFC projects (Numbers 59939190, 40272120 and 50229901), as well as the 7-5 National Programs on Science and Technology.

Much of the work presented in this book results from the efforts of the research groups headed by the authors. The accomplishment of this book would not have been possible without the great contributions by their colleagues and research students. The authors wish to thank Shiwei Bai, Kejun Wang, Rongming Pan, Xinping Li, Jingnan Xu, Guang Zhang, Ping Wang, Zuoyuan Liang, Bailin Wu, Suhua Li, Rui Ding, Haibin Xu, Haiying Bian, Jungang Cai, Shaogen Chen, Yuhui Zhao, Haibo Li, Yuyong Jiao, Xiaobao Zhao, Hongwei Song and Ashraf Hefny, for their contributions, comments and reviews. The authors also appreciate the encouragement and advice from Elsevier's editorial team James Sullivan, Vicki Wetherell and Lorna Canderton.

Weishen Zhu
Jian Zhao

# About the Authors

Weishen Zhu is a professor of rock engineering and director of the Geotechnical and Structural Engineering Centre at Shandong University in Jinan, China. Dr Zhu graduated in construction engineering from the Beijing Institute of Mining and Technology in 1956 and obtained a PhD from the Institute of Mining and Metallurgy, Cracow, Poland in 1962. Between 1962 and 2001, Dr Zhu was a research scientist in the Chinese Academy of Science Wuhan Institute of Rock and Soil Mechanics, and he was also a director of the Institute. Dr Zhu is Editor-in-Chief of the Chinese Journal of Rock Mechanics and Engineering, President of the Chinese Society of Underground Engineering and Underground Space, and an editorial board member of the international journal Rock Mechanics and Rock Engineering. Dr Zhu has published 4 books and over 150 technical papers. Dr Zhu's main research interests are mechanical properties of fractured rock masses, rock reinforcement by bolts, construction mechanics and stability analysis of large rock structures. Dr Zhu has more than 40 years experience in rock mechanics and engineering, and has been involved in numerous rock engineering and underground excavation projects in China.

Jian Zhao chairs the underground technology and rock engineering program at Nanyang Technological University of Singapore, and is an adjunct CKSP professor of geotechnical engineering at China University of Mining and Technology. Dr Zhao graduated in civil engineering from the University of Leeds in 1983 and obtained a PhD from Imperial College, London University, in 1987. Dr Zhao is an Editor of the international journal of Tunnelling and Underground Space Technology, editorial board member of the International Journal of Rock Mechanics and Mining Sciences and of 5 other technical journals. Dr Zhao's main research interests are rock joint properties, rock excavation and support, and rock structure stability under dynamic loads, and has published over 180 technical papers and 3 monographs. Dr Zhao also actively consults on rock engineering and tunnelling projects.

# Contents

Chapter 1

# Introduction

This book attempts to provide techniques for solving the problems in modelling and analysis of excavations in fractured rock masses, by taking into account the discontinuity influence, time-dependent behaviour and construction method dependent phenomenon. This chapter provides an overview on the rock mechanics issues related to underground excavation and stability.

## 1.1. ROCK MECHANICS AND UNDERGROUND EXCAVATION STABILITY

Rock structures are generally excavated within comparatively competent rock masses. However, rock masses usually consist of various discontinuity features, such as faults, joints and fractures. They significantly influence the stability of rock masses surrounding an excavation opening. The deformation and stability of the surrounding rock masses is controlled by the mechanics of the rock masses subjected to the change of conditions, which is a function of in situ condition before the excavation and disturbance due to the construction activities. The instability of underground works is mainly caused by the redistribution of stresses in the surrounding rock masses due to the excavation activity, due to excessive stress or excessive deformation. The process and result involves the interaction among the rock masses, the boundary conditions, and the engineering activity [1–3].

In order to understand the behaviour of the rock masses surrounding the excavations, it is necessary to understand the basic behaviour of the fractured rock masses subjected to loading and unloading conditions [4–7]. Common methods to study such behaviour are through theoretical analysis when the problems are simple, and through physical and numerical modelling when analytical solutions are not readily available [8–16].

Physical modelling is one of the basic tools to understand the behaviour and mechanism of jointed rock masses subjected to various boundary conditions, primarily loading conditions [17–19]. Whenever possible, physical modelling and testing should be conducted to provide the direct observation and basic understanding of engineering behaviour.

It is difficult, often impossible, to quantitatively predict the mechanical properties of jointed rock masses by physical modelling due to their complexity and large scale. Numerical modelling offers wide applications to simulate jointed rock mass behaviours including the effects of loading and time, and the behaviour of rock material, rock joints and rock masses [20–26].

The rationality and reliability of the results from numerical methods depend, to a great extent, upon the appropriate selection of computational model and mechanical and mathematical parameters [27–30]. Once the computational model is determined, the key to success hinges on the rational selection of the computing parameters. There are many factors and parameters that affect the rock mass behaviour and stability. One has to identify the order of importance of all the parameters [31,32]. In terms of computation, the limited resource may be the primary restriction. One of the common methods is the sensitivity analysis of various parameters within a system [33,34].

Rocks and rock masses often exhibit time-dependent behaviour, especially weak and soft rocks or highly fractured rock masses [35–45]. In underground excavation, the time-dependent phenomenon can be found that the loading on support elements gradually increases, leading to the final failure of the excavated structure [46–48].

Study of underground excavation in rock masses usually involves various modelling techniques and analysis approaches. Those methods can be physical tests, numerical modelling, observation and back analysis. The ultimate goal is to optimise excavation and support [8,49–52].

In recent years, the back analysis method has been widely applied in geotechnical engineering especially in the underground works [53–56]. Various related analytical and numerical techniques have been developed [57–60]. The method is based on the required input physical information, and can be divided into deformation back analysis method, stress back analysis method and coupled back analysis method. The physical information in the coupled back analysis method requires both deformation and stress. The back analysis has been applied to various rock engineering projects, particularly to underground excavation, to verify the support design and opening stability [57–62].

Construction of rock engineering projects usually requires a long duration, from a few months to a few years. The construction of these rock engineering projects will disturb the initial stable state of the rock masses. The various rock mass parameters interact in a dynamic interactive process until the rock mass reaches a new equilibrium state [63–65]. The construction is therefore a dynamic interactive process in time and in space. The success in constructing and managing a rock engineering project not only depends on the eventual state of the project, but also on the interim process and the construction methods adopted [66–68].

Construction of large-scale rock engineering projects is implemented by continual excavation of new working faces. Each newly excavated face interacts dynamically with the existing excavated space in time and in space. This dynamic interactive process of rock engineering constructions is non-inverse and non-linear. Its eventual state (or solution) is not unique but changeable with the interim process [1,63–65,69,70]. In other words, the eventual state is strongly dependent on the stress paths or stress histories. This leads to the possibility of the optimisation of construction process. Adopting a proper excavation sequence and installing effective rock reinforcement are engineering measures to stabilise large-scale underground excavations.

## 1.2. STRUCTURE OF THE BOOK

This book is divided into eight chapters. Chapter 1 provides an overview of rock mechanics issues in underground excavation in fractured rock masses, on the various topics covered by the main chapters of this book.

Chapter 2 addresses physical modelling of jointed rock masses. The observation and discussion are based on extensive laboratory studies on modelled rock masses with discontinuities. The physical modelling includes plane stress and plane strain tests of various jointed rock masses and bridged rock masses, and large-scale physical modelling of rock masses surrounding an excavation, and deformation analysis of excavation opening.

Chapter 3 deals with finite element based numerical modelling of the jointed rock masses together with computational examples. Deformation and strength equivalence formulations as well as treatment of rock joints are discussed. The numerical modelling is compared with physical modelling and examples of engineering application are given. Treatments of thick joints, collinear and multiple discontinuities are also dealt with in Chapter 3. It provides solutions on deformation and strength equivalence treatment for thick discontinuities, and collinear and multiple discontinuities.

Chapter 4 focusses on the sensitivity analysis of rock mass parameters. It covers the sensitivity analysis of common rock material and rock joint parameters, including methods of sensitivity analysis, and application of equivalent model for jointed rock mass. Examples on sensitivity analysis of rock mass parameters based on the extent of damage in the surrounding rock mass for a large underground cavern project is illustrated.

Weak and soft rocks with rheologic behaviour are often difficult mediums to model and analyse due to their time-dependent nature. Chapter 5 specifically addresses stability analysis of rheologic rock masses. It outlines rheological mechanical models for rocks and rock masses, visco-elastic and visco-plastic modelling, interaction of rock mass and support, and rheological damage and damage constitutive equations in numerical modelling.

Chapter 6 presents the principle of back analysis and observational methods commonly used in underground excavation stability assessment. The back analysis methods described include elastic back analysis, visco-elastic back analysis, back analysis and optimised methods in transverse isotropic rock masses. The method extends to plane and three-dimensional back analysis of jointed anisotropic rock masses, and stability analysis of excavation stability in such rock masses. Applications of various statistics models in deformation prediction in the stability assessment are discussed.

The principle of construction mechanics and excavation optimisation is presented in Chapter 7. It covers the principles of interactive construction mechanics and their applications to optimise cavern construction. It also outlines artificial intelligence techniques for construction optimisation and problem-solving algorithm in cavern construction optimisation. Engineering applications of artificial intelligence optimisation method and optimisation of excavation sequence are also presented in this chapter.

Chapter 8 specifically deals with reinforcement mechanism of rock bolts, including the effect of bolts on supporting the rock mass and reinforcement mechanism of rock bolts. Physical and numerical modelling of rock bolts are presented, with varying bolt spacing, length and layout, and with varying rock stress conditions. A scaled engineering model test is illustrated to verify the modelling results.

A comprehensive list of literatures on those topics is given at the end of this book, in the reference section.

Chapter 2

# Physical Modelling of Jointed Rock Mass

Rock structures are generally constructed within comparatively hard rock masses. The rock masses consist of various structural features, such as faults, joints and fractures. They significantly influence the stability of rock masses. In general, faults are treated as locally special fractures [71–73]. The characteristics of strength and deformation of the faults are studied specifically for the stability analysis. However, for joints and fractures, due to their abundant numbers, such special treatment generally does not apply. Instead they are regarded as basic rock mass elements in a considerable dimension [4–7,74–95]. However, the joints and fractures significantly affect the mechanical properties of the rock masses. Generally, distributions of joints are orderly in sets. In physical and numerical modelling, it is common that for a project, based on in situ geological investigation, two to three major joint sets are identified. The mean density, persistence, orientation and typical distribution patterns are generalised and modelled. Often, these joints are distributed intermittently. Therefore, the study presented here will emphasise the behaviour of rock masses where joints distribute orderly and intermittently.

If in situ tests are conducted to study the behaviour of rock mass containing sufficient joints, the dimension of the specimen needs to be a few tens of meters. However, these tests are not commonly conducted. Usually, the more practical method is the physical modelling with a reduced scale [78,96–102]. The mechanics of deformation and failure are observed, which forms the basis for further study.

## 2.1. MODELLING OF JOINTED ROCK MASSES UNDER PLANE STRESS STATE

### *2.1.1 Mechanical properties of equivalent materials*

For physical model tests, it is important to select a modelling material of which the properties are similar to those of the material modelled. Obviously, not all the properties can meet the law of similarity. Usually, the main parameters are made to meet the similarity conditions and secondary parameters are made to meet the conditions approximately [102].

A physical model test is performed to model a large-scale hydropower cavern project. The model is made of sand, barite powder and an organic polymer resin. Blocks are made by those materials through compaction and heating. The mechanical properties of the modelling material and the modelled rock masses are summarised in Table 2.1.

**Table 2.1.** Mechanical properties of the modelling material and the modelled rock mass.

| Mechanical properties | Modelling material and fracture | Modelled rock material and fracture | Actual rock material and fracture |
|---|---|---|---|
| $\sigma_c$ (MPa) | 0.545 | 218 | 212.4 |
| $E$ (MPa) | 68.3 | $2.73 \times 10^4$ | $3.0 \times 10^4$ |
| $c$ (MPa) | 0.79 | 31.16 | 16 |
| $\phi$ (°) | 44.4 | 44.4 | 56 |
| $c_j$ (MPa) | 0.0094 | 3.76 | 0.5 |
| $\phi_j$ (°) | 33.6 | 33.6 | 36.9 |

The physical model has the property similarity ratio of 400. The ratio of the rock strength to modelling material strength and the ratio of the rock modulus to the modelling material modulus are both 400.

Rock joints in the rock mass are simulated by the joints between blocks. To form intact portion, the blocks are cemented by the polymer resin. The testing results indicate that the physical model worked well in simulating the rock masses of different joint systems.

### 2.1.2 *Model tests of jointed rock mass*

In general, only two or three major joint sets have the governing influence on the engineering properties of the rock mass [102–106]. In most modelling studies, two representative joint sets are usually taken in the plane modelling. The two joint sets modelled usually intersect approximately orthogonally at an acute angle. Studies on both joint set arrangements were performed and they are presented in this section.

Rock masses that consist of two orthogonal joint sets are fairly common, e.g., the rock mass surrounding at Ertan and Xiaolangdi hydroelectric power stations. Physical modelling of the Ertan hydroelectric power cavern is conducted.

From the mapped joint data, three common joint distribution arrangements are modelled, to represent a number of possible combinations of the intersection between joint sets, as shown in Figure 2.1.

The modelling of mechanical characteristics of jointed rock masses should take into account the effect of joint quantity and joint properties. Studies by Muller and co-workers [107] have indicated that when the ratio of the rock mass dimension ($D_m$) over the joint spacing ($s_j$) is greater than 10 (i.e., $D_m/s_j > 10$), the strength and deformation characteristics of the rock mass model are consistent, and the size effect has little influence. The model has a size of $50 \times 50 \times 7$ cm and consists of blocks of $5 \times 10 \times 7$ cm, that satisfies the requirement of $D_m/s_j > 10$.

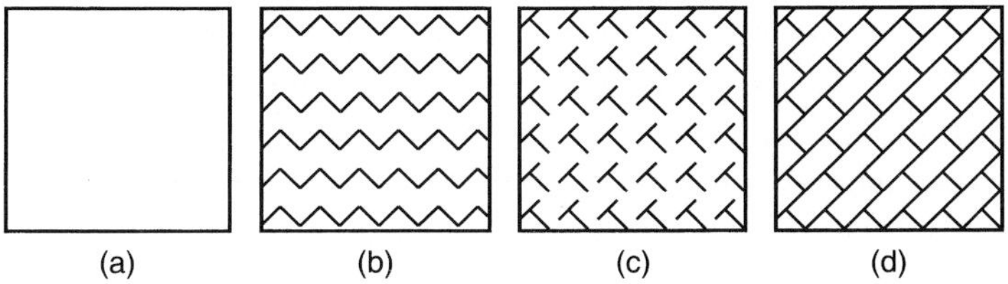


**Figure 2.1.** Patterns of joint set arrangements modelled. (a) intact (no joints rock); (b) two joint sets with the same persistence of 50% forming running-through in one direction; (c) two joint sets with the same persistence of 50% forming T-shapes; (d) two joint sets with persistence of 50% and 100% respectively.

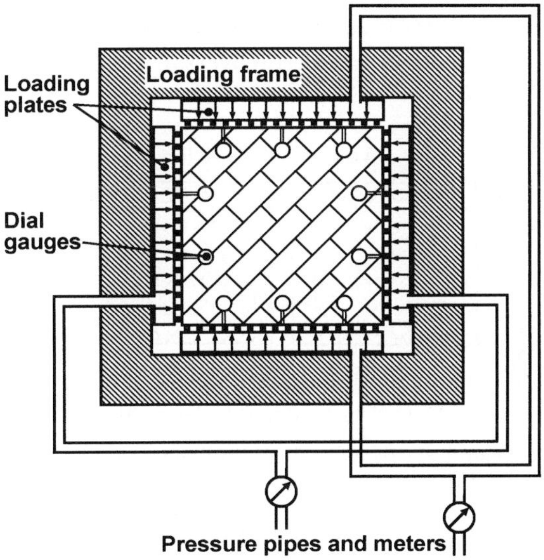


**Figure 2.2.** Loading and monitoring set-up of the test system.

The loading and monitoring set-up of the test system is shown in Figure 2.2, which consists of a loading frame, loading and monitoring devices. A 0.5 cm thick layer of sand is placed between the ends of the physical model and the rubber pockets to eliminate end friction.

Through a series of experiments, the failure mechanism of rock mass, the effect of joint distribution and effect of shearing are studied.

***2.1.2.1 Effect of joint distribution patterns.*** Three model tests are performed to represent rock masses of different joint distribution patterns as shown in Figure 2.1.

In Figure 2.1(b), major principal stress is applied at different directions. When the major principal stress is perpendicular to the direction of running-through (Figure 2.3(b)), the model has a low deformation module and a high compressive strength, because the joints can be completely closed and the rock masses become intact. In this case, the $\sigma_1$–$\varepsilon_1$ curve exhibits a concave shape. While for the same joint sets, when the major principal stress is applied parallel to the direction of running-through (Figure 2.3(a)), the lateral deformation is high and the tested model is liable to failure by splitting. In this case, the uniaxial compressive strength is lower than the former. The comparison between the two strain–stress curves is shown in Figure 2.4.

Results obtained from models shown in Figure 2.3(a) and Figure 2.3(c) are illustrated in Figure 2.5. As expected, because the joints in the model of Figure 2.3(c) are bridged, both longitudinal and transversal deformations are small, the

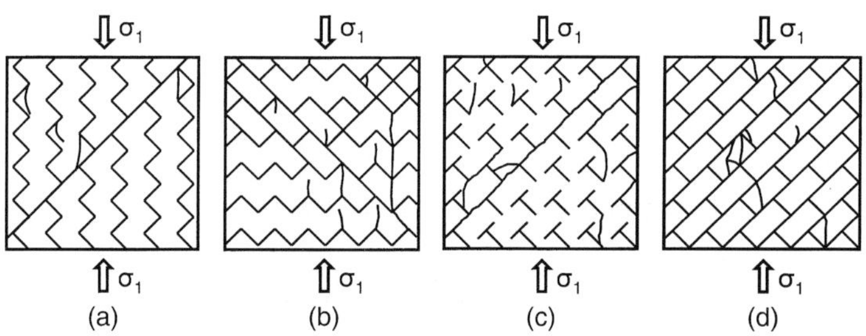


**Figure 2.3.** Failure patterns of jointed rock mass.

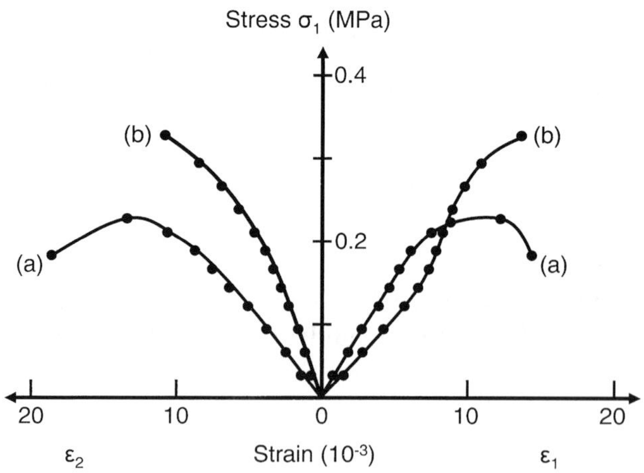


**Figure 2.4.** Stress–strain curves of different joint distribution patterns shown in Figure 2.3(a) and (b).

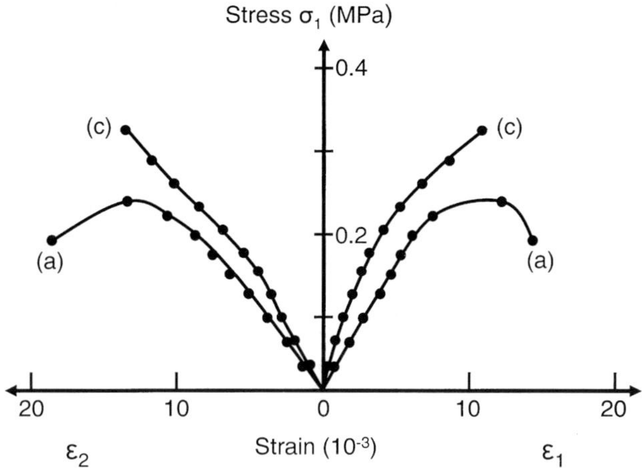


**Figure 2.5.** Stress–strain curves of different joint distribution patterns shown in Figure 2.3(a) and (c).

compressive strength is high and the $\sigma_1$–$\varepsilon_1$ curve is similar to that of intact rock material with no joints. In the model shown in Figure 2.3(b), because the running-through joint surface is roughly perpendicular to $\sigma_1$, its failure mostly exhibits the form in which longitudinal tensile-opening takes place at the sharp turning points of the joints, accompanied by local shearing failure. For the rock mass in Figure 2.3(c), T-shaped joints gradually run through each other longitudinally and lateral cracking occurs under compressive stresses. As a result, compound shear failure planes are finally formed. For the rock mass shown in Figure 2.3(d), the main failure mode is shear failure along joint planes that have been run through completely, accompanied by lateral tensile cracking of the materials. In summary, there are two basic failure patterns of the rock masses: lateral tensile cracking and shear sliding. Due to various combinations of joint set distributions and arrangements, failure mechanisms and development are different. In general, secondary tensile cracks occur at the joint tips first, then the cracks link up with the adjacent joints, followed by compound lateral tensile cracks and finally overall shear failure takes place along a plane shear.

***2.1.2.2 Strength of typical jointed rock masses.*** The effect of lateral shear on the rock mass strength is studied with the model shown in Figure 2.3(a), where major principal shear is applied vertically and linear principal shear ($\sigma_2$) is applied horizontally. Table 2.2 and Figure 2.6 give the typical results of tests under different $\sigma_2$. They show that the peak strength, $\sigma_1$, of the rock masses increases with the increasing $\sigma_2$. The deformation module also increases but the lateral deformation decreases rapidly.

**Table 2.2.** Testing results at different lateral pressure.

| Testing No. | $\sigma_2$ (MPa) | $\sigma_1$ (MPa) | $E$ (MPa) | $v$ |
|---|---|---|---|---|
| 1 | 0 | 0.28 | 29.75 | 0.1 |
| 2 | 0.05 | 0.41 | 65.45 | 0.1 |
| 3 | 0.15 | 0.59 | 61.85 | 0.25 |
| 4 | 0.25 | 0.69 | 117.3 | 0.31 |
| 5 | 0.34 | 0.96 | 122.9 | 0.43 |

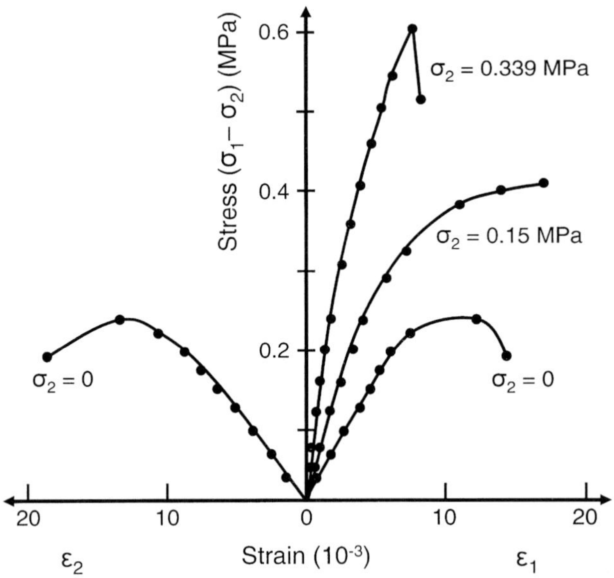


**Figure 2.6.** Change of stress and strain at different lateral stress for model in Figure 2.3(a).

The modelling results of the jointed rock mass in Figure 2.3(d) show that when the persistence of joint sets increases, the compressive strength of the rock mass decreases. The modelling results indicate that the deformation modules are governed by the persistence of joint sets, as shown in Figure 2.7.

Table 2.3 summarises the uniaxial compressive strength ($\sigma_c$) and deformation modules ($E$) of the four different rock masses shown in Figure 2.3.

In summary, the variation of joint set pattern strongly affects the strength and deformation characteristics of a rock mass. From the modelling studies, the following qualitative observation is noted for the rock mass represented by Figure 2.3(a), the peak strength and lateral stress are linearly proportional in general,

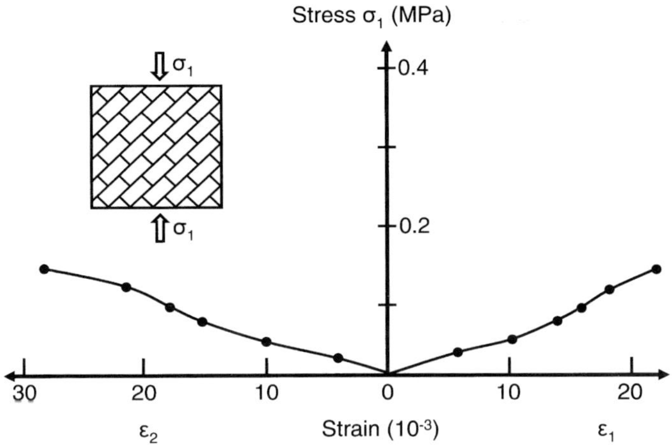


**Figure 2.7.** Stress–strain relationship of rock mass model of Figure 2.3(d).

**Table 2.3.** Strengths and elastic moduli of four models.

| Model type | $\sigma_c$ (MPa) | *E* (MPa) |
|---|---|---|
| 2.3 (a) | 0.29 | 29.75 |
| 2.3 (b) | 0.37 | 27.38 |
| 2.3 (c) | 0.35 | 43.60 |
| 2.3 (d) | 0.2 | 7.13 |

as shown in Figure 2.8, failures in most cases start with the lateral tensile cracking along joints and follow by the formation of a shear plane. Failure occurs finally along the inclined plane.

***2.1.2.3 Anisotropy of rock strength.*** The variation of the strength of the jointed rock mass in response to the change of the major principal stress direction is studied. Rock mass of joint pattern shown in Figure 2.3(a) is subjected to the fixed $\sigma_1$ and $\sigma_2$ with the direction of the shears.

The testing results are presented in Figure 2.9 and Table 2.4. The study shows that the lowest strength of the rock mass occurs when $\alpha_1$ is at $40 \sim 50^\circ$.

***2.1.2.4 Relation between rock mass strength and joint persistence.*** Studies on the effect of joint persistence on rock mass strength are carried out. The persistence of

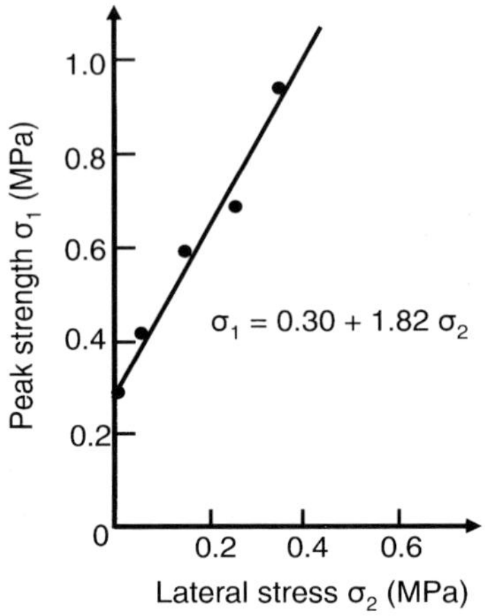


**Figure 2.8.** Relationship between peak strength and lateral stress for rock mass of Figure 2.3(a).

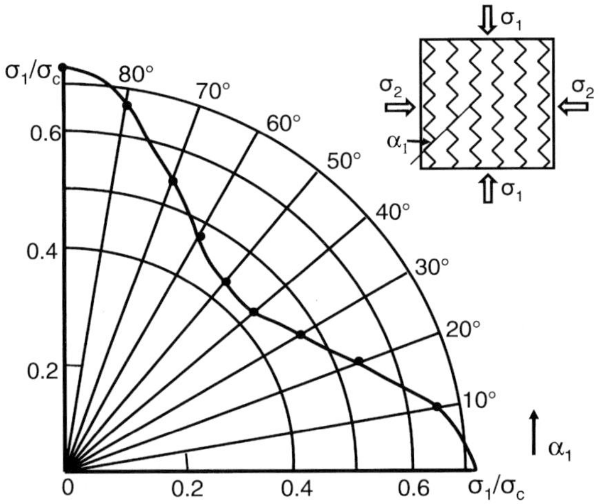


**Figure 2.9.** Change of rock mass strength with joint direction.

**Table 2.4.** Change of strength with joint direction.

| Strength | $\alpha_1$ | | | | |
|---|---|---|---|---|---|
| | 0° | 10° | 20° | 30° | 40° |
| $\sigma_1$ (MPa) | 0.4 | 0.37 | 0.31 | 0.27 | 0.24 |
| $\sigma_1/\sigma_c$ | 0.72 | 0.67 | 0.57 | 0.49 | 0.44 |

Note: $\sigma_c = 0.55$ MPa; $\sigma_2 = 0.05$ MPa.

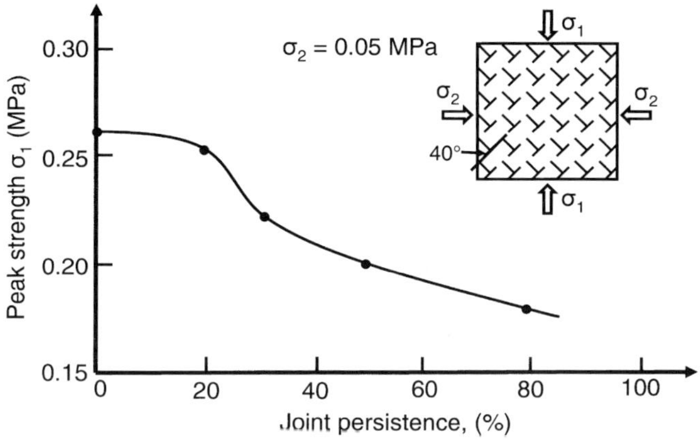


**Figure 2.10.** Change of rock mass strength with joint persistence.

**Table 2.5.** Effect of joint persistence on the rock mass strength.

| $n_2$ (%) | 0 | 20 | 30 | 50 | 80 |
|---|---|---|---|---|---|
| $\sigma_1$ (MPa) | 0.26 | 0.25 | 0.22 | 0.2 | 0.18 |

the second joint set ($n_2$) in Figure 2.3(c) is changed, while the loading conditions remain the same ($\sigma_2 = 0.05$ MPa). Modelling study reveals the relationship between the rock mass strength and the joint persistence, as shown in Figure 2.10 and Table 2.5.

It can be seen from Figure 2.10 that the strength of the rock mass decreases with the increasing joint persistence. The rate of the strength change is small when joint persistence is below 20%. A rapid strength decrease is observed when the persistence increases from 20% to 30%.

### *2.1.3 Large dimension plane strain model experiment*

In many underground rock engineering works, such as tunnels, the surrounding rock masses are in plane strain condition [108–111]. Therefore, the model study on mechanical behaviour of the surrounding rock masses under plane strain condition is of great importance. In plane strain modelling, the model has sufficient length to remove the influence of end friction. The results of plane strain model experiments for jointed rock mass are discussed in the following sections.

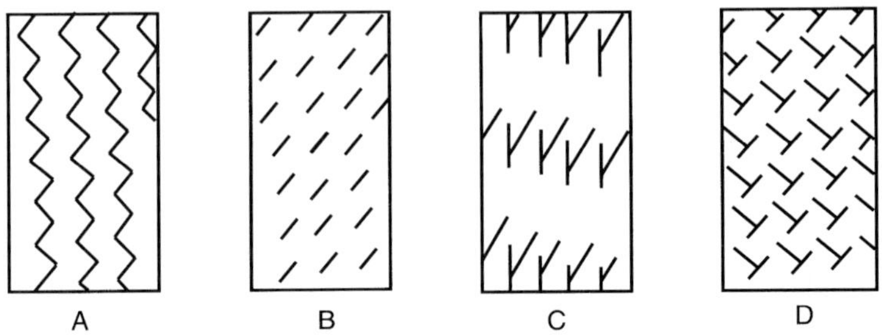


**Figure 2.11.** Four types of jointed rock mass modelled.

***2.1.3.1 Material properties and experimental methods.*** The preparation of the model material is similar to that of the plane stress model. Deformation characteristics of the joint plane are measured. The material has the following indexes: unit weight $\gamma = 2.2 \times 10^4\,\text{N/m}^3$, uniaxial compressive strength $\sigma_{c\,mm} = 0.87\,\text{MPa}$, tensile strength $\sigma_{t\,mm} = 0.11\,\text{MPa}$, elastic module $E_{mm} = 170\,\text{MPa}$, Poisson's ratio $\nu = 0.22$. Joint plane properties are: cohesion $c_j = 0.01\,\text{MPa}$, friction angle $\phi_j = 39°$; shear stiffness $K_s = 5\,\text{MPa/cm}$, and normal stiffness $K_n = 75\,\text{MPa/cm}$.

Large-size model of $100 \times 50 \times 14$ cm is used to eliminate size effect and end effect. The joints in the rock mass are simulated by the contact surfaces of the blocks. The shear strength and deformation stiffness of the joints are measured. To simulate the rock bridge, the contact surfaces of the blocks are cemented to the material strength. The shearing strength of the bridged joint is also measured. The loading is applied through a large steel frame under the condition of plane strain. The tests are performed on 20 rock mass models of four different joint patterns. These four patterns are shown in Figure 2.11.

***2.1.3.2 Testing results.*** In the tests, the deformations of the model in three directions are measured. Different failure stages are judged mainly by the deformation rate and the tendency of the deformation curves. Figure 2.12 and Table 2.6 show the typical testing results. The rock mass strength is calculated from various peak strengths ($\sigma_1$) under different lateral pressures ($\sigma_2$) using the Least Square Method, expressed by the overall equivalent cohesion, $c_{mm}$, and inner friction angle of $\phi_{mm}$. Figure 2.13 shows the typical deformation curves of the rock mass of joint distribution pattern of Figure 2.11(D) under different lateral pressures.

It can be seen from the modelling results that for the four different rock masses, their overall strength can be roughly related to the material strength and the joint strength, by the following equations: Rock mass cohesion = $(0.2 \sim 0.3) \times$ (Rock material cohesion + Rock joint cohesion), Rock mass friction angle = $0.5 \times$ (Rock material friction angle + Rock joint friction angle).

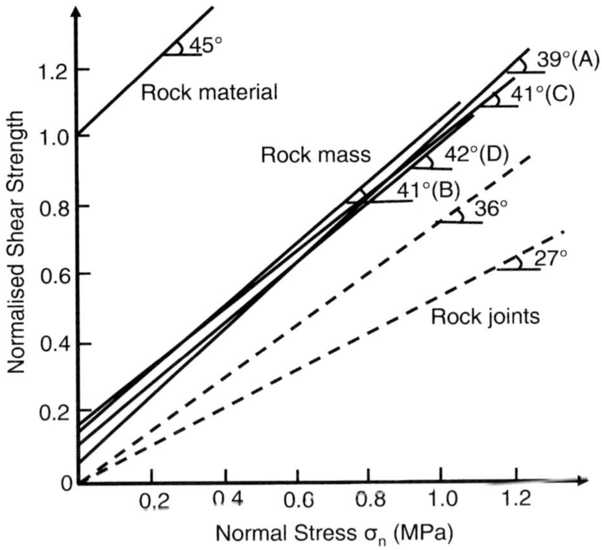


**Figure 2.12.** Relationship between specific strengths of rock mass, material and joint, for four rock masses with different joint distribution patterns.

**Table 2.6.** Results from large-size model tests of jointed rock masses.

| Model No | Model type | $\sigma_1$ (MPa) | $\sigma_2$ (MPa) | $\sigma_3$ (MPa) | $\sigma_0$ (MPa) | $\sigma_0/\sigma_1$ | $c_{mm}$ (MPa) | $\phi_{mm}$(°) |
|---|---|---|---|---|---|---|---|---|
| 90-1 | A | 0.31 | 0 | 0.02 | 0.19 | 0.61 | | |
| 90-2 | | 0.62 | 0.1 | 0.05 | 0.40 | 0.65 | 0.05 | 42 |
| 90-3 | | 0.71 | 0.15 | 0.06 | 0.52 | 0.73 | | |
| 90-4 | | 1.45 | 0.25 | 0.23 | 1.25 | 0.86 | | |
| 90-5 | | 1.28 | 0.31 | 0.11 | 1.11 | 0.87 | | |
| 91-1 | B | 0.18 | 0.02 | 0.06 | 0.09 | 0.50 | | |
| 91-6 | | 0.68 | 0.10 | 0.15 | 0.35 | 0.51 | 0.04 | 39 |
| 92-4 | | 1.42 | 0.30 | 0.62 | 0.95 | 0.67 | | |
| 92-3 | | 1.53 | 0.31 | 0.59 | 0.92 | 0.60 | | |
| 92-2 | C | 0.11 | 0 | 0.01 | 0.07 | 0.55 | | |
| 91-7 | | 0.52 | 0.08 | 0.07 | 0.28 | 0.54 | 0.02 | 40 |
| 91-8 | | 0.97 | 0.21 | 0.16 | 0.51 | 0.53 | | |
| 92-1 | | 1.44 | 0.27 | 0.39 | — | — | | |
| 91-2 | D | 0.12 | 0.01 | 0.03 | 0.09 | 0.73 | | |
| 91-5 | | 0.44 | 0.05 | 0.06 | 0.26 | 0.60 | 0.03 | 39 |
| 91-3 | | 0.55 | 0.09 | 0.12 | 0.36 | 0.66 | | |
| 91-4 | | 0.69 | 0.14 | 0.10 | 0.48 | 0.70 | | |

Note: $\sigma_0$ is the stress in the direction of $\sigma_1$ when the volumetric strain rate is zero; $\sigma_3$ is the normal stress applied on the model surfaces to keep the plane strain state.

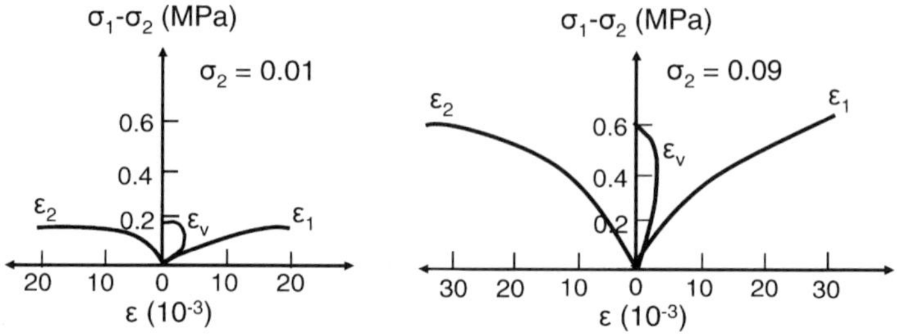


**Figure 2.13.** Stress–strain curves of rock mass model of joint distribution shown in Figure 2.11(D).

## 2.2. MODEL TEST OF ROCK MASS WITH ROCK BRIDGES

In this section, modelling is performed to study the failure of rock bridges, the relationship between the joint intensity and the strength of the rock, and the strength of the rock mass of the bridged joint. Results of two phases of modelling are presented in the following sections.

### *2.2.1 Experiment Phase I*

In this phase, the model materials are gypsum, diatomite and water mixture that can be conveniently poured to set. Fractures of given persistence are cast. The principal physico-mechanical parameters of the model are given in Table 2.7.

In fracture preparation, thin pieces of steel are buried in the specimen when the material is wet and then drawn out after set. The fractures are naturally closed. The specimens are subjected to drying and curing. In tests, the models are loaded by the plane stress, as shown in Figure 2.14. In addition to deformation transducers, special coating is applied at model surface around the fractures to observe crack initiation and propagation at the tips. The persistence of the fractures is at 0, 20%, 40%, 60% and 80% respectively. Seven normal stresses at 0.08, 0.2, 0.3, 0.35, 0.4, 0.5 and 0.6 MPa are applied. The sheared planes of all the specimens are examined after each test.

During the testing, normal stress is kept constant, and horizontal shear load is applied gradually until failure occurs. Typical stages of crack initiation and propagation before failure are observed, as shown in Figure 2.15:

(a) Small feather-shaped cracks appear around the existing fractures;
(b) Tensile cracks appears along the direction of shear load at one tip of the fracture first and then appear at the other tip;

**Table 2.7.** Materials and fracture parameters.

| | | |
|---|---|---|
| Model material | Dry density (g/cm$^3$) | 0.68 |
| | Porosity (%) | 68.9 |
| | Uniaxial compressive strength (MPa) | 2.05 |
| | Compressive modulus (MPa) | $1.3 \times 10^3$ |
| | Poisson's ratio | 0.2 |
| | Tensile strength (MPa) | 0.27 |
| | Tensile modulus (MPa) | $2.6 \times 10^3$ |
| | Fiction angle (°) | 35 |
| | Cohesion (MPa) | 0.53 |
| Fracture | Fracture length (mm) | 20, 40, 60, 80 |
| | Normal stiffness (MPa/cm) | 100 |
| | Shear stiffness (MPa/cm) | 20 |
| | Friction angle (°) | 10 |
| | Cohesion (MPa) | 0.1 |

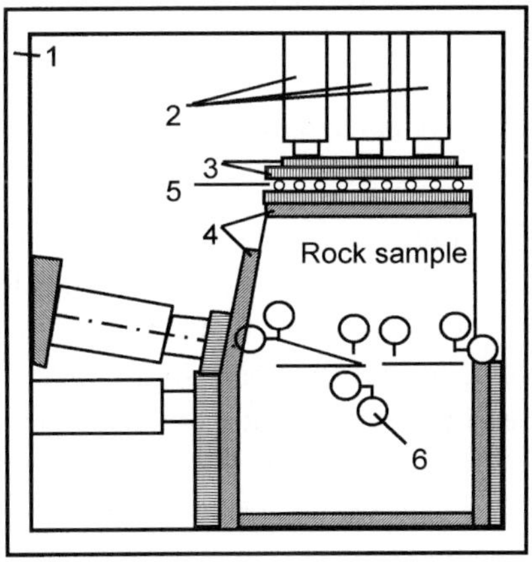


**Figure 2.14.** Model test arrangement.

(c) A pair of transversal compressive cracks appear along two approximately orthogonal directions between two adjacent tensile cracks;

(d) Rotation of lozenge block takes place, in association with shear dilation and small secondary cracks appearing around large cracks;

(e) Visible opening of joints takes place; and macro visible shear failure occurs where the block breaks along the lozenge diagonal.

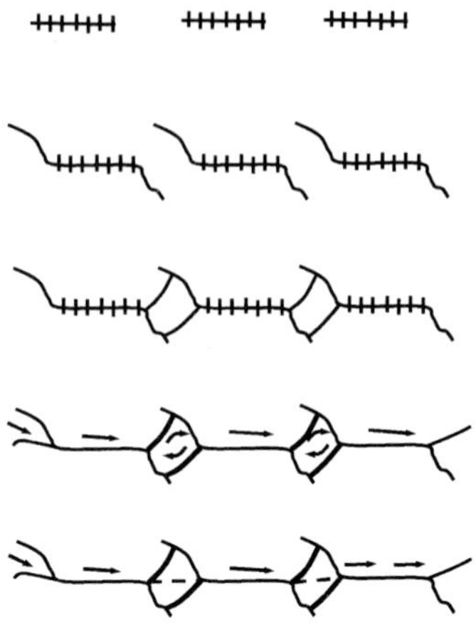


**Figure 2.15.** Stages of shear failure of the model.

For samples of different joint persistence and different vertical stresses, the shear failure has three categories [112–119]:

(a) For specimens with low joint persistence and medium vertical stress, considerable dilatancy of the specimen occurs. The failure mode is compressive torsional shear failure.
(b) For specimens with medium joint persistence and high vertical stress, the deformation and failure of the specimen undergoes the following stages: (i) micro cracks are generated at the fracture tips, (ii) cracks propagate and the secondary compressive torsional cracks occurs, (iii) lozenge blocks are formed, (iv) shear crack appears along the diagonal of the lozenge blocks and connects with tips of the fractures, and (v) the overall failure of the specimen occurs along the existing fractures and the new shear cracks, and the dilatancy takes place within the whole specimen. The failure mode is torsional tensile shear failure.
(c) For specimens with high joint persistence and high vertical stress, the deformation and failure of the specimen develop in the following stages: (i) small "rock bridge" is broken due to the vertical stress, (ii) with the increase of the shear stress, overall failure takes place along the existing fractures and the cracks crossing the rock bridges. Although a slight dilatancy can be observed in the shearing process, the specimen as a whole fails basically due to the compression. This is pure shear failure.

Results of over 20 specimens are summarised in Figure 2.16. It shows four shear strength envelopes for four groups of joint persistence. The apparent shear strength

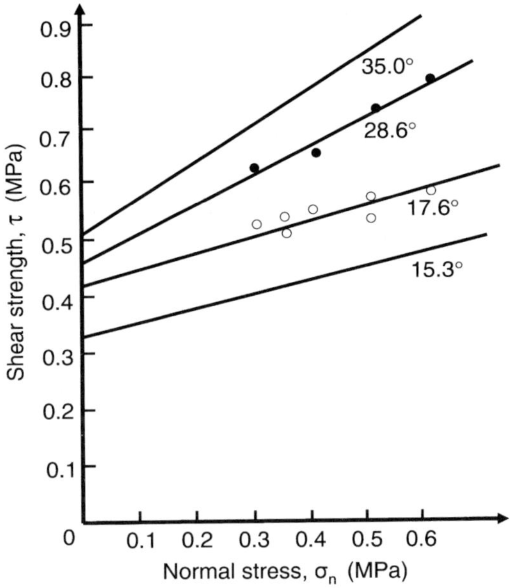


**Figure 2.16.** Change of shear strength with joint persistence.

falls with increasing persistence. When the persistence increases from 0 to 65%, internal friction angle decreases from 35° to about 15°, and the cohesion decreases from over 0.5 MPa to about 0.3 MPa.

### 2.2.2 *Experiment Phase II*

In order to have better control of joint persistence in the specimens, a new material mix is used in the second phase of experiments. The properties of the new material mix of the model are close to those of rock. Joint persistence is properly controlled. The material used is the mixture of sand, barite powder, colophony and alcohol. Fractures are created with polythene films of different lengths. Two types of fractures with different shear strength (friction coefficient) are created by two different methods: one by a single film and another one by two films with sandwiched grease. The designed joint persistence is the same as those in the phase I, i.e., 20%, 40%, 60% and 80%. Mechanical tests show that this material has similar properties as some sedimentary rocks and similar dilatancy character before failure. The tensile–compressive strength ratio is about 0.11. The mechanical parameters are summarised in Table 2.8.

**Table 2.8.** Properties of model material and fracture.

| | | |
|---|---|---|
| Model material | Dry density | 2.21 |
| | Uniaxial compressive strength (MPa) | 0.15 |
| | Compressive modulus (MPa) | $1.45 \times 10^3$ |
| | Poisson's ratio | 1.12 |
| | Tensile strength (MPa) | 0.122 |
| | Friction angle (°) | 38.5 |
| | Cohesion (MPa) | 0.25 |
| Single fracture | Fracture length (mm) | 20, 40, 60, 80 |
| | Normal stiffness (MPa/cm) | 43 |
| | Shear stiffness (MPa/cm) | 17 |
| | Friction angle (°) | 30 |
| | Cohesion (MPa) | 0.065 |
| Double fracture | Fracture length (mm) | 20, 40, 60, 80 |
| | Normal stiffness (MPa/cm) | 33 |
| | Shear stiffness (MPa/cm) | 11 |
| | Friction angle (°) | 24.3 |
| | Cohesion (MPa) | 0.041 |

The test equipment, installation and testing methods are the same as that adopted in Phase I. In the tests, failure processes are observed and noted. The crack initiation and propagation are similar to that in Phase I. From test data of over 60 specimens, a series of peak shear strength with corresponding normal stress are obtained by means of the Least Square Method, and are presented in Figures 2.17 and 2.18.

The results indicate a systematic decrease of shear strength of the specimen with increasing joint persistence. For specimens with double fracture the decreasing rate is greater. The strength envelopes show relatively good linearity. With increasing joint persistence, the rock mass cohesion and rock mass internal friction coefficient decrease accordingly.

The weighted mean method is commonly adopted for predicting the strength of the fractured rock mass. The shear resistance of the shear plane is obtained by summing the shear resistances contributed by the fracture and by the bridges. By assuming the normal stress on the shear plane before failure is uniformly distributed, the shear resistance of the shear plane can be calculated as,

$$\tau = [nc_j + (1-n)c_r] + \sigma_n[n\phi_j + (1-n)\phi_r] \tag{2.1}$$

where $n$ is the joint persistence, $c_j$ is the joint cohesion, $c_r$ is the cohesion of rock bridge, $\phi_j$ is the internal friction coefficient of joint plane, $\phi_r$ is the internal friction coefficient of rock bridge.

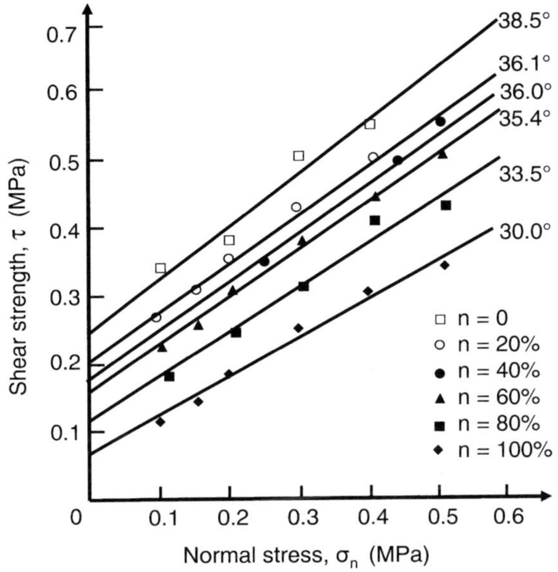


**Figure 2.17.** Shear strength at different normal stress and joint persistence for single fracture model.

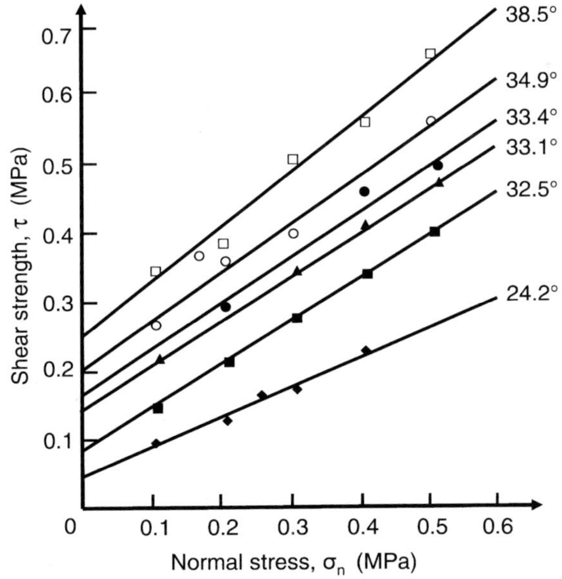


**Figure 2.18.** Shear strength at different normal stress and joint persistence for double fracture model.

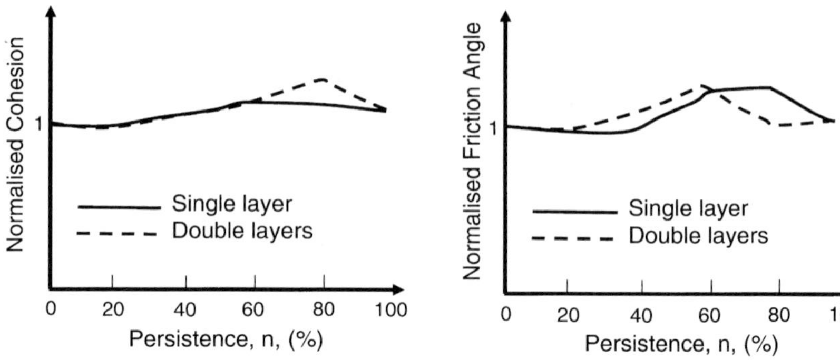


**Figure 2.19.** Change of normalised (a) cohesion and (b) friction angle with persistence.

However, the actual distribution of the normal stress on the shear plane before failure is not uniform. The failure mechanism of the fractured rock mass under the shear loading is complex, characterised by multi-stage development of fracturing [117–129]. It is often a combined tensile-shear failure [130–133]. Therefore the shear strength expressed in equation (2.1) differs from the actual strength. The differences in cohesion and friction angle are analysed by normalising the actual cohesion and friction to that expressed in equation (2.1), and are shown in Figure 2.19. It can be seen that the differences are little when the persistence is less than 40%. But with the increase of persistence the differences becomes greater. When the persistence is between 40–80%, the error is about 20%.

## 2.3. MODEL TESTS ON STABILITY OF SURROUNDING ROCK OF LARGE-SCALE CAVITY

Numerous model tests on large-scale underground projects have been conducted (e.g., [25,125,134–136]). However, model tests studying the stability of the surrounding jointed rock mass are limited. In this section, a large-scale model test on jointed rock mass stability is performed by modelling a rock cavern project [137]. Site investigation shows that there are three major joint sets: NE and NW sets with steep dip angles, and EW with a gentle dip angle. The surrounding rock mass of the cavern is cut into prisms and polyhedrons by these three joint sets. For simplification of the problem, only the NE and EW sets are modelled, as shown in Figure 2.20. The joint sets have cut the surrounding rock mass into cuboids of $8 \times 5 \times 10$ m in size. In the model, these joint distributions and block arrangements are simulated (Figure 2.21).

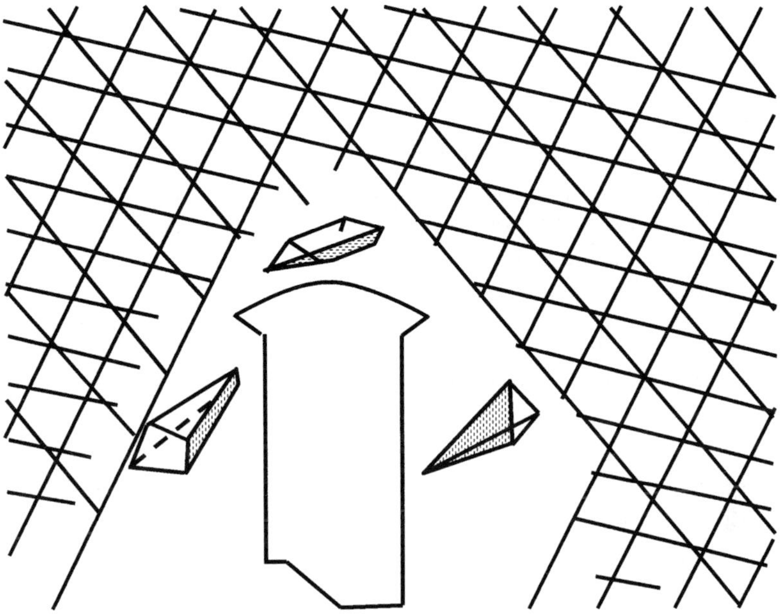

**Figure 2.20.** Polyhedron blocks produced by three joint sets cutting through the surrounding rock mass.

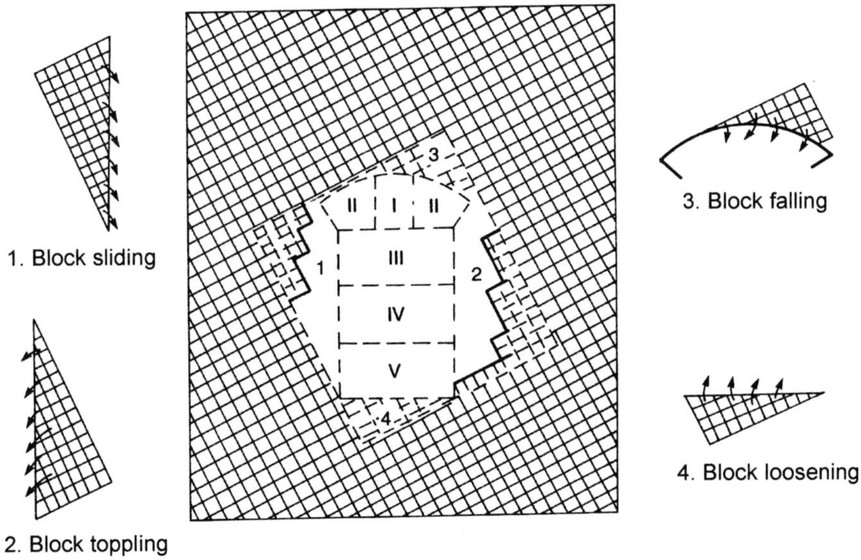


**Figure 2.21.** Potential failure patterns of the opening and sequence of excavation.

### 2.3.1 *Similarity conditions of modelling*

A physical simulation model should meet the similarity to the prototype in the following aspects:

(i) Size and configuration of the engineering project,
(ii) Geometry of the geological structures,
(iii) Physico-mechanical properties of the surrounding rock mass,
(iv) The initial stress state, and
(v) The construction sequences.

A scale of 1:200 is chosen between the model and the prototype. The rectangular model has a height of 1.7 m and a width of 1.6 m. The model material is made of gypsum, sands of various granular sizes and water. It is modelled by plane stress state condition. An initial horizontal in situ stress of 28.84 MPa and a vertical in situ stress of 13.35 MPa are applied, before the simulation of the excavation. In order to simulate the actual excavation sequence, a five-stage excavation sequence is adopted from the top to the bottom for the excavation, as shown in Figure 2.21.

### 2.3.2 *Deformation of the surrounding rock mass during excavation*

The model is applied with a vertical and a horizontal compressive stress first at boundaries to simulate the in situ stresses. The stresses are maintained at constant, while excavations in five stages are modelled. The deformation and failure are monitored during the excavation.

***2.3.2.1 Deformation of the surrounding rock mass during excavation.*** In the first to the third stage of excavation, both vertical and horizontal displacements are not remarkable.

From the beginning of the fourth stage through the fifth stage, large inward vertical displacements take place at the crown, with a maximum displacement of about 0.1 cm. The displacements at base are small. Large horizontal displacements are noted on two side walls, with maximum displacement of 0.3 cm, as shown in Figure 2.22.

After the fifth stage when the excavation has been completed, the displacement zones extends to 15 m above the crown and 53 m away from the sidewalls, as shown in Figure 2.22. The results show that the horizontal displacement zones are considerably large, about 8 times that of the excavation width.

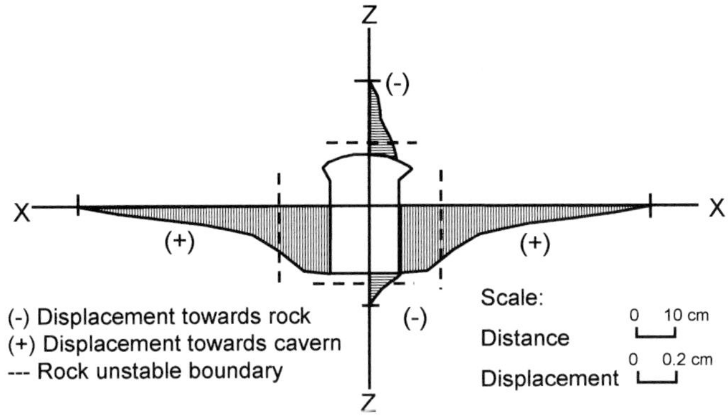


**Figure 2.22.** Displacement curve and affected range in rock mass surrounding the excavation.

***2.3.2.2 Failure of the surrounding rock mass during excavation.*** From the first to the third stage excavation, there is basically no failure taking place in the rock mass, only small tetrahedronal blocks are loosened at the crown.

From the fourth stage to the end of the fifth stage, the tetrahedronal blocks are sliding at the lower walls. In the upper walls, the blocks are falling along the steep dip joint into the opening. The rock mass at crown displays tensile rupture along joint planes and the rock mass at base starts to loose (Figure 2.21).

***2.3.2.3 Stability of the surrounding rock mass.*** Based on the deformation and failure characteristics of the surrounding rock mass, general conclusions on the stability of the rock mass can be drawn.

During the excavation from the first to the third stage, the surrounding rock mass is basically stable. From beginning of the fourth stage till the completion of the fifth, noticeable instability takes place.

The affected displacement zone in the surrounding rock mass is large. The affected displacement zone extends far into the sidewalls (50 m) and through the crown (15 m).

The instability is reflected in various patterns: tensile rupture in the crown, structural loosening in the bottom; and block sliding and falling on the sidewalls.

The instability occurs in the following sequences: first in the lower sidewall, followed by the upper sidewall and finally the rock mass at the top and bottom.

It should be noted that the model exaggerates the joint persistence, as 100% persistence is assumed. Simplification and approximation are applied to treat the three-dimensional problem by the two-dimensional model. The modelling is based on the law of scaling and the similarity principle, results obtained and observation made on the mechanism are qualitative, but have significant applications to research and engineering.

# Chapter 3
# Numerical Modelling of Jointed Rock Mass

It is difficult to quantitatively predict the mechanical properties of jointed rock masses due to their complexity, even by expensive, large-scale in situ tests. The parameters obtained from laboratory or field tests are usually for joint planes or rock blocks to a very limited scale. In most cases, the scale is limited to a few meters. The overall equivalent properties of rock masses of large size are almost not available through direct measurement. Prediction of jointed rock mass behaviours by numerical modelling is useful to study the effects of loading, the behaviour of rock material, rock joints and rock masses. Therefore, it has wide and promising applications. However, the validity of numerical modelling should be supported by physical simulation and correlation.

For jointed rock masses, various types of numerical models have been developed [12,14,15,138–162]. For rocks containing a small number of joints, joint element [25,26] can be adopted to represent the discontinuous planes. Modelling of densely jointed rock masses can be mainly realised by two methods. One is the approach of continuum mechanics, or the equivalence approach, such as material parameter equivalence, energy equivalence, deformation equivalence, composite equivalence, fracture mechanics and damage mechanics [77,83,114,163–167]. The other approach is to take the rock blocks as particles of a discontinuum, to study the mechanical properties of the assembly of these particles, including stress, strain and stability in light of discontinuum theory, and to derive mechanical law for the blocks. Examples of this approach are the rigid block method, discrete element method [15,150,152–154,168,169], discontinuity deformation analysis method [20,92,156,159]. However, all the methods have their own limitations, for example, incompatibility of joint elements with adjacent continuum elements, the scale effect in equivalence, definition of damage tensor, equation of damage evolution and discretion error. This chapter attempts to develop a new numerical approach for jointed rock masses and a modelling technique for rapid and accurate prediction of their mechanical behaviour.

An equivalent continuum model for jointed rock mass based joint element assembly concept, coupled with damage–fracture mechanics and analytical approach is introduced in the following sections.

## 3.1. EQUIVALENT CONTINUUM MODEL FOR JOINTED ROCK MASSES

### *3.1.1 Basic principles*

The basic principle of an "equivalence continuum method" is: (a) to have an overall consideration of the influence of joints on rock mass properties based on equivalence principles; (b) to make the jointed rock mass homogenous and continuous so as to derive a set of constitutive relations and then to obtain mechanical properties of the jointed rock mass by means of numerical analysis. These equivalence models are usually elastic. Elasto-plastic constitutive relations are seldom adopted.

As shown in Figure 3.1, in order to model the surrounding rock mass consisting of two joint sets, typical elements should be identified. The typical elements should be large enough to include the two joint sets and their interaction characteristics (Figure 3.2). It is desired that the size of a typical element should be sufficiently small compared with the engineering dimension. Therefore numerical analysis of this typical element can lead to understanding of the strength and deformation properties of the typical rock mass containing joints. It further leads to the development of constitutive relation of the jointed rock mass and the analysis and modelling of rock mass stability.

The modelling procedure discussed here involves several steps: (a) the typical jointed rock mass is discretised into intact rock elements and rock joint elements; (b) equivalent constitutive relations and strength-deformation properties of the

**Figure 3.1.** Jointed rock mass surrounding an opening.

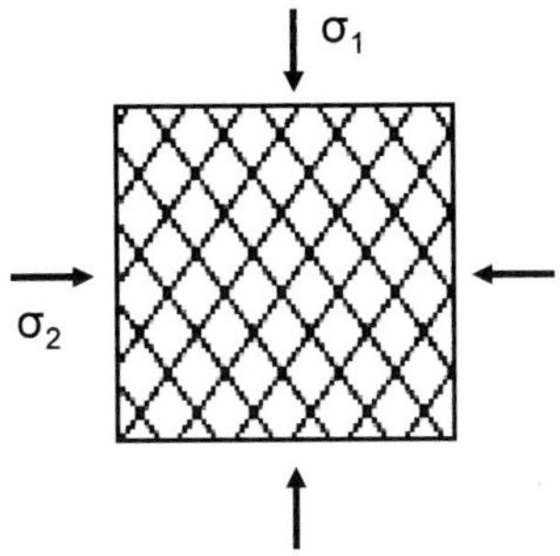


**Figure 3.2.** Typical element representing the jointed rock masses.

**Table 3.1.** Principle of the equivalent continuum model.

| | Original rock model | Equivalent continuum model |
|---|---|---|
| Geometry mode | Discontinuous | Equivalent continuous body |
| Deformability | Overall deformation comprises rock deformation and joint deformation | Deformation equivalence: deformation equal to that of original jointed model under the same loading |
| Strength characteristics | Joint strength is dominant when failure takes place along the joint plane, and rock strength is used when failure takes place in the intact rock | Strength equivalence: failure of model takes place as the original jointed model. |

jointed rock elements are established; (c) continuum modelling of the typical jointed rock mass element is performed and the overall mechanical properties of the element are obtained. This approach has the 'generality' of the equivalence method and the 'particularity' of the joint element method. The modelling method has significant applications in modelling jointed rock masses.

The principle of the equivalent continuum model is illustrated in Table 3.1. The equivalent constitutive relation is established based on deformation and strength. Both approaches are discussed in the following sections.

### *3.1.2 Deformation equivalence*

The principle of deformation equivalence approach assumes that the equivalent continuum element and the jointed rock mass element deform exactly the same under the same loading. Based on this principle, the relation of material constants between the equivalent continuum element and jointed rock mass element is derived. The equivalent continuum element can be treated as an anisotropic medium.

Assuming the joint strike is along $z$ axis, then in the plane-stress condition, the anisotropic material has a stress–strain relation of

$$\begin{pmatrix} \sigma_x \\ \sigma_y \\ \tau_{xy} \end{pmatrix} = \begin{pmatrix} c_{11} & c_{12} & c_{13} \\ c_{21} & c_{22} & c_{23} \\ c_{31} & c_{32} & c_{33} \end{pmatrix} \begin{pmatrix} \varepsilon_x \\ \varepsilon_y \\ \gamma_{xy} \end{pmatrix} \tag{3.1}$$

where $\sigma_x$, $\sigma_y$ and $\tau_{xy}$ are stresses acting on $x$, $y$ and $xy$ plane; $\varepsilon_x$, $\varepsilon_y$ and $\gamma_{xy}$ are strains in $x$, $y$ and $xy$ direction; $c_{ij}$ are elastic constants, called elastic stiffnesses.

Because $c_{ij} = c_{ji}$ $(i, j = 1, 2, 3)$, there are six independent elastic constants. Since these six parameters can be determined, elastic finite element method analysis will not be difficult. However, in situ measurement of these constants is not easy. As a jointed rock mass can be regarded as a composition of isotropic intact rock material and rock joints, the deformation parameters of the equivalent continuum medium can be obtained from the properties of the rock material and the joints, such as the elastic constants of the rock material and rock joints, or joint stiffnesses and the joint geometrical parameters (spacing, persistence, orientation, aperture and roughness) [4–7,170–197]. Methods of obtaining the equivalence are discussed in the following sections.

***3.1.2.1 Deformation equivalence with no joint dilation.*** Assume that an elastic medium is generally anisotropic. The elastic constitutive relation can be written in the form of:

$$\begin{pmatrix} \varepsilon_x \\ \varepsilon_y \\ \gamma_{xy} \end{pmatrix} = \begin{pmatrix} s_{11} & s_{12} & s_{13} \\ s_{21} & s_{22} & s_{23} \\ s_{31} & s_{32} & s_{33} \end{pmatrix} \times \begin{pmatrix} \sigma_x \\ \sigma_y \\ \tau_{xy} \end{pmatrix} \tag{3.2}$$

where $s_{ij}$ are the elastic constants, called elastic moduli.

For a rock mass containing a single joint shown in Figure 3.3, the stress on the joint plane can be obtained from the equilibrium equation:

$$\left.\begin{aligned} \sigma_n &= \sigma_x \sin^2\alpha + \sigma_y \cos^2\alpha - \tau_{xy} \sin 2\alpha \\ \tau &= \sigma_y \sin\alpha\cos\alpha - \sigma_x \sin\alpha\cos\alpha + \tau_{xy} \cos 2\alpha \end{aligned}\right\} \tag{3.3}$$

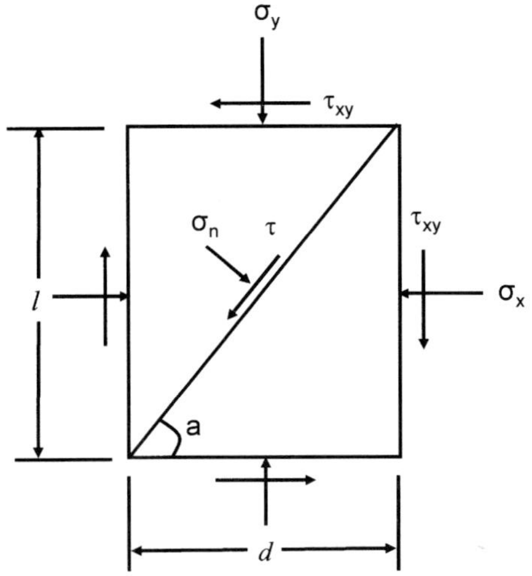


**Figure 3.3.** Mechanical analysis of a jointed rock element.

From the principle of superposition, the deformation components of the rock mass are,

$$\left.\begin{aligned}\delta_x^m &= s_{11}\sigma_x d + s_{12}\sigma_y d + s_{13}\tau_{xy} d + \frac{\sigma_n}{k_n}\sin\alpha - \frac{\tau}{k_s}\cos\alpha \\ \delta_y^m &= s_{21}\sigma_x l + s_{22}\sigma_y l + s_{23}\tau_{xy} l + \frac{\sigma_n}{k_n}\cos\alpha + \frac{\tau}{k_s}\sin\alpha\end{aligned}\right\} \tag{3.4}$$

After substituting equation (3.3) into equation (3.4) and rearranging,

$$\left.\begin{aligned}\delta_x^m = &\left(s_{11}d + \frac{1}{k_n}\sin^2\alpha\sin\alpha + \frac{1}{k_s}\cos^2\alpha\sin\alpha\right)\sigma_x \\ &+ \left(s_{12}d + \frac{1}{k_n}\cos^2\alpha\sin\alpha - \frac{1}{k_s}\cos^2\alpha\sin\alpha\right)\sigma_y \\ &\left(s_{13}d - \frac{1}{k_n}\sin 2\alpha\sin\alpha - \frac{1}{k_s}\cos 2\alpha\sin\alpha\right)\tau_{xy}\end{aligned}\right\} \tag{3.5a}$$

$$\left.\begin{aligned}\delta_y^m = &\left(s_{21}l + \frac{1}{k_n}\sin^2\alpha\cos\alpha - \frac{1}{k_s}\sin^2\alpha\cos\alpha\right)\sigma_x \\ &+ \left(s_{22}l + \frac{1}{k_n}\cos^2\alpha\cos\alpha + \frac{1}{k_s}\sin^2\alpha\cos\alpha\right)\sigma_y \\ &\left(s_{23}l - \frac{1}{k_n}\sin 2\alpha\cos\alpha - \frac{1}{k_s}\cos 2\alpha\cos\alpha\right)\tau_{xy}\end{aligned}\right\} \tag{3.5b}$$

The elastic constitutive relation of the equivalent continuum medium is:

$$\begin{pmatrix} \varepsilon_x^e \\ \varepsilon_y^e \\ \gamma_{xy}^e \end{pmatrix} = \begin{pmatrix} s_{11}^e & s_{12}^e & s_{13}^e \\ s_{21}^e & s_{22}^e & s_{23}^e \\ s_{31}^e & s_{32}^e & s_{33}^e \end{pmatrix} \times \begin{pmatrix} \sigma_x \\ \sigma_y \\ \tau_{xy} \end{pmatrix} \tag{3.6}$$

Under the same loading, the deformation of the equivalent medium is:

$$\left.\begin{aligned} \delta_x^e &= s_{11}^e \sigma_x d + s_{12}^e \sigma_y d + s_{13}^e \tau_{xy} d \\ \delta_y^e &= s_{21}^e \sigma_x l + s_{22}^e \sigma_y l + s_{23}^e \tau_{xy} l \end{aligned}\right\} \tag{3.7}$$

Following the deformation equivalence principle, i.e. $\delta_x^m = \delta_x^e$, $\delta_y^m = \delta_y^e$ then

$$\left.\begin{aligned} s_{11}^e &= s_{11} + \left(\frac{1}{k_n d}\sin^2\alpha + \frac{1}{k_s d}\cos^2\alpha\right)\sin\alpha \\ s_{12}^e &= s_{12} + \left(\frac{1}{k_n d}\cos^2\alpha - \frac{1}{k_s d}\cos^2\alpha\right)\sin\alpha \\ s_{13}^e &= s_{13} - \left(\frac{1}{k_s d}\cos 2\alpha\cos\alpha + \frac{1}{k_n d}\sin 2\alpha\sin\alpha\right) \\ s_{21}^e &= s_{21} + \left(\frac{1}{k_n l}\sin^2\alpha - \frac{1}{k_s l}\sin^2\alpha\right)\cos\alpha \\ s_{22}^e &= s_{22} + \left(\frac{1}{k_n l}\cos^2\alpha + \frac{1}{k_s l}\sin^2\alpha\right)\cos\alpha \\ s_{23}^e &= s_{23} + \left(\frac{1}{k_s l}\cos 2\alpha\sin\alpha - \frac{1}{k_n l}\sin 2\alpha\cos\alpha\right) \end{aligned}\right\} \tag{3.8}$$

From Figure 3.3, $l = d\tan\alpha$ and hence $s_{12}^e = s_{21}^e$. From the symmetry of the constitutive relation, it is easy to see that $s_{31}^e = s_{13}^e$ and $s_{32}^e = s_{23}^e$. However, $s_{33}$ is very difficult to derive. To simplify the analysis, it can be assumed that $s_{33}^e = s_{33}$. Therefore, all six parameters needed for building the constitutive relations are obtained.

For rock mass element with two joints, $s_{ij}^e$ of the rock mass element containing one joint can be obtained from equation (3.8). By replacing $s_{ij}$ with $s_{ij}^e$ in equation (3.8) and repeating the analysis for the second joint, $s_{ij}^e$ and the constitutive relation of the equivalent medium with two joints can be obtained. As for the case of multiple joints, the method is similar by repeating the above procedure.

***3.1.2.2 Deformation equivalence with joint dilation.*** Let the joint dilation (normal displacement caused by shearing) be $\delta_d$, the dilation angle be $i$, then $\delta_d = \tau/(k_s \tan i)$, by considering dilation, equation (3.4) becomes

$$\left.\begin{aligned} \delta_x^m &= s_{11}d\sigma_x + s_{12}d\sigma_y + s_{13}d\tau_{xy} + \frac{\sigma_n}{k_n}\sin\alpha - \frac{\tau}{k_s}\cos\alpha - \frac{\tau}{k_s}\tan i \sin\alpha \\ \delta_y^m &= s_{21}l\sigma_x + s_{22}l\sigma_y + s_{23}l\tau_{xy} + \frac{\sigma_n}{k_n}\cos\alpha + \frac{\tau}{k_s}\sin\alpha - \frac{\tau}{k_s}\tan i \cos\alpha \end{aligned}\right\} \tag{3.9}$$

Let the two terms on the right-hand side of each of the equations (3.9) be:

$$\left.\begin{aligned} -\frac{\tau}{k_s}\cos\alpha - \frac{\tau}{k_s}\tan i \sin\alpha &= -\frac{\tau}{k'_s}\cos\alpha \\ \frac{\tau}{k_s}\sin\alpha - \frac{\tau}{k_s}\tan i \cos\alpha &= -\frac{\tau}{k''_s}\sin\alpha \end{aligned}\right\} \tag{3.10}$$

where

$$\left.\begin{aligned} k'_s &= k_s/(1 + \tan i \tan\alpha) \\ k''_s &= k_s/(1 - \tan i \ \mathrm{cotan}\alpha) \end{aligned}\right\}$$

By replacing $k_s$ with $k'_s$ in the first three equations in equation (3.8), and replacing $k_s$ with $k''_s$ in the last three equations of equation (3.8), then $s_{11}^e$, $s_{12}^e$, $s_{13}^e$, $s_{21}^e$, $s_{22}^e$ and $s_{23}^e$ can be obtained. Similarly, as an approximation, $s_{33}^e = s_{33}$, the deformation equivalence formula with joint dilation is then established.

### *3.1.3 Formula of strength equivalence*

The strength of jointed rock mass is governed by the strengths of rock material and of rock joint. The jointed rock mass may undergo two types of failure: the failure of the rock material and the failure of the rock joint.

***3.1.3.1 Strength equivalence in the case of a single joint.*** Assuming that the strength of the intact rock, rock joint and the equivalent rock mass element follow the Mohr–Coulomb criterion and the parameters are ($c_r$, $\phi_r$), ($c_j$, $\phi_j$), ($c_e$, $\phi_e$), respectively. The strength conditions of the rock material element and equivalent continuum rock mass element are:

$$\frac{\sigma_1 - \sigma_3}{2} - \frac{\sigma_1 + \sigma_3}{2}\sin\varphi_r = c_r\cos\varphi_r \tag{3.11}$$

$$\frac{\sigma_1 - \sigma_3}{2} - \frac{\sigma_1 + \sigma_3}{2}\sin\varphi_e = c_e\cos\varphi_e \tag{3.12}$$

and it is evident that

$$\left.\begin{aligned} c_e &= c_r \\ \varphi_e &= \varphi_r \end{aligned}\right\} \tag{3.13}$$

For jointed rock mass element, let $\beta$ be the angle between the joint planes and the plane of the major principal stress, as shown in Figure 3.4.

When $\beta < \beta_{min}$ or $\beta > \beta_{max}$, the strength of the jointed rock mass is dominated by the strength of the rock material. The strength of the jointed rock mass element, in this case, is the same as equations (3.13). When $\beta_{min} \le \beta \le \beta_{max}$, the failure of jointed rock mass element takes place along the joint plane, the strength is governed by:

$$\frac{\sigma_1 - \sigma_3}{2}\sin(2\beta - \varphi_j) - \frac{\sigma_1 + \sigma_3}{2}\sin\varphi_j = c_j\cos\varphi_j \tag{3.14}$$

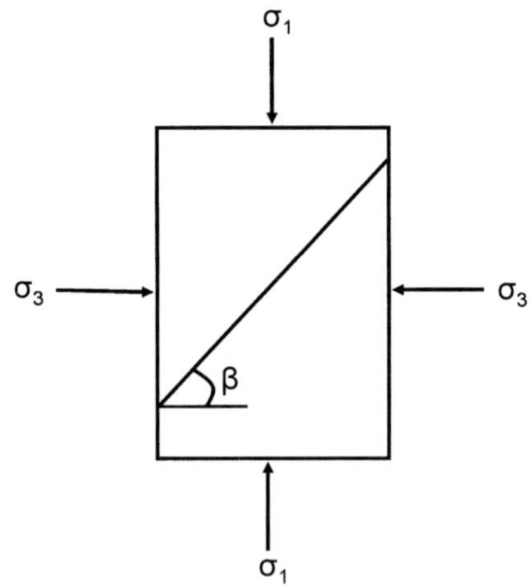


**Figure 3.4.** Jointed rock element containing single joint.

when $\beta_{min} \leq \beta \leq \beta_{max}$, by loading the jointed rock mass element, the strength envelope ($c$ and $\phi$ ) can be obtained. If this strength envelope is regarded as the envelope of the equivalent medium, then $c_e$ and $\phi_e$ can be obtained. Rewriting equation (3.11) in the following form:

$$\frac{\sigma_1 - \sigma_3}{2} - \frac{\sigma_1 + \sigma_3}{2} \frac{\sin \varphi_j}{\sin(2\beta - \varphi_j)} = c_j \frac{\cos \varphi_j}{\sin(2\beta - \varphi_j)},$$

and compared to equation (3.12), leading to:

$$\left.\begin{aligned} \varphi_e &= \sin^{-1} \frac{\sin \varphi_j}{\sin(2\beta - \varphi_j)} \\ c_e &= c_j \frac{\cos \varphi_j}{\sin(2\beta - \varphi_j) \cos \varphi_e} \end{aligned}\right\} \tag{3.15}$$

Unlike the intact rock material, when the failure of jointed rock mass is governed by the joint plane, the strength of jointed rock mass is not constant but varies with joint inclination ($\beta$).

***3.1.3.2 Strength equivalence in the case of two joint planes.*** Figure 3.5 shows a jointed rock mass element containing two joints. The failure of the rock mass element is dependent on the geometric combination of the two joints, the stress distribution and the shear strength of the joints [187–207]. Failure usually takes place

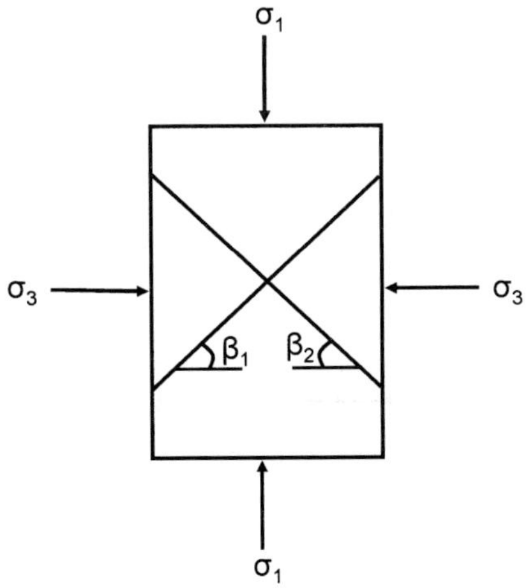


**Figure 3.5.** Jointed rock element containing two joints.

along one of the joint planes. The equivalent strength is therefore dependent on the strength of that joint plane where failure occurs.

Let the strength criterion of the joint plane be written in another form:

$$\sigma_1 = \frac{2c_j + 2\tan\varphi_j\sigma_3}{(1 - \tan\varphi_j \text{cotan}\beta)\sin 2\beta} + \sigma_3 \tag{3.16}$$

Let $\beta_1$ and $\beta_2$ be the inclination of the joint plane 1 and the joint plane 2, and both $\beta_1$ and $\beta_2$ meet the condition of $\beta_{min} \leq \beta \leq \beta_{max}$. In addition, let

$$A_1 = \frac{1}{(1 - \tan\varphi_j \text{cotan}\beta_1)\sin 2\beta_1},$$

$$A_2 = \frac{1}{(1 - \tan\varphi_j \text{cotan}\beta_2)\sin 2\beta_2}.$$

When $A_1 > A_2$, joint 1 will be the dominant plane governing the strength of the rock mass. Similarly when $A_2 > A_1$, joint 2 will be the dominant plane governing the strength. It is easy to obtain the equivalent strength following the method similar to equation (3.15).

For a rock mass element containing three joints, the parameters of equivalent rock mass strength can be determined in a similar way. Hoek and Brown [194] pointed out that the strength and deformation properties of a rock mass element containing four joint sets or more can be considered to be isotropic.

***3.1.3.3 Analysis of tensile failure.*** When a jointed rock mass element is subjected to tension, its tensile strength, to a large extent, depends on the tensile strength of joints. Since the rock joint is normally considered having zero tensile strength, the tensile strength of the jointed rock mass is usually taken as zero [186,194–196]. Therefore, non-tension analysis should be adopted in this circumstance.

### *3.1.4 Treatment of elements with non-persistent joint*

In a jointed rock mass, the joints are often not persistent throughout the joint planes [175,178,197]. Therefore, relevant modification should be made on the fundamental formula of deformation and strength equivalence. For simplicity, let us define the joint projection length ratios in the $x$- and $y$-directions, respectively as:

$$R_x = \frac{L_{jx}}{L_x} \quad \text{and} \quad R_y = \frac{L_{jy}}{L_y}, \tag{3.17}$$

where $L_{jx}$ and $L_{jy}$ are the joint projection lengths in the $x$- and $y$-directions. $L_x$ and $L_y$ are the element sizes in the $x$- and $y$-directions. It is obvious that deformation of elements is positively correlative to $R_x$ and $R_y$. Thus, the following modification should be made for equation (3.8): when $0 \leq \alpha \leq 45°$, the terms containing $K_n$ and $K_s$ are multiplied by $R_y$ and $R_x$; when $45° < \alpha \leq 90°$, the terms with $K_n$ and $K_s$ are multiplied by $R_x$ and $R_y$.

The modified formula is verified in terms of joint length, joint inclination and scale effect. First, deformation $\delta_1$ (or strain $\varepsilon_1$) in the direction of $\sigma_1$ can be calculated using finite element method (e.g., Goodman's joint element). With $\delta_1$ as the equivalent deformation, elastic constant $s_{11} = \varepsilon_1/\sigma_1$ can be obtained, and $D_{11}$ is defined as $s_{11}^r/s_{11}$. Alternatively, $s_{11}$ and $D_{11}$ can be derived from the modified equation (3.8) according to the joint distribution. The values obtained from FEM modelling and from equation (3.8) can be compared. The relative error of the value of $D_{11}$ can be estimated by treating the FEM results as the exact solutions.

***3.1.4.1 Variation of joint length.*** Assuming that the jointed rock element is square in shape at the XOY plane with a size of 5 × 5 cm, joint inclination is at 45°, Young's modulus of material is 68 MPa and Poisson's ratio of material is 0.25. The joint has a normal stiffness of 75.0 MPa/cm and a shear stiffness of 5.0 MPa/cm. Six cases, as illustrated in Figure 3.6, are modelled and discussed. The modelling results are summarised in Table 3.2.

***3.1.4.2 Variation of joint inclination.*** As shown in Figure 3.7, three models are studied to examine the effects of joint orientation. In the models, a joint with a length

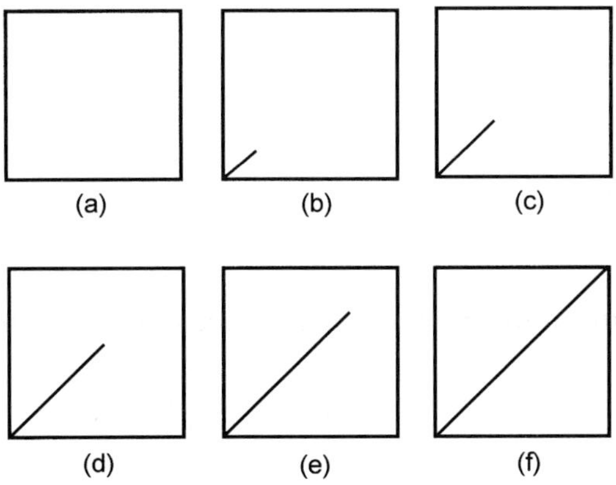


**Figure 3.6.** Verification of models with different joint persistence.

**Table 3.2.** Influence of joint length distribution on elastic constant $D_{11}(D_{11} = s^r_{11}/s_{11})$.

| Case | $R_x$ | $R_y$ | $s^r_{11}/s_{11}$ (Analytical) | $s^r_{11}/s_{11}$ (FEM) | Relative error (%) |
|---|---|---|---|---|---|
| a | 0.0 | 0.0 | 1.00 | 1.00 | 0 |
| b | 0.2 | 0.2 | 0.83 | 0.94 | 12 |
| c | 0.4 | 0.4 | 0.71 | 0.83 | 14 |
| d | 0.6 | 0.6 | 0.62 | 0.72 | 14 |
| e | 0.8 | 0.8 | 0.55 | 0.62 | 11 |
| f | 1.0 | 1.0 | 0.49 | 0.49 | 0 |

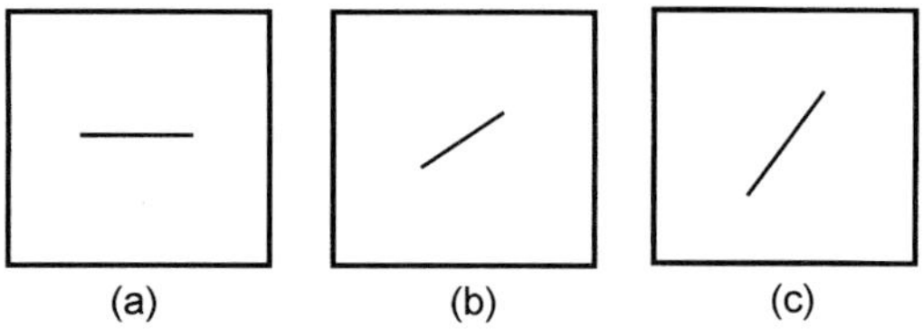


**Figure 3.7.** Verification models with different joint inclination.

**Table 3.3.** Influence of joint angle on elastic constant $D_{11}(D_{11} = s^r_{11}/s_{11})$.

| Case | $R_x$ | $R_y$ | $s^r_{11}/s_{11}$ (Analytical) | $s^r_{11}/s_{11}$ (FEM) | Relative error (%) |
|---|---|---|---|---|---|
| a | 0.2 | 0.0 | 0.96 | 0.90 | 7 |
| b | 0.2 | 0.2 | 0.83 | 0.88 | 6 |
| c | 0.2 | 0.3 | 0.81 | 0.93 | 13 |

of 1.0 cm at the centre of the element dips at 0, 45 and 71.56 degrees, respectively. Results obtained from the three models are given in Table 3.3.

***3.1.4.3 Scale effect of element.*** Scale effects are studied using the jointed element models shown in Figure 3.6(d), with the joint dip angle at 45°, $R_x$ and $R_y$ at 0.6. Six different model sizes are considered, and the results are presented in Table 3.4.

The results show that when the jointed rock element size is greater than 10 cm, the error decreases to 5% or less. In discretisation of the jointed rock mass in engineering practice, the size of the finite elements is generally much greater than 10 cm. Therefore, the treatment of the deformation equivalence using modified equation (3.8) is reasonably accurate and easy to be realised in modelling.

**Table 3.4.** Scale effect on elastic constant $D_{11}(D_{11} = s^r_{11}/s_{11})(\alpha = 45^\circ,\ R_x = R_y = 0.6)$.

| Joint length (cm) | $s^r_{11}/s_{11}$ (Analytical) | $s^r_{11}/s_{11}$ (FEM) | Relative error (%) |
|---|---|---|---|
| 5 | 0.62 | 0.72 | 14 |
| 10 | 0.77 | 0.81 | 5 |
| 20 | 0.87 | 0.88 | 1 |
| 30 | 0.90 | 0.91 | 1 |
| 40 | 0.93 | 0.93 | 0 |
| 50 | 0.94 | 0.94 | 0 |

### *3.1.5 Verification of the numerical model by physical modelling*

The overall mechanical properties of jointed rock masses can be reflected when the proposed jointed rock elements are incorporated as the basic elements into the FEM analysis. Verifications have been performed through correlation with physical modelling [8,9].

The size of the physical model is 50 × 50 × 7 cm, containing two sets of orthogonal joints with the same spacing and length of 5 cm. Figure 3.8 shows three types of joint arrangements and the FEM meshes.

Due to the complexity of the physical model test, comparisons are only made on strength and deformation properties under the uniaxial stress state for all the three types of joint patterns (Figure 3.8a, b and c) and the properties under the biaxial stress state for one joint pattern (Figure 3.8a). The comparisons are shown in Figures 3.9 and 3.10. In Figure 3.9, $\varepsilon_1^*$ and $\varepsilon_2^*$ are the strains in the direction of $\sigma_1$ and $\sigma_2$ under uniaxial loading, i.e., $\sigma_2 = 0$. In Figure 3.10, $\varepsilon_1^*$ and $\varepsilon_2^*$ are the strains under biaxial loading, i.e., when $\sigma_1 = \sigma_2$. Figures 3.9 and 3.10 indicate that the stress–strain relations obtained from the physical and the numerical modelling under the uniaxial stress state match well. In the biaxial stress state, the major principal stress–strain relations, peak strengths and strains match well between the physical and the numerical modelling results. But the major principal stress–lateral strain relations dismatch. The reason may be that the boundary friction in physical modelling leads to smaller measured deformations than the expected values.

### *3.1.6 Examples of engineering applications*

***3.1.6.1 Prediction of strength of jointed rock mass.*** The above method is used to model the strength of the rock mass at the Ertan hydropower cavern project in China. The rock masses at the Ertan project site have three major joint sets, two of which dip steeply and the other dips gently. The two orthogonal main joint sets (Figure 3.11) are considered in the modelling.

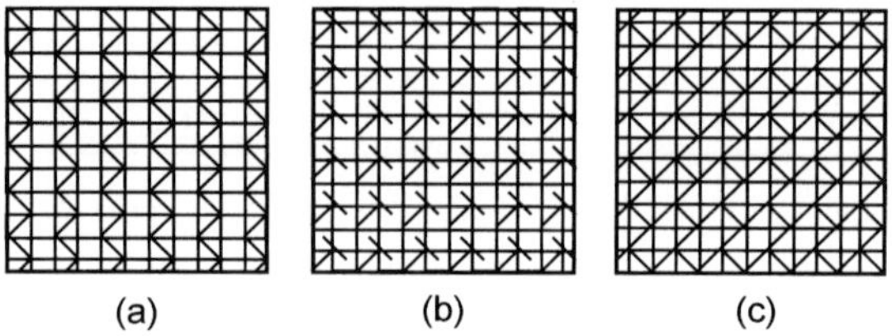


**Figure 3.8.** Models of different joint arrangement and FEM mesh.

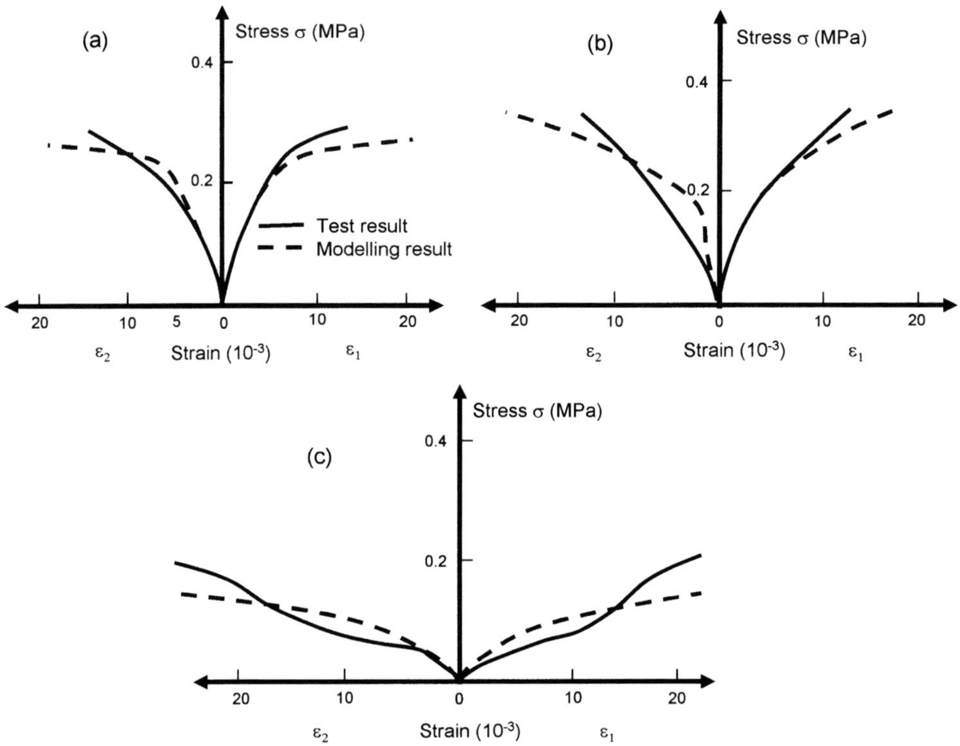


**Figure 3.9.** Comparison of σ–ε relations of the physical and the numerical models under uniaxial loading.

The measured length of the joints is between 4 and 7 m, and is taken as 5 m in average in the modelling, with a joint spacing of 1 m, based on the geological investigation. The joint persistence is 50% and 30%, respectively, for the two joint sets. The model dimension is 17.8 × 17.8 m, which satisfies the requirement that each boundary of the rock mass model contains about 10 joints, in order to represent

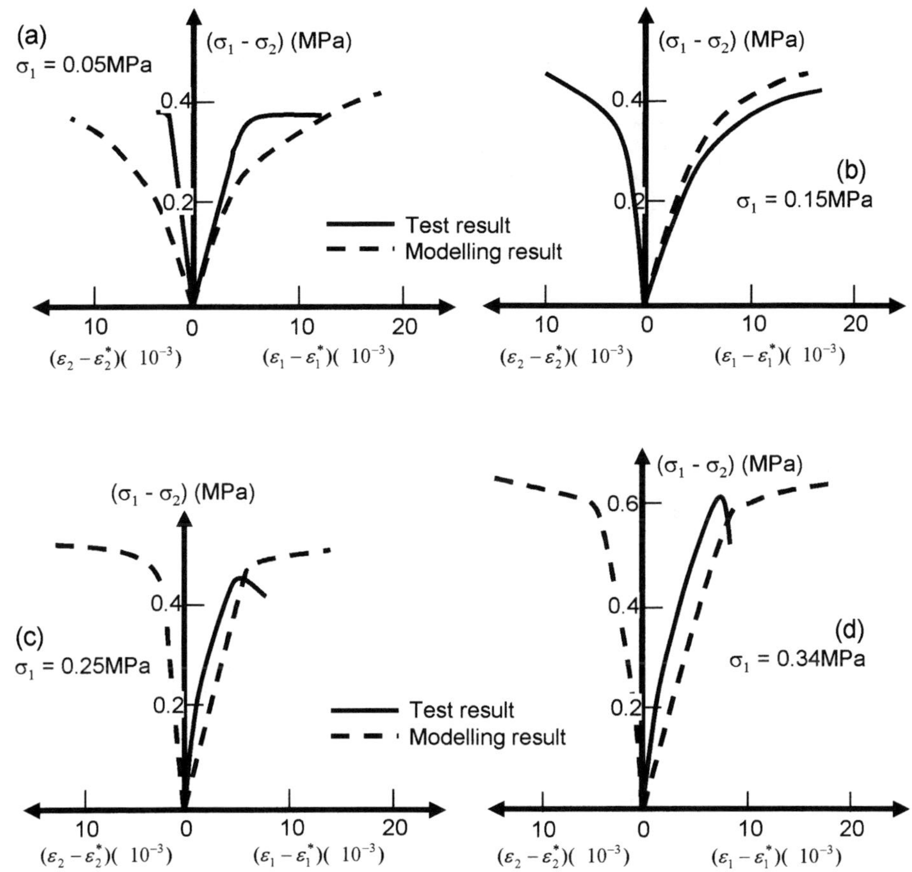


**Figure 3.10.** Comparison of σ–ε relations of the physical and the numerical models under biaxial loading.

the rock mass adequately. The distribution of joints and the FEM meshes are shown in Figure 3.11. There are 625 elements, 242 of which are jointed rock elements. The number of nodes is 676. The joint arrangement belongs to the type shown in Figure 3.8a. The mechanical parameters of rock and joints are obtained from the laboratory tests as given in Table 3.5.

In the study, only strength equivalence is considered. FEM analysis involves three types of elements: intact rock element, and two types of jointed rock elements, as shown in Figure 3.11. The intact rock element has a size of 71.2 × 71.2 cm. The equivalence of strength property is automatically treated by the program.

Loading simulated by the numerical modelling gives the yield values of $\sigma_1$ with respect to different lateral pressures $\sigma_2$. Strength parameters of the rock masses are

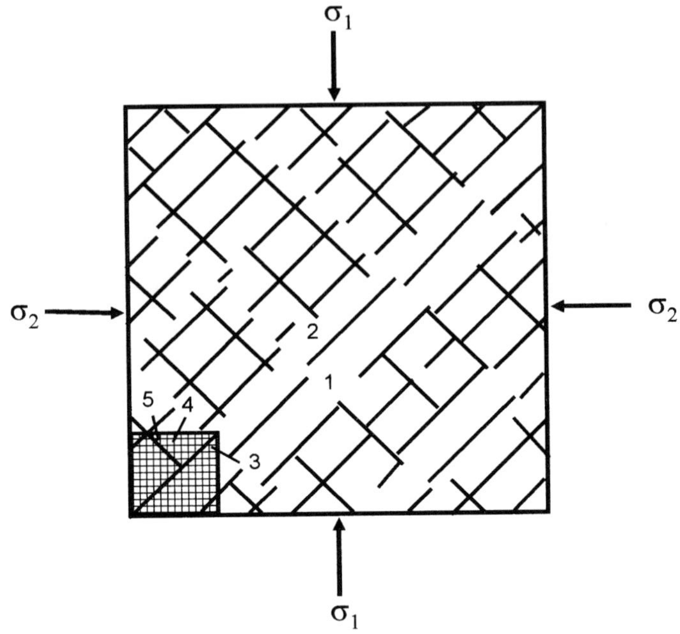


**Figure 3.11.** Distribution of joints and meshing of jointed rock.

**Table 3.5.** Parameters of rock material and rock joints.

| Material parameters | $E$ (MPa) | ν | $c$ (MPa) | φ (°) |
|---|---|---|---|---|
| Intact rock | 35,000 | 0.30 | 14.58 | 65.20 |
| Joint plane | – | – | 0.5 | 36.89 |

obtained through back analysis, using the Drucker–Prager criterion [208,209]. Results are summarised in Figure 3.12, showing the strength envelopes of the rock masses.

***3.1.6.2 Stability analysis of jointed rock masses.*** Numerical modelling is performed to study the stability of the jointed rock mass surrounding a large opening 400 m below the ground surface. The surrounding rock mass is granite. The separation between the powerhouse and the transformer chamber, the cavern dimension and the excavation sequence are shown in Figure 3.13. During the excavation, deformation measurement with extensometers was taken on several sections of the caverns. The distribution of two major joint sets and three faults is shown in Figure 3.14.

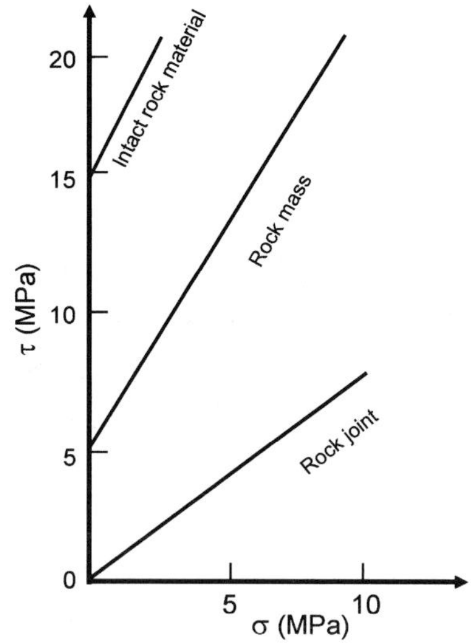


**Figure 3.12.** Rock mass strength envelope obtained from the modelling.

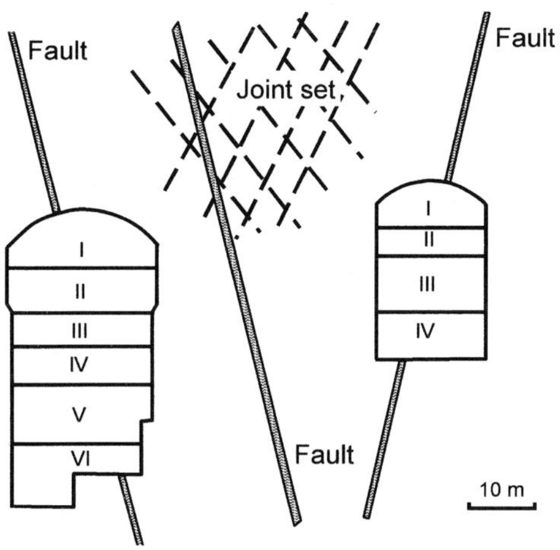


**Figure 3.13.** Layout of the powerhouse, the transformer chamber and excavation sequence.

Deformation properties of the jointed rock mass and the constitutive relations are obtained by the deformation equivalence method. In modelling the jointed rock element is generally assumed anisotropic and the joints have zero aperture and are evenly spaced across the model. The persistence of the joints is taken as 50% based on site investigation. Constitutive relation of the rock mass with two joint sets is derived with the method stated in Section 3.1.2. Mechanical parameters are listed in Table 3.6. The in situ stresses are $\sigma_x = 10.19$ MPa, $\sigma_y = 8.51$ MPa, and $\tau_{xy} = 1.12$ MPa.

The size of model was 525 × 320 m, containing 654 elements and 630 nodes. The initial stresses were obtained from back analysis of the displacement data of the fourth excavation stage of the powerhouse and the data of the second excavation stage of the transformer chamber.

Using the above mechanical parameters and stress conditions, non-linear FEM modelling has been performed on the stability of the caverns. The Drucker–Prager criterion is adopted. The extent of failure zone around the caverns and displacements are obtained as shown in Figure 3.14. Table 3.7 gives the comparison of the displacements obtained by modelling of the jointed rock mass and by elastic analysis of the rock without considering joints.

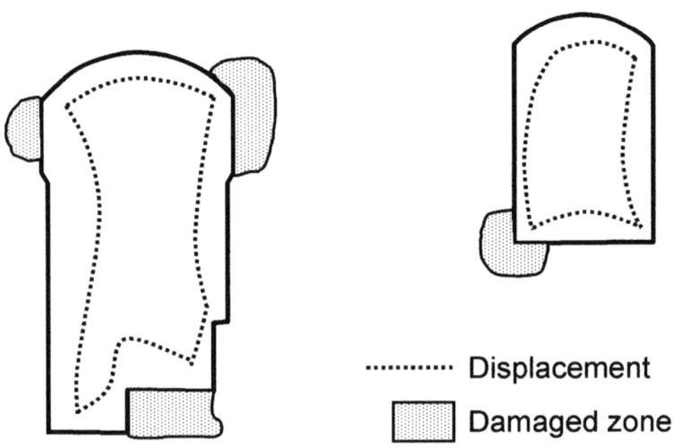


**Figure 3.14.** Over-stressed areas of the surrounding rock mass.

**Table 3.6.** Mechanical parameters of the rock material and rock joints.

| Materials | $E$ (MPA) | $\nu$ | $c$ (MPa) | $\phi$ (°) | $\sigma_1$(MPa) | $K_n$ (MPa/cm) | $K_s$ (MPA/cm) |
|---|---|---|---|---|---|---|---|
| Intact rock | 37,000 | 0.20 | 22.27 | 48.1 | 5.0 | – | – |
| Rock joint | – | – | 0.5 | 35 | 0.0 | 600,000 | 7500 |

**Table 3.7.** Predicted displacement convergence values (mm).

| Measurement | 1 | 2 | 3 | 4 | 5 | 6 | 7 | 8 | 9 |
|---|---|---|---|---|---|---|---|---|---|
| Modelling of jointed rock mass | 11.4 | 18.5 | 19.5 | 6.0 | 6.9 | 7.3 | 10.8 | 7.4 | 6.9 |
| Elastic analysis of non-jointed rock mass | 9.7 | 15.5 | 16.3 | 5.0 | 5.8 | 6.1 | 9.0 | 6.3 | 5.8 |

## 3.2. EQUIVALENT ANALYSIS FOR ROCK MASSES CONTAINING THICK JOINTS

In this section an equivalent continuous model is proposed for rock mass consisting of thick joint sets. The method is based on strength analysis with the same principle discussed in Section 3.1.

### *3.2.1 Equivalent deformation principle and method*

When a rock mass is cut by two joint sets in direction $\alpha_1$ and $\alpha_2$ respectively, and the rock material is homogeneous, then there are two parameters describing the rock material ($E_r$ and $\sigma_r$) and four parameters describing each joint set ($E_j$, $\sigma_j$, $\beta$ and $\alpha$). $E$ is the elastic modulus, $\sigma$ is the strength, $\beta$ is the ratio of joint aperture to rock mass width and $\alpha$ is the orientation of joint set. Only the basic formulae for plane condition are given here, which are also illustrated in Figure 3.34. In the process, the effect of one joint set is taken into account first to produce an 'equivalent medium'. The other joint set is subsequently taken into consideration to derive the equivalent continuous medium of the rock mass containing two joint sets. As shown in Figure 3.34, considering the first joint set, the joint stress–strain relation is as follows:

$$\{\sigma\}_j = [\sigma_{xj} \quad \sigma_{yj} \quad \tau_{xyj}]^T \qquad \{\varepsilon\}_j = [\varepsilon_{xj} \quad \varepsilon_{yj} \quad \gamma_{xyj}]^T \tag{3.18}$$

The stress–strain relation of the rock material is

$$\{\sigma\}_R = [\sigma_{xR} \quad \sigma_{yR} \quad \tau_{xyR}]^T \qquad \{\varepsilon\}_R = [\varepsilon_{xR} \quad \varepsilon_{yR} \quad \gamma_{xyR}]^T \tag{3.19}$$

If the stress–strain state of the above combined joints and material is replaced by the stress–strain state of the jointed rock mass as shown in Figure 3.15, namely

$$\{\sigma\}_1 = [\sigma_{x1} \quad \sigma_{y1} \quad \tau_{xy1}]^T \qquad \{\varepsilon\}_1 = [\varepsilon_{x1} \quad \varepsilon_{y1} \quad \gamma_{xy1}]^T \tag{3.20}$$

then they are related by the following equations:

#### *3.2.1.1 Equilibrium condition.*

$$\left.\begin{aligned} \sigma_{x1} &= (1-\beta_1)\sigma_{xR} + \beta_1\sigma_{xj} \\ \sigma_{y1} &= \sigma_{yR} = \sigma_{yj} \\ \sigma_{xy1} &= \tau_{xyR} = \tau_{xyj} \end{aligned}\right\} \tag{3.21}$$

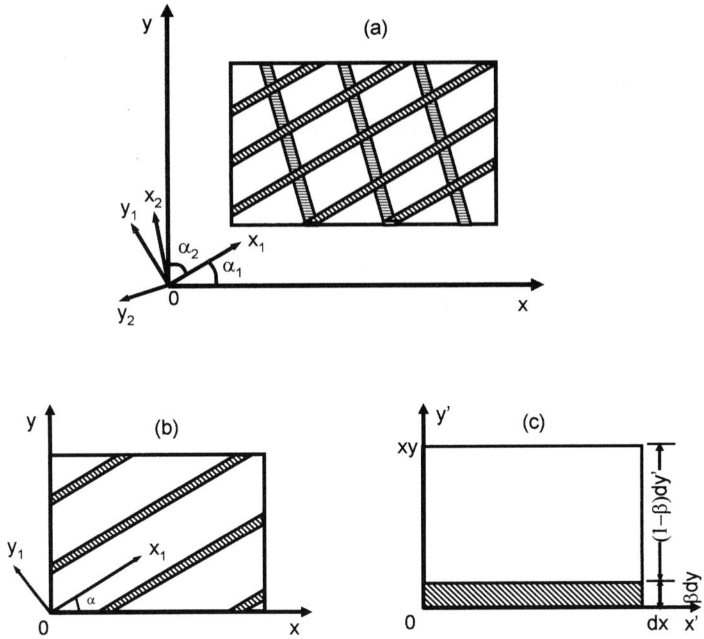


**Figure 3.15.** Representation of rock joints and rock material.

### *3.2.1.2 Displacement compatibility condition.*

$$\left.\begin{aligned} \varepsilon_{x1} &= \varepsilon_{xR} = \varepsilon_{xj} \\ \varepsilon_{y1} &= (1-\beta_1)\varepsilon_{yR} + \beta_1\varepsilon_{yj} \\ \gamma_{xy1} &= (1-\beta_1)\gamma_{xyR} + \beta_1\gamma_{xyR} \end{aligned}\right\} \tag{3.22}$$

***3.2.1.3 Physical equations.*** Suppose that both the rock material and joint material are homogeneous, isotropic and elastic, then for joint

$$\left\{\begin{aligned} \sigma_{xj} &= a_{j1}\varepsilon_{xj} + b_{j1}\varepsilon_{yj} \\ \sigma_{yj} &= b_{j1}\varepsilon_{xj} + c_{j1}\varepsilon_{yj} \\ \tau_{xyj} &= d_{j1}\gamma_{xyj} \end{aligned}\right. \tag{3.23}$$

for rock material:

$$\left\{\begin{aligned} \sigma_{xR} &= a_R\varepsilon_{xR} + b_R\varepsilon_{yR} \\ \sigma_{yR} &= b_R\varepsilon_{xR} + c_R\varepsilon_{yR} \\ \tau_{xyR} &= d_R\gamma_{xyR} \end{aligned}\right. \tag{3.24}$$

where $a_R$, $b_R$, $c_R$, $d_R$ are defined by $E_r$ and $\sigma_r$, the elastic constants of rock material. $a_{j1}$, $b_{j1}$, $c_{j1}$, $d_{j1}$ are defined by $E_{j1}$ and $\mu_{j1}$, the elastic constants of joint material. In plane strain condition,

$$\begin{cases} a = c = E(1-\mu)/(1+\mu)(1-2\mu) \\ b = E\mu/(1+\mu)(1-2\mu) \\ d = E/2(1+\mu) \end{cases} \tag{3.25}$$

Based on further derivation, the relation between $\{\sigma\}_1$ and $\{\varepsilon\}_1$ in $X_1OY_1$ coordinate system can be obtained

$$\{\sigma\}_1 = [D]_{0j}[H]_1^{-1}\{\varepsilon\}_1 = [\overline{D}]_1\{\varepsilon\}_1 \tag{3.26}$$

where

$$[H]_1 = \begin{bmatrix} 1 & 0 & 0 \\ \beta_1 \dfrac{b_R - b_{j1}}{c_{j1}} & 1 - \beta_1 + \beta_1 \dfrac{c_R}{c_{j1}} & 0 \\ 0 & 0 & 1 - \beta_1 + \beta_1 \dfrac{d_R}{d_{j1}} \end{bmatrix} \tag{3.27}$$

$$[D]_{0j} = \begin{bmatrix} (1-\beta_1)a_R + \beta_1 a_{j1} + \beta_1 \dfrac{b_{j1}}{c_{j1}}(b_R - b_{j1}) & (1-\beta_1)b_R + \beta_1 \dfrac{b_{j1}}{c_{j1}} c_R & 0 \\ b_R & c_R & 0 \\ 0 & 0 & d_R \end{bmatrix} \tag{3.28}$$

In the $X_2OY_2$ coordinate system, the elastic matrix of the jointed rock mass is

$$[D]_1 = [T]_1[\overline{D}]_1[T]_1^T \tag{3.29}$$

where

$$[T]_1 = \begin{bmatrix} \cos^2(\alpha_1 - \alpha_2) & \sin^2(\alpha_1 - \alpha_2) & -\sin^2(\alpha_1 - \alpha_2) \\ \sin^2(\alpha_1 - \alpha_2) & \cos^2(\alpha_1 - \alpha_2) & \sin^2(\alpha_1 - \alpha_2) \\ \frac{1}{2}\sin 2(\alpha_1 - \alpha_2) & -\frac{1}{2}\sin 2(\alpha_1 - \alpha_2) & \cos 2(\alpha_1 - \alpha_2) \end{bmatrix} \tag{3.30}$$

It is followed by the second joint set subsequently taken into consideration. Similarly, the elastic matrices of the bi-directional jointed rock mass in $XOY$

coordinate system are given as:

$$[D]_2 = [T]_2[D]_{ij}[H]_2^{-1}[T]_2^T \tag{3.31}$$

where

$$[T]_2 = \begin{bmatrix} \cos^2\alpha_2 & \sin^2\alpha_2 & -\sin^2\alpha_2 \\ \sin^2\alpha_2 & \cos^2\alpha_2 & \sin^2\alpha_2 \\ \frac{1}{2}\sin 2\alpha_2 & -\frac{1}{2}\sin 2\alpha_2 & \cos 2\alpha_2 \end{bmatrix} \tag{3.32}$$

$$[H]_2 = \begin{bmatrix} 1 & 0 & 0 \\ \beta_2\dfrac{d_{21}^{1} - b_{j2}}{c_{j2}} & 1 - \beta_2 + \beta_2\dfrac{d'_{22}}{c_{j2}} & \beta_2\dfrac{d'_{23}}{c_{j2}} \\ \beta_2\dfrac{d'_{31}}{d_{j2}} & \beta_2\dfrac{d'_{32}}{d_{j2}} & 1 - \beta_2 + \beta_2\dfrac{d'_{33}}{d_{j2}} \end{bmatrix} \tag{3.33}$$

$$[D]_{ij} = \begin{bmatrix} (1-\beta_2)d'_{11} + \beta_2 a_{j2} + \beta_2\dfrac{b_{j2}}{c_{j2}}(d'_{21} - b_{j2}) & (1-\beta_2)d'_{12} + \beta_2\dfrac{b_{j2}}{c_{j2}}d'_{22} & (1-\beta_2)d'_{13} + \beta_2\dfrac{b_{j2}}{c_{j2}}d'_{23} \\ d'_{21} & d'_{22} & d'_{23} \\ d'_{31} & d'_{32} & d'_{33} \end{bmatrix} \tag{3.34}$$

and $d'_{ij}$ $(i, j = 1, 2, \ldots, 6)$ are elements of $[D]_1$ as expressed in equation (3.29).

Generalised in the spatial condition, the general matrix for anisotropic elasticity is

$$\begin{bmatrix} \sigma_x \\ \sigma_y \\ \sigma_z \\ \tau_{xy} \\ \tau_{yz} \\ \tau_{zx} \end{bmatrix} \begin{bmatrix} d_{11} d_{12} \cdots\cdots d_{16} \\ d_{21} d_{22} \cdots\cdots d_{26} \\ \vdots \quad \vdots \\ \vdots \quad \vdots \\ \vdots \quad \vdots \\ d_{61} d_{62} \cdots\cdots d_{66} \end{bmatrix} \begin{bmatrix} \varepsilon_x \\ \varepsilon_y \\ \varepsilon_z \\ \gamma_{xy} \\ \gamma_{yz} \\ \gamma_{zx} \end{bmatrix} \tag{3.35}$$

Suppose that $[D]_{i-1}$ is the elastic matrix of rock mass in the $i$th local coordinate system $OX_iY_iZ_i$ of the rock mass cut by the ($i$–1)-th group of joint, the elements of $[D]_{i-1}$ are the coefficients ($k$, $l = 1, 2, \ldots, 6$) in the above expression.

Similarly, suppose that the rock mass is cut by the $i$th joint set and the joints are parallel to plane $OX_iY_i$, then

$$[H]_i = \begin{bmatrix} 1 & 0 & 0 & 0 & 0 & 0 \\ 0 & 1 & 0 & 0 & 0 & 0 \\ \beta_i \dfrac{d_{31}^{i-1} - b_{ij}}{c_{ij}} & \beta_i \dfrac{d_{32}^{i-1} - b_{ij}}{c_{ij}} & 1 - \beta_i + \beta_i \dfrac{d_{33}^{i-1}}{c_{ij}} & \beta_i \dfrac{d_{34}^{i-1}}{c_{ij}} & \beta_i \dfrac{d_{35}^{i-1}}{c_{ij}} & \beta_i \dfrac{d_{36}^{i-1}}{c_{ij}} \\ 0 & 0 & 0 & 1 & 0 & 0 \\ \beta_i \dfrac{d_{51}^{i-1}}{d_{ij}} & \beta_i \dfrac{d_{52}^{i-1}}{d_{ij}} & \beta_i \dfrac{d_{53}^{i-1}}{d_{ij}} & \beta_i \dfrac{d_{54}^{i-1}}{d_{ij}} & 1 - \beta_i + \beta_i \dfrac{d_{55}^{i-1}}{d_{ij}} & \beta_i \dfrac{d_{56}^{i-1}}{d_{ij}} \\ \beta_i \dfrac{d_{61}^{i-1}}{d_{ij}} & \beta_i \dfrac{d_{62}^{i-1}}{d_{ij}} & \beta_i \dfrac{d_{63}^{i-1}}{d_{ij}} & \beta_i \dfrac{d_{64}^{i-1}}{d_{ij}} & \beta_i \dfrac{d_{65}^{i-1}}{d_{ij}} & 1 - \beta_i + \beta_i \dfrac{d_{66}^{i-1}}{d_{ij}} \end{bmatrix} \tag{3.36}$$

$$[D]_{i-1j}^{T} = \begin{bmatrix} (1-\beta_i)d_{11}^{i-1} + \beta_i d_{ij} + \beta_i \dfrac{b_{ij}}{c_{ij}}(d_{31}^{i-1} - b_{ij}) & (1-\beta_i)d_{21}^{i-1} + \beta_i d_{ij} + \beta_i \dfrac{b_{ij}}{c_{ij}}(d_{31}^{i-1} - b_{ij}) & d_{31}^{i-1} & (1-\beta_i)d_{41}^{i-1} & d_{51}^{i-1} & d_{61}^{i-1} \\ (1-\beta_i)d_{12}^{i-1} + \beta_i d_{ij} + \beta_i \dfrac{b_{ij}}{c_{ij}}(d_{32}^{i-1} - b_{ij}) & (1-\beta_i)d_{22}^{i-1} + \beta_i d_{ij} + \beta_i \dfrac{b_{ij}}{c_{ij}}(d_{32}^{i-1} - b_{ij}) & d_{32}^{i-1} & (1-\beta_i)d_{42}^{i-1} & d_{52}^{i-1} & d_{62}^{i-1} \\ (1-\beta_i)d_{13}^{i-1} + \beta_i \dfrac{b_{ij}}{c_{ij}} d_{33}^{i-1} & (1-\beta_i)d_{23}^{i-1} + \beta_i \dfrac{b_{ij}}{c_{ij}} d_{33}^{i-1} & d_{33}^{i-1} & (1-\beta_i)d_{43}^{i-1} & d_{53}^{i-1} & d_{63}^{i-1} \\ (1-\beta_i)d_{14}^{i-1} + \beta_i \dfrac{b_{ij}}{c_{ij}} d_{34}^{i-1} & (1-\beta_i)d_{24}^{i-1} + \beta_i \dfrac{b_{ij}}{c_{ij}} d_{34}^{i-1} & d_{34}^{i-1} & (1-\beta_i)d_{44}^{i-1} & d_{54}^{i-1} & d_{64}^{i-1} \\ (1-\beta_i)d_{15}^{i-1} + \beta_i \dfrac{b_{ij}}{c_{ij}} d_{35}^{i-1} & (1-\beta_i)d_{25}^{i-1} + \beta_i \dfrac{b_{ij}}{c_{ij}} d_{35}^{i-1} & d_{35}^{i-1} & (1-\beta_i)d_{45}^{i-1} & d_{55}^{i-1} & d_{65}^{i-1} \\ (1-\beta_i)d_{16}^{i-1} + \beta_i \dfrac{b_{ij}}{c_{ij}} d_{36}^{i-1} & (1-\beta_i)d_{26}^{i-1} + \beta_i \dfrac{b_{ij}}{c_{ij}} d_{36}^{i-1} & d_{36}^{i-1} & (1-\beta_i)d_{46}^{i-1} & d_{56}^{i-1} & d_{66}^{i-1} \end{bmatrix} \tag{3.37}$$

Having been cut by the $i$th joint set, the rock mass has the elastic matrix

$$[\overline{D}]_i = [D]_{i-1j}[H]_i^{-1} \tag{3.38}$$

The direction cosines between system $OX_{i+1}Y_{i+1}Z_{i+1}$ and $OX_iY_iZ_i$ are as follows:

| New coordinates | Original coordinates | | |
|---|---|---|---|
| | $X_i$ | $Y_i$ | $Z_i$ |
| | direction cosines | | |
| $X_{i+1}$ | $l_{i1}$ | $m_{i1}$ | $n_{i1}$ |
| $Y_{i+1}$ | $l_{i2}$ | $m_{i2}$ | $n_{i2}$ |
| $Z_{i+1}$ | $l_{i3}$ | $m_{i3}$ | $n_{i3}$ |

In the $(i+1)$-th local system $OX_{i+1}Y_{i+1}Z_{i+1}$, the elastic matrix of the jointed rock mass is

$$[D]_i = [T]_i[\overline{D}]_i[T]_i^T \tag{3.39}$$

$$[T]_i = \begin{bmatrix} l_{i1}^2 & m_{i1}^2 & n_{i1}^2 & 2l_{i1}m_{i1} & 2m_{i1}n_{i1} & 2n_{i1}l_{i1} \\ l_{i2}^2 & m_{i2}^2 & n_{i2}^2 & 2l_{i2}m_{i2} & 2m_{i2}n_{i2} & 2n_{i2}l_{i2} \\ l_{i3}^2 & m_{i3}^2 & n_{i3}^2 & 2l_{i3}m_{i3} & 2m_{i3}n_{i3} & 2n_{i3}l_{i3} \\ l_{i1}l_{i2} & m_{i1}m_{i2} & n_{i1}n_{i2} & l_{i1}m_{i2}+l_{i2}m_{i1} & m_{i1}n_{i2}+m_{i2}n_{i1} & n_{i1}l_{i2}+n_{i2}l_{i1} \\ l_{i2}l_{i3} & m_{i2}m_{i3} & n_{i2}n_{i3} & l_{i2}m_{i3}+l_{i3}m_{i2} & m_{i2}n_{i3}+m_{i3}n_{i2} & n_{i2}l_{i3}+n_{i3}l_{i2} \\ l_{i3}l_{i1} & m_{i3}m_{i1} & n_{i3}n_{i1} & l_{i3}m_{i1}+l_{i1}m_{i3} & m_{i3}n_{i1}+m_{i1}n_{i3} & n_{i3}l_{i1}+n_{i1}n_{i3} \end{bmatrix} \tag{3.40}$$

$i = 1, 2, \ldots, 6.$

### 3.2.2 *Basic principle of strength equivalence*

Suppose that the strengths of the rock material, the joint and the equivalent continuum body all follow the Mohr–Coloumb criterion, the strengths are $c_r$, $\phi_r$, $c_j$, $\phi_i$, $c_e$, $\phi_e$, respectively. The strength of jointed rock mass element consists of the rock material strength and the joint strength. Adopting the equivalent strength method described in Section 3.1, the strength condition is

$$\frac{\sigma_1 - \sigma_2}{2} - \frac{\sigma_1 + \sigma_2}{2} \sin \varphi_e = C_e \cos \varphi_e \tag{3.41}$$

With different angles of the joint plane to the major principal stress plane $\beta$, $c_e$ and $\phi_e$ are different:
when $\beta < \beta_{\min}$ or $\beta > \beta_{\max}$,

$$C_e = C_R \qquad \varphi_e = \varphi_R \tag{3.42}$$

when $\beta_{\min} \leq \beta \leq \beta_{\max}$

$$\left.\begin{aligned} C_e &= C_j \frac{\cos \varphi_j}{\sin(2\beta - \varphi_j)} \cdot \frac{1}{\cos \varphi_e} \\ \varphi_e &= \arcsin(\sin \varphi_j / \sin(2\beta - \varphi_j)) \end{aligned}\right\} \tag{3.43}$$

If the rock mass contains more than two joint sets, the related results can be obtained by superposition. But it should be noted that the smaller value of $c$, $\phi$ should be adopted if two values exist.

For different stress states, if all the rock material and rock joint parameters are known, the strength of the rock mass can be derived by the above procedure. The equivalent rock mass strength parameters $c_e$ and $\phi_e$ can be determined on the basis of Drucker–Prager criterion as shown before.

With two stress states ($a$ and $b$) known, the equivalent rock mass strength parameters ($c_e$ and $\phi_e$) can be determined through the Drucker–Prager criterion.

$$\begin{aligned} \alpha I_1^a + \sqrt{J_2^a} &= K \\ \alpha I_1^b + \sqrt{J_2^b} &= K \end{aligned} \tag{3.44}$$

From equation (3.44), we have

$$\alpha = \frac{\sqrt{J_2^b} - \sqrt{J_2^a}}{I_1^a - I_1^b} \tag{3.45}$$

Because

$$\alpha = \frac{\sqrt{3} \sin \varphi}{3\sqrt{3 + \sin^2 \varphi}},$$

the expression below is obtained,

$$\varphi_e = \varphi = \arcsin\left(\frac{3\alpha}{\sqrt{1-3\alpha^2}}\right) \tag{3.46}$$

From

$$K = \frac{\sqrt{3}C\cos\varphi}{\sqrt{3+\sin^2\varphi}}$$

we have

$$C_e = C = \frac{\left(\sqrt{3+\sin^2\varphi}\right)K}{\sqrt{3}\cos\varphi} \tag{3.47}$$

The values of $c_e$ and $\phi_e$ for the rock mass can be obtained from equations (3.46) and (3.47).

### *3.2.3 Engineering applications*

The above equivalent analysis of jointed rock masses is applied to model a large underground cavern complex of a hydropower project, as shown in Figure 3.16. The main work includes the selection of separations between caverns, the excavation sequences and the support schemes. Two-dimensional non-linear analysis and three-dimensional anisotropic analysis are conducted.

In the FEM modelling, continuous rock mass is simulated with the commonly used isoparametric elements. Dominant faults are simulated with the isoparametric joint elements. The fractured rock mass containing joint sets is simulated with equivalent medium. In two-dimensional modelling, primarily 4-node quadrilateral elements and 3-node triangle elements are used. The number of elements is 661, and the number of nodes is 603. Six schemes for cavern separation optimisation, three schemes for excavation sequence optimisation and two schemes for shotcrete–rockbolt support optimisation are modelled. In three-dimensional modelling, 8-node brick-shaped elements and 6-node pentahedron elements are commonly used. The number of elements is 1094, and the node is 1368. Two schemes of cavern separation optimisation and two schemes of excavation and support optimisation are modelled.

A large number of modelling cases have been conducted. Some important conclusions are outlined here. The maximum displacements in plane elastic analysis

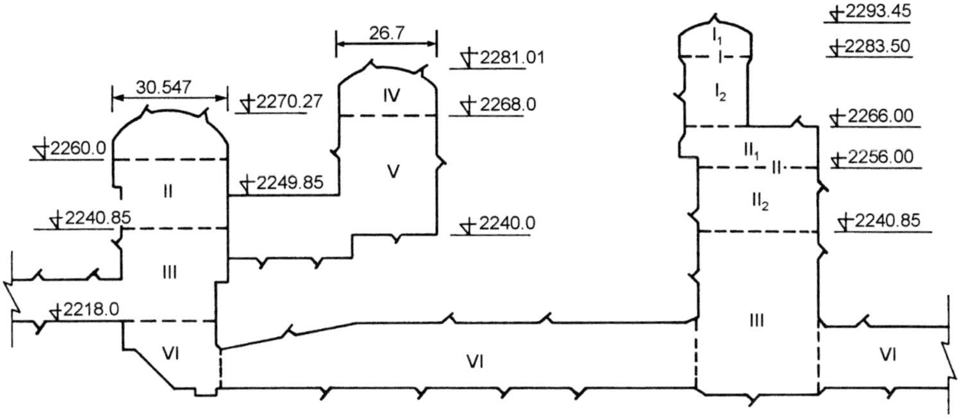


**Figure 3.16.** Systematic view of cavern complex of a hydropower station.

and in non-linear analysis occur at different locations. The former appears at the middle of the left wall in the powerhouse (Figure 3.16). In non-linear analysis the yield plastic zones between the main powerhouse and the transformer room is joined together, and the maximum displacement appears at the middle of the right wall of the powerhouse with a magnitude of about 6.7 cm. The maximum compressive stresses also appear at the vault and the floor. The maximal compressive stress at the left vault is nearly 47.0 MPa. With the increase of spacing between the caverns, the areas of yield zones significantly decrease, and eventually the yield zones become isolated. The cavern separation has significant influence on the area of yield zone.

Non-linear analysis is conducted for three schemes of excavation sequence:

Scheme 1: I → II → III → IV → V (five excavation steps)

Scheme 2: I → II → [III → / IV →] → V (four excavation steps)

Scheme 3: I → [II → / IV →] → [III → / V →] (three excavation steps)

Modelling results indicate that the maximum displacement is decreased by 32% in Scheme 1, by 35% in Scheme 2 and by 12% in Scheme 3. As for yield zone, Scheme 1 has the least yield zone compared with Schemes 2 and 3. Through comprehensive consideration, Scheme 1 is recommended for the actual construction.

In the non-linear modelling, the effects of two different bolt schemes are compared. New concept and method are developed for modelling the bolt effects. From the model tests, the results show that the main effect of the bolts is to increase the strength and the stiffness of the surrounding rock mass. Its effect to produce support reaction is much less significant. The results show that the peak strength and modulus of the bolted rock mass increase approximately by 10–20%.

However, the residual strength of the bolted rock mass increases considerably as compared with unbolted rock mass. Under the uniaxial compression, the residual strength increases by 60–80%, and is influenced by the bolt spacing and the stiffness ratio between the bolt and the rock mass. The study on bolt is discussed in detail in Chapter 8.

### 3.3. STRENGTH CHARACTERISTICS OF FRACTURED ROCK MASS UNDER COMPRESSIVE SHEAR STRESS

The natural rock masses consist of faults, joints, fractures and other discontinuities. Engineering projects often generate compressive or shearing stress on the rock mass [186,194–196,210–218]. It is important to understand the interaction of the rock discontinuities and the strength and failure mechanisms of the rock masses under compressive and shear stresses.

Joints and fractures in a rock mass generally distribute intermittently and the strength of the joint and the bridge area in the rock mass differ greatly in different rock masses [205,206]. Researchers usually pay close attention to the overall shear strength of a jointed rock mass and to the secondary crack initiation. Many researchers suggested that the compressive strength of collinear fractures and rock bridges can be determined by means of the weighted mean method [219–221]. Others believed that the different deformation characteristics of the joint and the bridge in a rock mass should be taken into consideration to assess the rock mass strength. Because of different stress distributions in the rock bridge and on the joint surfaces, the friction on the joint surfaces must be multiplied by a mobilising factor [199–200]. Tests conducted on gypsum show that the strength properties of a rock mass with multi-rowed intermittent fractures aligning along the same direction [205]. Brown [206] studied the strength of jointed rock mass by means of fracture mechanics. Horii and Nemat-Nasser [219] gave analytical solution and stress intensity factor of the secondary crack trajectory, and discussed the formation of a failure plane. Reyes and Einstein [220], through tests and damage analyses, studied the stress distribution and failure mode of a rock bridge between two fractures in en-echelon crack arrays.

Analytical models for fracturing mechanisms of the rock bridge between adjacent fractures in a bilateral compressive stress field, based upon the phenomena observed in the experimental studies are proposed and discussed in this section. The models can be used to predict the overall shear strengths of the rock mass. The analytical solutions have been verified through experiments.

### 3.3.1 *Strength of rock mass containing collinear cracks*

***3.3.1.1 Fracture propagating at tight cracks.*** Stress concentration phenomena will take place at crack tips under loading. Although the average stress of the crack is generally less than the threshold stress of crack propagating, the stresses at the tips are by far higher than the threshold stress. As a result, secondary cracks will occur near the tips. The initiating point of secondary cracks can be determined from the fracture toughness of the rock material. For collinear cracks (Figure 3.17) loaded by compound stresses of compression and shear, stress intensity factor at the crack tips can be derived as follows:

$$k_{\mathrm{I}} = 0 \tag{3.48}$$

$$k_{\mathrm{II}} = (\tau - \sigma \tan \varphi_j - c_j)\sqrt{2b \tan \frac{\pi a}{4b}} \tag{3.49}$$

Where compressive stress is defined as positive and tensile stress as negative; $c_j$, $\phi_j$ are the cohesion and the friction angle of the crack, respectively. The condition for the crack to propagate is $K_{\mathrm{II}} \geq K_{\mathrm{II}c}$ where $K_{\mathrm{II}c}$ is type II fracture toughness of the crack.

The crack propagation does not indicate that the stress reaches the rock strength. The strength is reflected by the stress that makes the cracks join together to form a failure plane. The crack initiation and the propagating path become very important

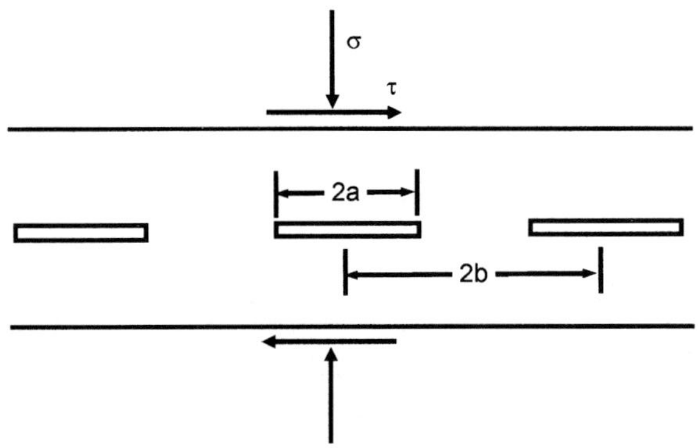


**Figure 3.17.** Collinear cracks under compressive shear stress.

for estimating the strength. As shown in Figure 3.18, during the test, rupture initiates at crack tips, it tends to propagate gradually along the direction of maximum principal stress, $\sigma_1$. When a crack propagates to a certain extent and meets the adjacent crack, rock failure occurs. Analysis on the above-observed phenomenon has shown that stress intensity factors at the tip (point A) of a crack under compressive and shear stresses are $k_I^A = 0$ and $k_{II}^A \neq 0$. When $|k_{II}^A| \geq k_{IIc}$, the crack rupture initiates at point A with an initial rupture angle of approximately 70° (the angle between initial rupture line and line of the crack). However, with the increment of external loading, the direction of the secondary crack gradually turns to the direction of $\sigma_1$. The secondary crack path with direction of turning is schematically described in Figure 3.19. The cracking starts along $\overline{A}\,\overline{B}$. When the crack propagates to point B, the shear stress between $\overline{B}\,\overline{C}$ exceeds the maximum shear strength of the rock material, resulting in shearing failure on plane $BC$ to form crack $\overline{B}\,\overline{C}$. Therefore, the fracture ABC becomes the final fracture plane.

***3.3.1.2 Determination of shearing strength of crack body.*** At present, the commonly used method to analyse the strength of a jointed rock mass takes into consideration the joint persistence $\eta$, the shear strength parameters of the crack, $c_f$ and $f_f$, the shear parameter of rock bridge material, $c_r$ and $f_r$ [221].

$$\left.\begin{aligned} \bar{f} &= \eta f_j + (1-\eta) f_r \\ \bar{c} &= \eta c_j + (1-\eta) c_r \end{aligned}\right\} \tag{3.50}$$

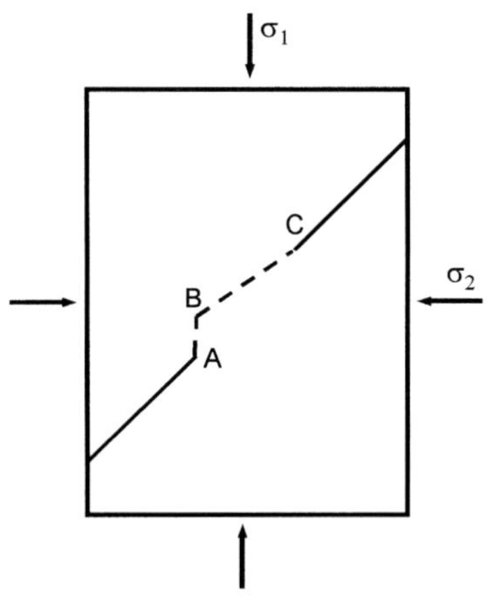


**Figure 3.18.** Fracture through bridged rock materials.

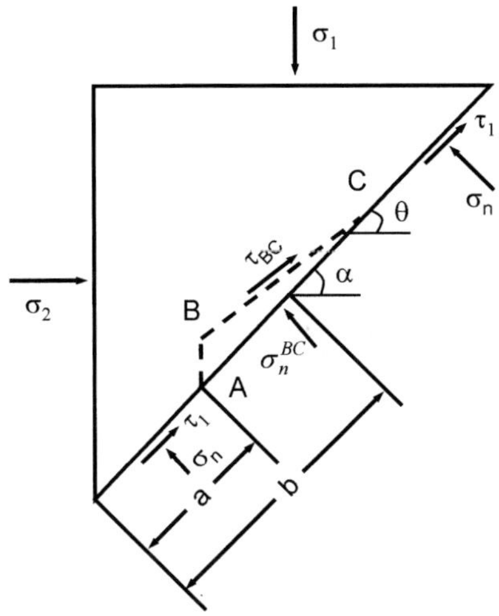


**Figure 3.19.** Stress analysis on the fracture plane.

$$\tau = \eta(\sigma f_j + c_j) + (1 - \eta)(\sigma f_r + c_r) \tag{3.51}$$

Taking the stress difference between rock bridge and crack plane into account, the following equation is also used.

$$\tau = \eta[(1 - \beta)\sigma f_j + c_j] + (1 - \eta)[\beta\sigma f_r + c_r] \tag{3.52}$$

where $\beta$ is a factor concerning the normal stress distribution on failure planes and is to be determined through experiments or numerical modelling.

As shown in Figure 3.10, $DA$ and $CE$ are the existing crack surfaces and $AB$ is the secondary crack. Shear strength can be expressed as:

$$\sigma_n = \sigma_1 \cos^2 \alpha + \sigma_2 \sin^2 \alpha \tag{3.53}$$

then

$$\tau_j = \sigma_n H(\sigma_n) f_j + c_j \tag{3.54}$$

where

$$H(x) = \begin{cases} 1 & x > 0 \\ 0 & x \leq 0 \end{cases} \tag{3.55}$$

The equilibrium equations of the crack element (Figure 3.19) are

$$F_x = 0 \qquad F_y = 0 \tag{3.56}$$

Consequently,

$$2b\sigma_2 \sin\alpha + 2a(\tau_j \cos\alpha - \sigma_n \sin\alpha) + 2c(\tau_{BC}\cos\theta - \sigma_n^{BC}\sin\theta) = 0 \tag{3.57}$$

$$-2b\sigma_1 \cos\alpha + 2a(\tau_j \sin\alpha + \sigma_n \cos\alpha) + 2c(\tau_{BC}\sin\theta - \sigma_n^{BC}\cos\theta) = 0 \tag{3.58}$$

where $c = (b - a)(\cos\alpha)/(\cos\theta)$

Rewriting (3.57) and (3.58), then

$$\tau_{BC} - \sigma_n^{BC}\tan\theta = \frac{a(\sigma_n \tan\alpha - \tau_j) - b\sigma_2 \tan\alpha}{b - a} \tag{3.59}$$

$$\tau_{BC}\tan\theta + \sigma_n^{BC} = \frac{b\sigma_1 - a(\tau_j \tan\alpha + \sigma_n)}{b - a} \tag{3.60}$$

Solving simultaneously, equations (3.59) and (3.60) gives

$$\sigma_n^{BC} = \frac{B_0 - A_0 \tan\theta}{1 + \tan^2\theta} \tag{3.61}$$

$$\tau_{BC} = \frac{A_0 + B_0 \tan\theta}{1 + \tan^2\theta} \tag{3.62}$$

where

$$A_0 = \frac{a(\sigma_n \tan\alpha - \tau_j) - b\sigma_2 \tan\alpha}{b - a} \tag{3.63}$$

$$B_0 = \frac{b\sigma_1 - a(\tau_j \tan\alpha + \sigma_n)}{b - a} \tag{3.64}$$

Let

$$F = |\tau_{BC}| - \sigma_n^{BC} H(\sigma_n^{BC}) f_r - c_r$$

Hence, when $F \geq 0$, shear failure occurs at plane $BC$ and a failure plane is formed. At this moment, $\partial F/\partial\theta = 0$, thus

$$\frac{\partial \tau_{BC}}{\partial\theta} - H(\sigma_n^{BC}) f_r \frac{\partial \sigma_n^{BC}}{\partial\theta} = 0 \tag{3.65}$$

based on

$$\frac{\partial \tau_{BC}}{\partial \theta} = \frac{B_0(1 - \tan^2 \theta) - 2A_0 \tan \theta}{(1 + \tan^2 \theta)^2 \cos^2 \theta} \tag{3.66}$$

$$\frac{\partial \sigma_{BC}}{\partial \theta} = \frac{A_0(\tan^2 \theta - 1) - 2B_0 \tan \theta}{(1 + \tan^2 \theta)^2 \cos^2 \theta} \tag{3.67}$$

Substituting equations (3.66) and (3.67) into (3.65) gives

$$\theta = \frac{1}{2}\operatorname{arcctg}\frac{A_0 - B_0 f_r H(\sigma_r^{BC})}{B_0 + A_0 f_r H(\sigma_n^{BC})} \tag{3.68}$$

***3.3.1.3 Verification through model tests.*** The modelling material made in the physical model is a mixture of gypsum and ceyssatile, which has the following mechanical properties: uniaxial tensile strength = 0.27 MPa, compressive elastic modulus = $1.3 \times 10^3$ MPa, Poisson's ratio = 0.20, internal friction angle = 35°, cohesion = 0.53 MPa. The cracks are closed and have the following properties: length = 20, 40, 60, 80 mm, internal friction angle = 10° and cohesion = 0.1 MPa.

Layout of the direct shear test is shown in Figure 2.14. The rock mass model has crack persistence ranging from 0 to 62%. In the test, the normal stress on shear plane is kept constant and the lateral loads are gradually increased till the model fails. The test results are analysed by least square method, giving shear strength curves as shown in Figure 3.20. It can be seen from the Figure 3.20 that the experimental results agree well with the analytical solution. The comparison is summarised in Table 3.8.

The above comparison indicates that the strength criterion proposed reasonably represents the properties of the rock mass containing collinear cracks.

### *3.3.2 Strength of rock mass containing multiple cracks*

Joints, fractures and cracks in a rock mass may be distributed parallel but not necessarily co-linear, as shown in Figure 3.21. The failure of such rock mass is indicated by dotted lines. Because the crack is in compressive shear stress state, the stress intensity factor at the tip is $K_I = 0$ and $K_{II} \neq 0$. Hence under the compressive shear state the condition of the crack initiation should be $K_{II} \geq K_{IIc}$.

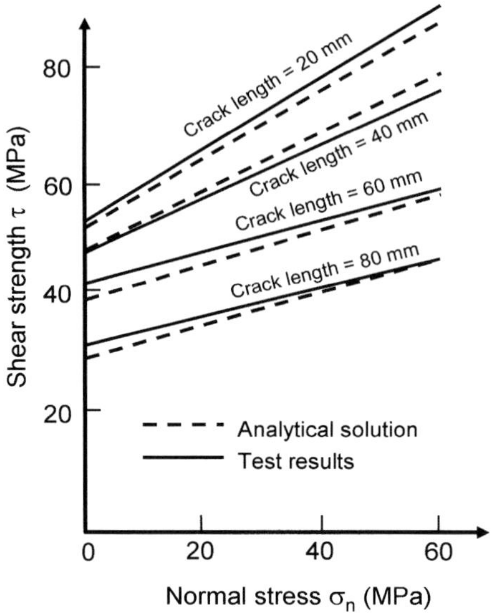


**Figure 3.20.** Comparison of strength envelopes between measured and calculated results.

As shown in Figure 3.23, crack initiates at fracture tips, and the secondary cracks *BC* and *DE* are formed at the tips B and D. The secondary cracks expand in the direction of maximum compressive stress, and at a certain stage, the cracks join each other to form *CD*. The rock mass containing multiple cracks then fails.

With understanding the failure mechanism after multi-fractured rock mass, the following assumptions are used to develop a strength criterion:

(i) The secondary crack propagates in the direction of the maximum compressive stress, and along a straight line,
(ii) The failure of multi-fractured rock mass is due to shear.

As shown in Figure 3.23, *BC* and *ED* are the secondary crack surfaces. Stresses can be expressed as:

$$\sigma_n = \sigma_1 \cos^2 \alpha + \sigma_2 \sin^2 \alpha$$

$$\tau_j = \sigma_n H(\sigma_0) f_j + c_j$$

For the element in Figure 3.23, the force equilibrium condition is

$$F_x = 0$$

$$F_y = 0$$

**Table 3.8.** Comparison between measured and calculated results.

| Persistence (%) | Normal stress (0.1 MPa) | Shear strength (0.1 MPa) | | Theoretical error (%) |
|---|---|---|---|---|
| | | Measured | Calculated | |
| 0 | 3 | 7.43 | 7.4 | 0.4 |
| 0 | 4 | 8.56 | 8.1 | 5.4 |
| 0 | 5 | 8.99 | 8.8 | 2.1 |
| 9 | 0.8 | 5.76 | 5.44 | 5.56 |
| 9 | 4 | 7.86 | 7.53 | 4.2 |
| 13 | 3 | 6.43 | 6.64 | −3.3 |
| 13 | 4 | 6.80 | 7.27 | −6.9 |
| 13 | 6 | 8.06 | 8.54 | −6.00 |
| 20 | 5 | 7.49 | 7.42 | 1 |
| 42 | 5 | 5.91 | 5.9 | 0.17 |
| 43 | 0.8 | 4.39 | 3.84 | 12.5 |
| 43 | 3.5 | 5.43 | 5.12 | 5.7 |
| 44 | 3 | 5.36 | 4.82 | 10 |
| 45 | 3.5 | 5.03 | 4.99 | 0.8 |
| 46 | 4 | 5.63 | 5.16 | 8.3 |
| 46 | 6 | 6.10 | 6.08 | 0.3 |
| 50 | 5 | 5.35 | 5.34 | 0.18 |
| 60 | 0.8 | 3.52 | 3.03 | 14 |
| 60 | 2 | 3.65 | 3.50 | 4.1 |
| 62 | 3 | 4.34 | 3.76 | 13.36 |
| 62 | 4 | 4.14 | 4.14 | 0 |
| 62 | 6 | 4.95 | 4.89 | 1.2 |

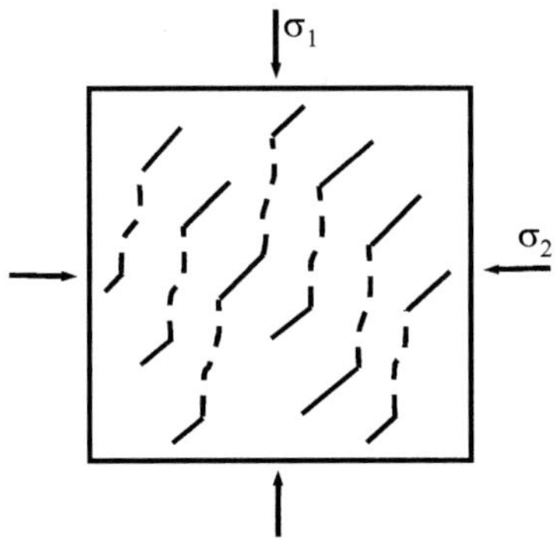


**Figure 3.21.** Rock mass with a group of parallel cracks.

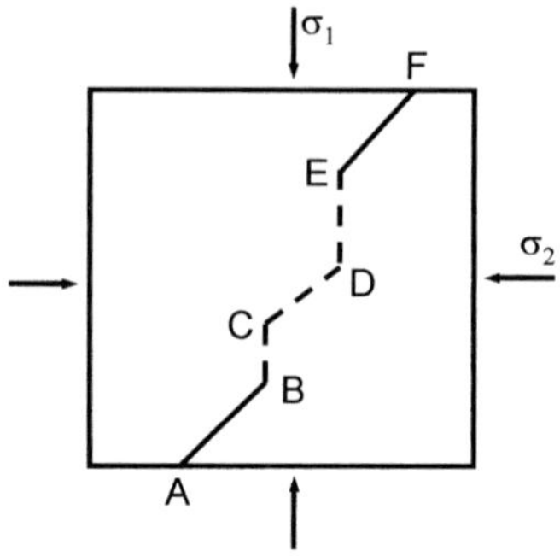


**Figure 3.22.** Enechelon cracking trajectory.

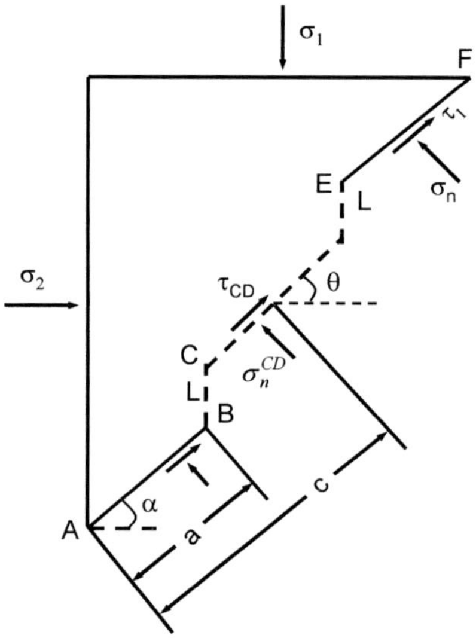


**Figure 3.23.** Force analysis.

Namely,

$$\left.\begin{aligned} 2c\sigma_2 \sin\alpha + 2a(\tau_j \cos\alpha - \sigma_n \sin\alpha) + 2m(\tau_{CD}\cos\theta - \sigma_n^{CD}\sin\theta) = 0 \\ -2c\sigma_2 \cos\alpha + 2a(\tau_j \sin\alpha + \sigma_n \cos\alpha) + 2m(\tau_{CD}\sin\theta - \sigma_n^{CD}\cos\theta) = 0 \end{aligned}\right\} \quad (3.69)$$

where

$$m = (c - a)\frac{\cos\alpha}{\cos\theta}$$

Arrangement of equation (3.69) leads to

$$\tau_{CD} - \sigma_n^{CD} \tan\theta = \frac{a(\sigma_n \tan\alpha - \tau_j) - c\sigma_2 \tan\alpha}{c - a} \tag{3.70}$$

$$\tau_{CD} \tan\theta + \sigma_n^{CD} = \frac{c\sigma_1 - a(\tau_j \tan\alpha + \sigma_n)}{c - a} \tag{3.71}$$

Solution of equations (3.70) and (3.71) gives

$$\sigma_n^{CD} = \frac{B_0 - A_0 \tan\theta}{1 + \tan^2\theta} \tag{3.72}$$

$$\tau_{CD} = \frac{A_0 + B_0 \tan\theta}{1 + \tan^2\theta} \tag{3.73}$$

in which

$$A_0 = \frac{a(\sigma_n \tan\alpha - \tau_j) - c\sigma_2 \tan\alpha}{c - a}$$

$$B_0 = \frac{c\sigma_1 - a(\tau_j \tan\alpha + \sigma_n)}{c - a}$$

It follows from Figure 3.23 by geometrical analysis that

$$\theta = \arctan\left[c\tan\alpha - 4d\tan\alpha - \frac{4L}{3c - 2a} \cdot \frac{1}{\sin\alpha}\right] \tag{3.74}$$

According to the above analysis, the failure criterion for multi-fractured rock mass under the action of compressive shear stress is

$$F \geq 0 \tag{3.75}$$

where

$$\begin{aligned} F &= \tau_{CD} - \sigma_n^{CD} H(\sigma_n^{CD}) f_r - c_r \\ \tau_{CD} &= \frac{A_0 + B_0 \tan\theta}{1 + \tan^2\theta} \\ \sigma_n^{CD} &= \frac{B_0 - A_0 \tan\theta}{1 + \tan^2\theta} \end{aligned} \tag{3.76}$$

$$A_0 = \frac{a(\sigma_n \tan\alpha - \tau_j) - c\sigma_2 \tan\alpha}{c - a} \tag{3.77}$$

$$B_0 = \frac{c\sigma_j - a(\tau_j \tan\alpha + \sigma_n)}{c - a} \tag{3.78}$$

$$\theta = \arctan\left[c \tan\alpha + 4d \tan\alpha - \frac{4L}{3c - 2a} \cdot \frac{1}{\sin\alpha}\right] \tag{3.79}$$

For uniaxial loading;

$$\sigma_1 \neq 0 \qquad \sigma_2 = 0$$
$$L = \frac{6.703T^2}{\pi K_{1c}^2} \tag{3.80}$$

For biaxial loading;

$$\sigma_1 \neq 0 \qquad \sigma_2 \neq 0$$
$$L = \frac{1}{\pi}\left[\frac{\sqrt{K_{1C}^2 + 11.614\sigma_2} - K_{1C}}{2.234\sigma_2}\right]^2 \tag{3.81}$$

$$T = 2aF_n \cos\alpha \tag{3.82}$$

$$F_n = |(\sigma_1 - \sigma_2)\sin\alpha\cos\alpha| - f_j H(\sigma_n)\sigma_n - c_j \tag{3.83}$$

$$H(x) = \begin{cases} 1 & x > 0 \\ 0 & x \leq 0 \end{cases} \tag{3.84}$$

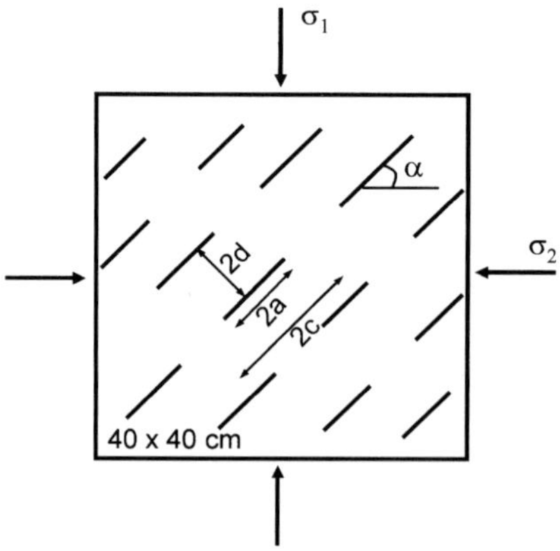


**Figure 3.24.** Diagram of the model test.

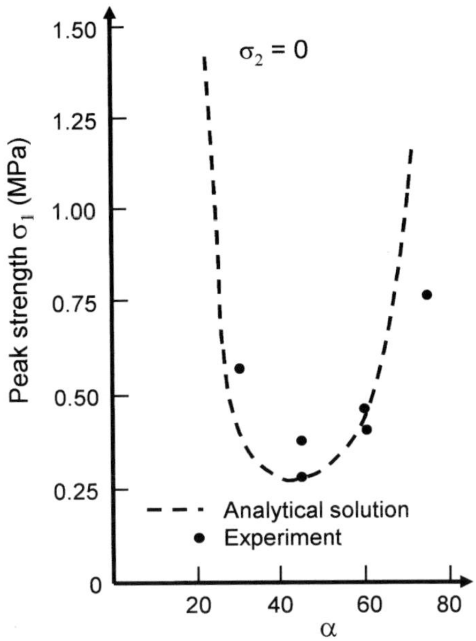


**Figure 3.25.** Comparison of predicted values with the tested values when $\sigma_2 = 0$.

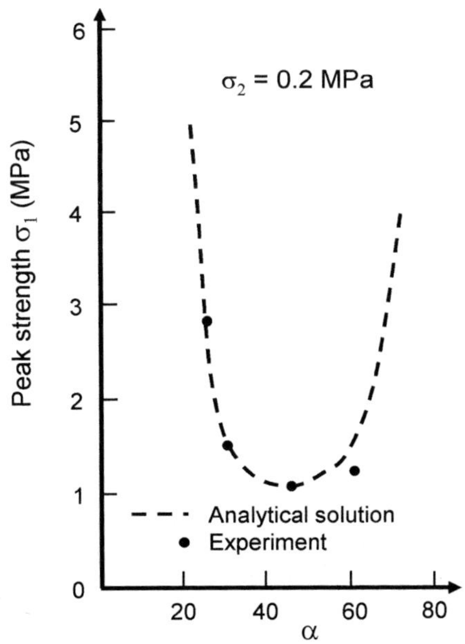


**Figure 3.26.** Comparison of predicted values with the tested values when $\sigma_2 = 0.2\,\text{MPa}$.

In addition, when $L \leq 2d/\cos\alpha$, connection takes place among the propagated cracks, and at the same time, the rock mass strength reaches the peak. The present criterion for multi-fractured rock mass is verified by physical model tests [8].

Figure 3.24 shows the schematic layout of the physical model tests. The dimension of the existing fracture are: $a = 4.8$ cm, $b = 8.0$ cm, $d = 1.0$ cm. The model material is gypsum and its main physical parameters are: uniaxial compressive strength = 1.84 MPa, uniaxial tensile strength = 0.38 MPa, modulus of elasticity = $0.238 \times 10^4$ MPa, Poisson's ratio = 0.20, internal friction angle = 45°, cohesion = 0.38 MPa and the fracture toughness of type I $K_{IC} = 74.06\ \mathrm{N/cm^{3/2}}$. The main physical parameters of the fracture are: the friction coefficient = 0.365 and the cohesion = 0.

The criterion is also implemented with numerical modelling code. A computer program using the new strength criterion is written in FORTRAN and was used to simulate the physical model tests. Comparisons of the results obtained from the numerical modelling and the physical modeling are shown in Figures 3.25 and 3.26.

From the comparisons, it can be seen that the numerical modelling based on the new strength criterion generally agree well with the physical test results.

Chapter 4

# Sensitivity Analysis of Rock Mass Parameters

At present, a variety of numerical methods are available for stability analysis of rock masses. Among them, the most widely used are the finite element method (FEM) [138–140,222–232], the boundary element method (BEM) [14,233–241], the discrete element method (DEM) [15,148–154,242–246] and coupled methods [16,247–253]. The rationality and reliability of the results from those methods depend, to a great extent, upon the appropriate selection of computational model and mechanical and mathematical parameters. Once the computational model is determined, the key to success hinges on the rational selection of the computing parameters. There are many factors and parameters that affect the rock mass stability. One has to identify the order of importance of all the parameters [1–3]. In terms of computation, the limited resource may be the primary restriction. This section presents some specific assessments on the effects of various parameters by means of the sensitivity analysis in systematic approaches.

## 4.1. SENSITIVITY ANALYSIS OF COMMONLY USED PARAMETERS

### *4.1.1 Method of sensitivity analysis*

Sensitivity analysis method is the method used to analyse the stability of a system [254]. Given a system whose character, $P$, is governed mainly by $n$ factors of $\alpha=\{\alpha_1,\alpha_2,\ldots,\alpha_n\}$, $P=f\{\alpha_1,\alpha_2,\ldots,\alpha_n\}$. Under a reference state of $\alpha^*=\{\alpha_1^*,\alpha_2^*,\ldots,\alpha_n^*\}$, the character is described by $P^*$. The sensitivity analysis is to, let the above individual factors vary within respective possible range and then analyse both tendency and extent to which the character of the system, $P$, departs from the base state due to the variation of the factors.

The first step of sensitivity analysis is to establish the system model, i.e., the functional relation between the system character and the factors, $P=f\{\alpha_1,\alpha_2,\ldots,\alpha_n\}$. This relation should be, if possible, described by analytical expression. In the case of a complex system, it can be expressed by numerical method or through presentation of graphic chart. It is a key step of the effective analysis on parameter sensitivity to establish a model that reflects the system with the reality as fully as possible.

The basic parameter set should be given after the system model has been established. The basic parameter set should be established to reflect the subjects considered. For example, when the sensitivity of the stability of an underground project to the variation of rock mass parameters is to be studied, the suggested rock mass parameters of the in situ rock mass can be used as the basic parameter set. Once the basic parameter set is determined, the sensitivity analysis can be performed on each parameter. When analysing the effect of $\alpha_k$ on the characteristic of the system, $P$, the parameter $\alpha_k$ in the set is varied within a possible range while the remaining parameters are kept constant. In this case, the system's character displays the following relation:

$$P = f(\alpha_1^*, \ldots, \alpha_{k-1}^*, \alpha_k, \alpha_{k+1}^*, \ldots, \alpha_n^*) = \varphi_k(\alpha_k) \tag{4.1}$$

The character curve of $P$–$\alpha_k$ is plotted from equation (4.1), which roughly describes the sensitivity of $P$ to the disturbance from $\alpha_k$. For example, the rapid change of the curve around $\alpha_{k1}$ shows a high sensitivity of $P$ to $\alpha_k$, i.e., a slight change of $\alpha_k$ will cause a great change of $P$. On the contrary, the curve is gently around $\alpha_{k2}$, then the system character of $P$ is less sensitive to the parameter of $\alpha_k$, i.e., $P$ varies slightly with a large change of $\alpha_k$. In other words, when the parameter is near $\alpha_{k1}$, it is a parameter of high sensitivity; when it is near $\alpha_{k2}$, then is of low sensitivity (Figure 4.1).

The above analysis only gives the sensitivity behaviour of the system character $P$ to a single factor. The character of a real system is generally governed by many factors of different physical quantities and units. Therefore, it is difficult to compare the sensitivity of the various factors by the above method. To solve this problem, dimensionless treatment and analysis can be applied.

In dimensionless analysis, the sensitivity function and the sensitivity factor are defined in dimensionless terms. The ratio of the relative error ($\delta_p$), of the system character $P$ ($\delta_p = |\Delta P|/P$) to the relative error of parameter $\alpha_k$ ($\delta_{\alpha k} = |\Delta\alpha_k|/\alpha_k$) is defined as the sensitivity function, $S_k(\alpha_k)$, of the parameter $\alpha_k$.

$$S_k(\alpha_K) \triangleq \left(\frac{|\Delta P|}{P}\right) \Big/ \left(\frac{|\Delta\alpha_k|}{\alpha_k}\right) = \left|\frac{\Delta P}{\Delta\alpha_k}\right| \frac{\alpha_k}{P} \quad k = 1, 2, \ldots, n \tag{4.2}$$

When $|\Delta\alpha_k|/\alpha_k$ is small, the function of $S_k(\alpha_k)$ can be expressed approximately as

$$S_k(\alpha_k) = \left|\frac{d\varphi_{k(\alpha_k)}}{d\alpha_k}\right| \frac{\alpha_k}{P} \quad k = 1, 2, \ldots, n \tag{4.3}$$

From Equation (4.3), the sensitivity function curve of $\alpha_k$ can be obtained, which is shown in Figure 4.2. Given $\alpha_k = \alpha_k^*$, the sensitivity factor $S_k^*$ of the parameter $\alpha_k$ is obtained as

$$S_k^* = S_k(\alpha_k^*) = \left(\frac{d\varphi_{k(\alpha_k)}}{d\alpha_k}\right)_{\alpha_k = \alpha_k^*} \frac{\alpha_k^*}{P^*} \quad k = 1, 2, \ldots, n \tag{4.4}$$

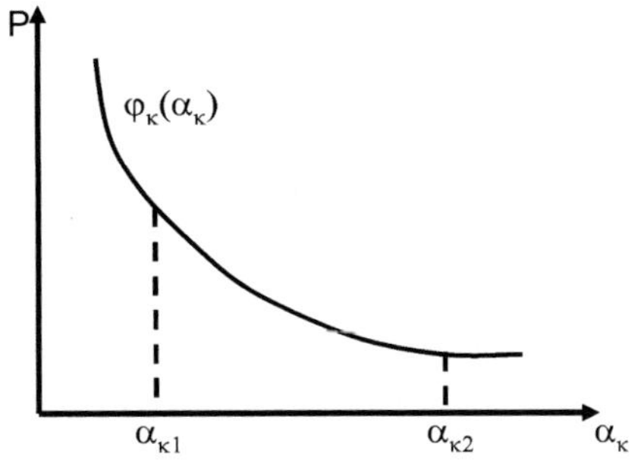


**Figure 4.1.** Characteristic curve of system $P$–$\alpha_k$.

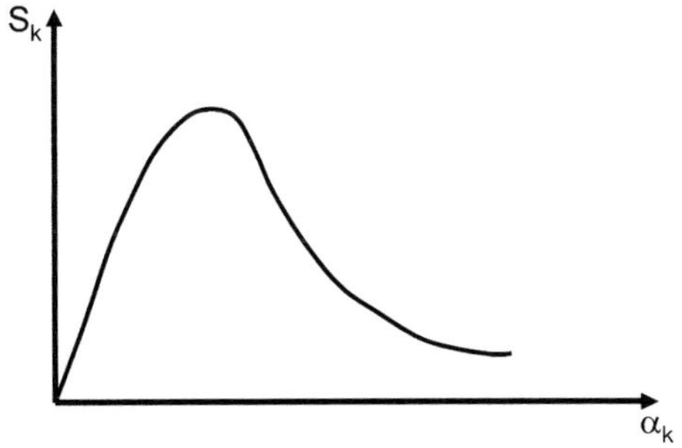


**Figure 4.2.** Curve of sensitivity function $S_k$–$\alpha_k$.

where $S_k^*$ and $k = 1, 2, \ldots, n$ are a group of non-negative dimensionless real numbers. The higher $S_k^*$ is, the more sensitive $P$ is to $\alpha_k$. Based on comparison between different $S_k^*$ values, one can give synthetic assessment on the sensitivity of various factors.

When $\varphi_k\ (\alpha_k)$ is a sectional function, its derivative may be discontinuous at section boundary point of $\alpha_{k0}$, which makes Equation (4.3) fail to give the sensitivity of $S_k^0$ at $\alpha_{k0}$. In this case, one of the following methods can be used:

(i) Choose the higher value of the left and the right limits $S_k(\alpha_k)$ at $\alpha_k = \alpha_{k0}$ as the sensitivity $S_k^0$, i.e.,

$$S_k^0 = \max\{S_k(\alpha_k-), S_k(\alpha_k+)\}$$

(ii) Smooth the function of $\varphi_k\ (\alpha_k)$ using cubic spline function fitting method or other techniques to eliminate the discontinuous point of the derivative; and

(iii) Set a common relative error $\varepsilon$ for every parameter, i.e., let $|\Delta\alpha_k|/\alpha_k = \varepsilon$, $k = 1, 2, \ldots, n$ and then calculate $S_k^0$ using equation (4.2).

#### 4.1.2 Sensitivity analysis of stability of underground works

Examples of using the sensitivity analysis method described in the previous section are given in this section. It analyses various factors that affect the stability of a rock cavern complex of a hydropower project.

The project comprises main powerhouse, transformer house and tailrace surge chamber, and the sectional layout is shown in Figure 4.3. The powerhouse measures 64.4 m in height and 27 m in width. The deformations of the side wall and the crown are of interest to the stability analysis. The project is located in the region of high in situ stress field where horizontal stresses are greater than the vertical stresses. Due to the high overlying depth above the chamber complex, the in situ stress field is assumed to be uniform and the directions of its two principal stresses are oriented horizontally and vertically at $\sigma_x = 13.3$ MPa and $\sigma_y = 9.5$ MPa respectively.

***4.1.2.1 Computational model.*** A two-dimensional non-linear FEM programme has been used to model and analyse the stability of the underground structures. The following simplifications and assumptions have been made:

(i) The surrounding rock mass is homogeneous and continuous with the effect of faults neglected, but the joint effect is considered using the equivalent elastic module, $E_e$, from the in situ measurements.

(ii) The initial in situ stress is uniformly distributed within the computational domain and the two principal stresses act in horizontal and vertical directions.

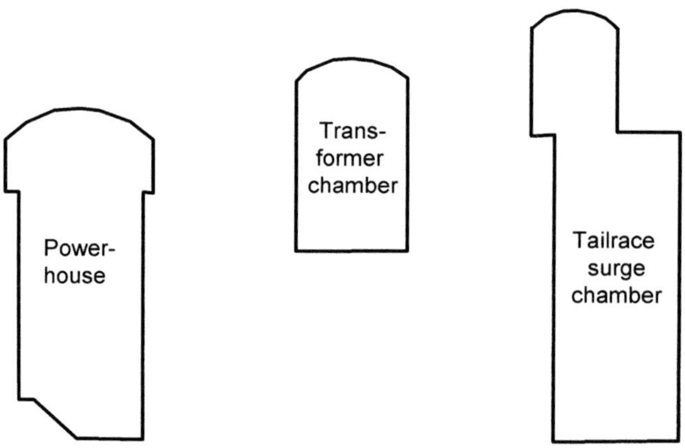


**Figure 4.3.** Sectional diagrammatic sketch of Laxiwa hydroelectric power station.

The excavation is simulated by *unloading process* and the effect of the excavation of the tailrace surge chamber on the weakening of the rock is treated with the *stiffness-reduction method.*

The system character, namely, the stability of the underground structure complex is reflected by the maximum horizontal deformation of the left sidewall of the main powerhouse (or by other indexes, e.g., damage areas). The parameters used for the sensitivity analysis include elastic module ($E$), Poisson's ratio ($\nu$), cohesion (c), internal friction angle ($\phi$), horizontal and vertical principal stresses ($\sigma_x$ and $\sigma_y$). For the above exercise, the basic parameter set is given in Table 4.1.

***4.1.2.2 Analysing the results.*** All parameters have been analysed one by one using the method stated earlier. The procedure to analyse the horizontal stress $\sigma_x$ and the elastic module of $E$ is described below.

From experiences, the possible varying range of those two parameters are determined. $E$ is in the range of 1.0–4.5 × $10^4$ MPa, and $\sigma_x$ is in the range of 4.75–19.95 MPa. The values of $E$ and $\sigma_x$ are adjusted step by step to calculate the maximum horizontal displacement, $u$, of the left sidewall of the main powerhouse. Curves representing $u$–$E$ and $u$–$\sigma_x$ are plotted from the computing results, as shown in Figures 4.4 and 4.5. The function relations of $u$–$E$ and $u$–$\sigma_x$ have been obtained from the curves and can be expressed as:

$$u = \varphi_E(E) = 8.70/\varphi \tag{4.5}$$

$$u = \varphi_{\sigma_x}(\sigma_x) = 0.225\sigma_x - 0.2425 \tag{4.6}$$

From the equation (4.3) we have two sensitivity functions of $S_E(E)$ and $S_{\sigma_x}(\sigma_x)$:

$$S_E(E) = \frac{8.70}{Eu} \equiv 1 \tag{4.7}$$

$$S_{\sigma_x}(\sigma_x) = 0.225\frac{\sigma_x}{u} = \frac{0.225\sigma_x}{0.225\sigma_x - 0.2425} \tag{4.8}$$

**Table 4.1.** Basic parameter set.

| $E$ (×$10^4$ MPa) | $\nu$ | $c$ (MPa) | $\phi$ (°) | $\sigma_x$ (MPa) | $\sigma_y$(MPa) |
|---|---|---|---|---|---|
| 3.2 | 0.21 | 25 | 48 | 13.3 | 9.5 |

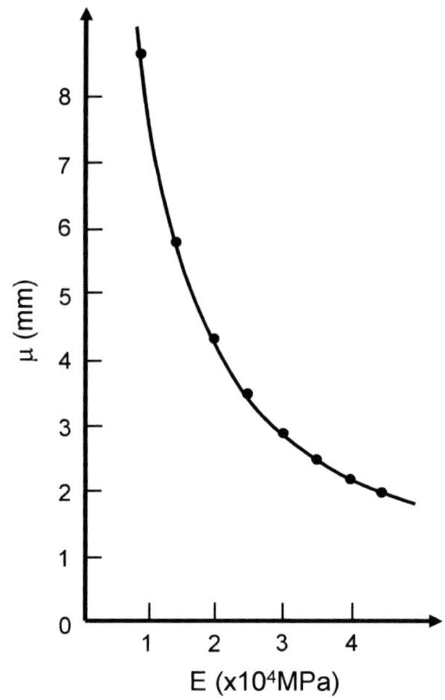


**Figure 4.4.** *u–E* curve.

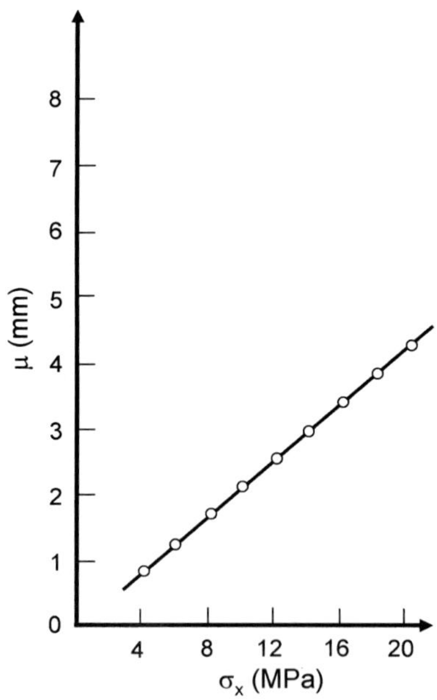


**Figure 4.5.** $u$–$\sigma_x$ curve

The corresponding sensitivity curves of $S_E$–$E$ and $S_{\sigma_x}$–$\sigma_x$ are shown in Figures 4.6 and 4.7 respectively.

As shown in Figures 4.6 and 4.7, the sensitivity function of $S_E(E) \equiv 1$ means that the sensitivity factor, $S_E^*$, is constantly equal to 1, and is not influenced by the basic value of the elastic module $E$. $S_{\sigma_x}(\sigma_x)$ is a decreasing function, and the sensitivity is high when $\sigma_x$ is low and decreases with increasing $\sigma_x$. The limit of $S_{\sigma_x}$ is 1. By substituting $\sigma_x^* = 1.33$ MPa into Equation (4.8), the sensitivity factor, $S_{\sigma_x}^{*} \cdot S_{\sigma_x}^*$, of the parameter $\sigma_x$ is 1.088.

Similar analysis can be performed on other parameters. Table 4.2 summarises the sensitivity factor of other parameters.

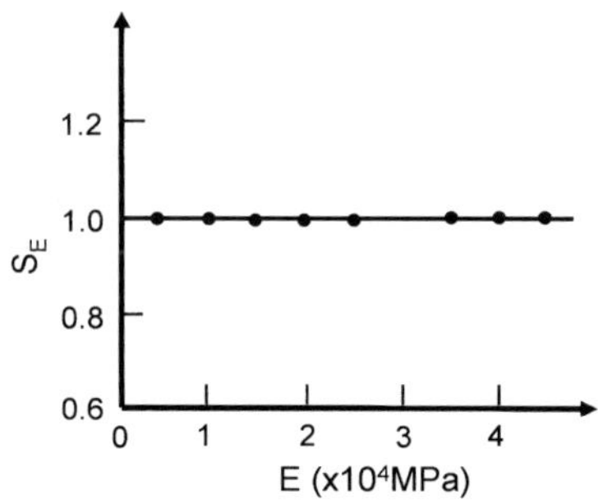


**Figure 4.6.** $S_E$–$E$ curve.

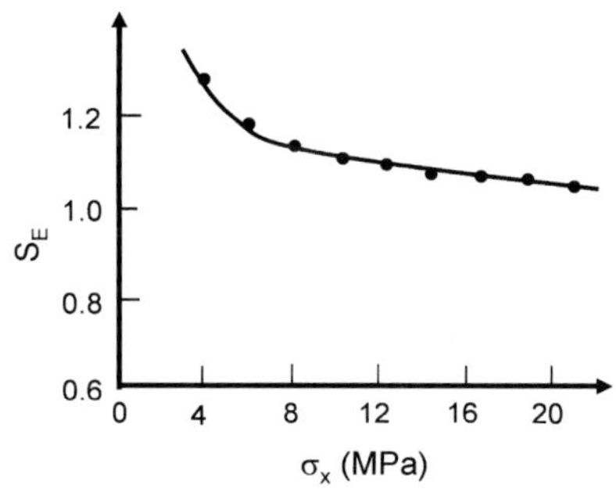


**Figure 4.7.** $S_{\sigma_x}$–$\sigma_x$ curve.

**Table 4.2.** Sensitivity factor of various parameters.

| $S_E^*$ | $S_\mu^*$ | $S_c^*$ | $S_\varphi^*$ | $S_{\sigma_x}^*$ | $S_{\sigma_y}^*$ |
|---|---|---|---|---|---|
| 1.0 | 0.077 | 0.039 | 0.020 | 1.088 | 0.077 |

Table 4.2 indicates that when the horizontal displacement of the sidewall of the powerhouse is used to judge the stability of the cavern, then the most sensitive factor that affects the stability is the horizontal stress. It has sensitivity as high as 1.088. In other words, if there exists an error of 15% in $\sigma_x$, the relative error ($\delta u$) of $u$ is $1.088 \times 15\% = 16.32\%$. The internal friction angle $\phi$ is the least sensitive factor ($S^*_\varphi = 0.02\%$). An error of 15% in $\phi$ only results in a 0.3% ($0.2 \times 15\%$) error of $u$. As seen from the analysis, the horizontal stress and the elastic module have high sensitivity factors and should be treated with great care.

It should be noted that the above conclusion is drawn from a given set of basic parameters that are directly involved in a specific engineering project. The conclusion differs from problem to problem and project to project. As shown in Figure 4.7, the sensitivity factor, $S^*_{\sigma_x}$, of $\sigma_x$ varies, depending upon the selection of the basic value of $\sigma^*_x$. In addition, the sensitivity of some parameters is affected by the interaction of other parameters. For example when a structure is in a perfect elastic state, cohesion and friction have no effect on the deformation of the structure. But when the structure is in an elasto-plastic state, and cohesion and friction affect the deformability, such effect becomes more remarkable with increasing plastic area. The sensitivity of $c$ and $\phi$ is therefore dependent not only on their basic value but also on the values of $\sigma_x$ and $\sigma_y$. Further analysis of such interaction between parameters requires more rock mechanics knowledge of the problem. One can use the Rock Engineering System [1–3] or the Grey System [254] to study the interaction between the parameters. For such cases, when sensitivity analysis is performed on a parameter that has active interaction with others, it is desirable to make the latter varying within their possible ranges.

### *4.1.3 Application to optimisation of test schemes*

In order to analyse the stability of the underground rock structures, the mechanical and engineering parameters of the rock mass must be known. Accurate rock mass properties can only be obtained from large in situ tests. Such tests are seldom carried out as they are very expensive and time consuming. Sensitivity analysis of parameters can be applied for the optimisation of testing schemes.

Test scheme optimisation is to obtain the rock properties that meet the engineering requirement with the least amount of work for large in situ tests. Studies of sensitivities of all the parameters would identify the parameters of high sensitivities that should be measured during the in situ tests. In the example in Section 4.1.2, the horizontal deformation of the high wall, the horizontal stress and the rock mass module are the key parameters. They must be determined through field tests, while the other parameters are less critical and can be obtained through less expensive tests or other methods.

Sensitivity analysis helps to avoid mistakes due to subjective conjecture. Different basic parameter sets have different orders of sensitivity. For this reason, the sensitivity analysis of parameters should be aimed at specific engineering problems so as to distinguish key parameters from the rocks. The key parameters may vary from project to project.

The amount of field tests for rock parameters can be rationalised according to their sensitivity factors. In principle, parameters of higher sensitivity should be subjected to more tests.

Selections of appropriate test methods shall be made in accordance with the requirements of the engineering project and based on the sensitivity factors. For example, when there are two options available in measuring the horizontal stress $\sigma_x$: one is cheap but has a relative error of $\delta_{\sigma_{x1}} = 15\%$, while another is expensive but has a relative error of $\delta_{\sigma_{x2}} = 10\%$. As the sensitivity of $S^*_{\sigma_x}$ is 1.088, the relative errors of the two methods results in measurement errors of $\delta_u$ are 16.32% and 10.88%, respectively. If a relative error $\delta_u < 15\%$ is required, the second method should be adopted although the first one is cheap. If a larger error of $\delta_u < 20\%$ is allowed, the first method can be used.

Although the sensitivity analysis provides, in view of accuracy, scientific basis for the optimisation of testing schemes, the overall optimisation should consider other factors, such as testing cost, duration and availability.

## 4.2. ANALYSIS OF THE EFFECT OF JOINT PARAMETERS ON ROCK MASS DEFORMABILITY

In the previous section, the sensitivity analysis is performed by assuming rock masses as isotropic and homogeneous media. However, natural rock masses contain sets of joints, which have influence on the strength and the deformation characteristics of the rock mass. Chapter 3 has suggested several mechanical models for different kinds of jointed rock masses. This section examines the sensitivity of the joint mechanical and geometric parameter on the rock mass stability.

### *4.2.1 Application of equivalent model for jointed rock mass*

The sensitivity analysis of the effect of joints on underground rock structures involves a considerable amount of work. The current study therefore adopts the same analytical model and method described in Section 3.3.

The rock mass is assumed to contain two joint sets with different orientations (Figure 4.8). The following notations are used in the analysis. For the rock material: $E_r$ - elastic module, $\nu_r$ - Poisson's ratio; and for joint sets (subscripts 1 and 2 for set I

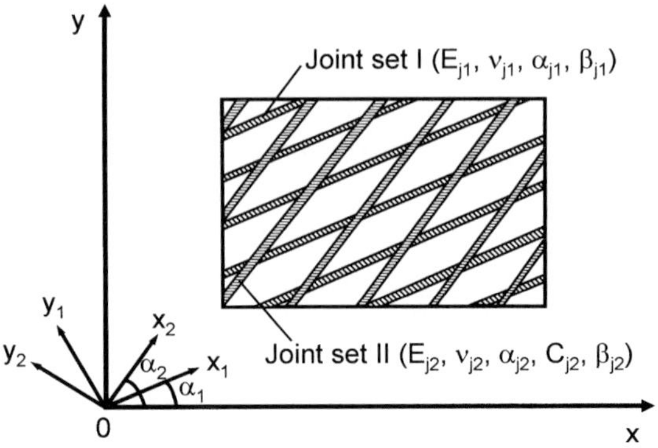


**Figure 4.8.** Typical rock mass containing two joint sets.

and II respectively): $E_{j1}$ and $E_{j2}$ - elastic moduli, $\nu_{j1}$, $\nu_{j2}$ - Poisson's ratio, $\alpha_{j1}$ and $\alpha_{j2}$ - dip angle, $\eta_{j1}$ and $\eta_{j2}$ - persistence, $\beta_{j1}$ and $\beta_{j2}$ - specific width (total joint width percentage in the rock mass).

The model is treated under the plane strain condition, and the strike of the joint sets coincides with the $z$ axis, and the joint persistence $C_{j1}$ and $C_{j2}$ are both equal to 1.

Equivalent properties of the rock mass are obtained according to the properties of individual joint sets and the rock material, as discussed in Section 3.3. The stress states of the joint and the rock material are respectively (refer to Section 3.3).

$$\begin{aligned} \{\sigma\}_{j1} &= \{\sigma_x^{j1}, \sigma_y^{j1}, \tau_{xy}^{j1}\}^r \quad \{\varepsilon\}_{j1} = \{\varepsilon_x^{j1}, \varepsilon_y^{j1}, \gamma_{xy}^{j1}\}^r \\ \{\sigma\}_r &= \{\sigma_x^r, \sigma_y^r, \tau_{xy}^r\}^r \quad \{\varepsilon\}_r = \{\varepsilon_x^r, \varepsilon_y^r, \gamma_{xy}^r\}^r \end{aligned} \tag{4.9}$$

### *4.2.2 Basic parameter for sensitivity analysis*

The sensitivity analysis of the joint parameter is conducted for the same case as those in the previous section (Section 4.1). Sensitivity of various parameters to the maximum horizontal displacement of the cavern sidewall and to the maximum settlement of the roof is studied.

The project was described in Section 4.1.2 and shown in Figure 4.3. There exists two prevailing joint sets in the region and the elastic constants of the rock material are: elastic module $E_r = 5.8 \times 10^4$ MPa and Poisson's ratio $\nu_r = 0.21$. The parameters of the two joint sets are obtained from the site investigation and are summarised in Table 4.3.

**Table 4.3.** Basic parameter values of joint sets used in sensitivity analysis.

| | | | | | |
|---|---|---|---|---|---|
| Joint set I | $E_{j1} = 0.5 \times 10^4$ MPa | $\nu_{j1} = 0.23$ | $\alpha_{j1} = 70$ | $\eta_{j1} = 0.6$ | $\beta_{j1} = 3.2 \times 10^{-2}$ MPa |
| Joint set II | $E_{j1} = 1.4 \times 10^4$ MPa | $\nu_{j1} = 0.23$ | $\alpha_{j2} = 110$ | $\eta_{j2} = 0.6$ | $\beta_{j2} = 3.2 \times 10^{-2}$ MPa |

The values of parameter in Table 4.3 are also suggested for the sensitivity analysis. Accordingly, given that each parameter varies within a certain range, the side wall displacement ($u_x$) and the roof settlement ($u_y$) are calculated. The change and trend of $u_x$ and $u_y$ with the variation of the parameters are examined. A FEM program is used for the analysis.

### 4.2.3 Computational results

***4.2.3.1 Effects of joint elastic moduli on displacement.*** Figure 4.9 shows the relation curves between the joint elastic modulus ($E_j$) and the surrounding rock mass displacement (μ), for each joint set. $u_x$ is the deformation of the high wall and $u_y$ is the deformation of the crown. Relative convergence ($u/B$) and the relative ratio of the joint elastic modulus to the rock mass modulus ($E_j/E_r$) are also presented in Figure 4.9. $B$ is the cavern width in calculating $u_x$, and $B$ is the cavern height in calculating $u_y$.

From Figure 4.9, it can be seen that the $u$–$E_j$ curve can be fitted with the function of $Y = A/X + B$. The regression analysis determines the $u$–$E_j$ relationship for

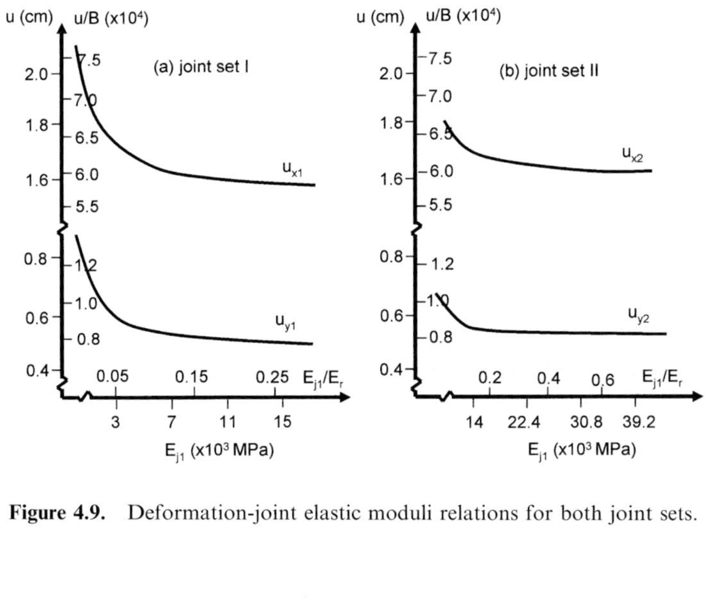


**Figure 4.9.** Deformation-joint elastic moduli relations for both joint sets.

the two joints as:
For joint set I,

$$u_x = 527.6/E_{j1} + 1.5505 \tag{4.10a}$$

$$u_y = 326.4/E_{j1} + 0.4772 \tag{4.10b}$$

For joint set II,

$$u_x = 675.8/E_{j2} + 1.6098 \tag{4.10c}$$

$$u_y = 286.2/E_{j2} + 0.5251 \tag{4.10d}$$

where $u_x$ and $u_y$ (cm) stand for the maximum displacement of the sidewall and the maximum vertical deformation of the cavern, respectively. $E_{j1}$ and $E_{j2}$ (MPa) are the modulus of joint sets I and II, respectively.

The following conclusions can be drawn from Figure 4.9 and equation (4.10):

(i) If the actual elastic moduli of the rock joints are lower than the suggested elastic moduli used in the computation, then the actual deformation of the surrounding rock mass is greater than that computed.

(ii) For the main powerhouse cavern, the total deformation of the sidewall is greater than that of the roof, for the same change in the joint modulus. However, the relative deformation of the sidewall is less than that of the roof. For example, when the elastic modulus of joint set I changes from $E^*_{j1} = 5000$ to 2500 MPa (given that the maximum relative error of $E_j$ is 50%), then the absolute values of $u_x$ and $u_y$ are 0.106 and 0.065 cm, while the corresponding relative changes are 6.37 and 12.04%, respectively. The results show that the displacement of roof is more sensitive than the displacement of sidewall to the change of joint elastic modulus.

(iii) The same changes of the elastic moduli of the joint sets with different initial moduli have different effects. Given the same relative errors of 50% of both joint sets whose initial moduli are 5000 and 14,000 MPa, the errors caused are 12.04 and 3.76% at the roof, respectively. It shows that lower elastic moduli has higher sensitivity to the rock mass deformation.

***4.2.3.2 Effect of joint Poisson's ratio on rock mass deformation.*** Figure 4.10 shows the relation curves between the joint Poisson's ratio ($\nu_j$) and the rock mass displacement ($u$). The Poisson's ratios of the joint sets ($\nu_{j1}$ and $\nu_{j2}$) vary between 0.184 and 0.276, and the deformations ($u_x$ and $u_y$) remain almost the same. This shows that the rock mass deformation is by far less sensitive to the Poisson's ratio.

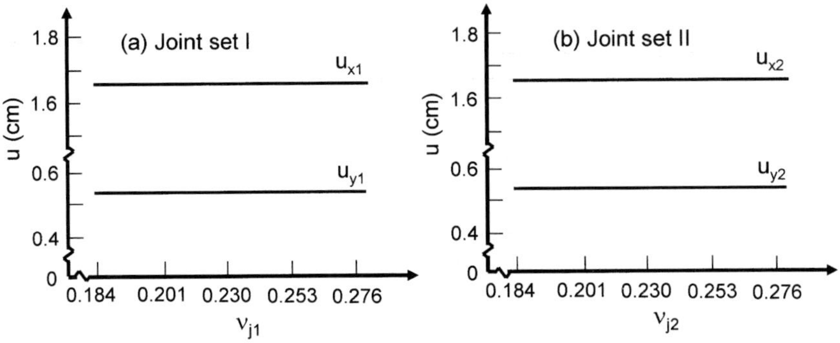


**Figure 4.10.** Effects of joint Poisson's ratio on rock mass deformation.

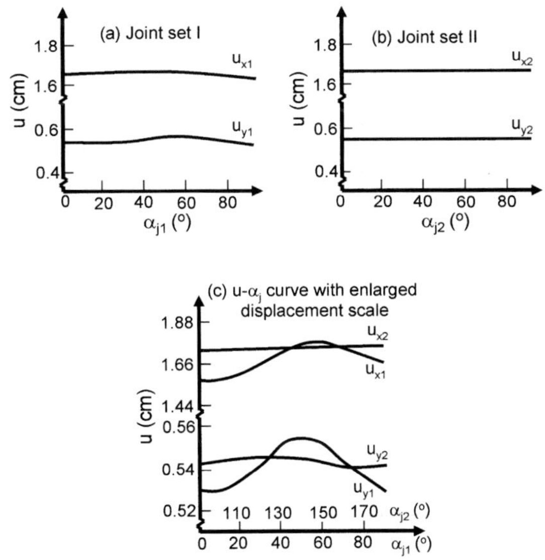


**Figure 4.11.** Relationship between displacement and dip angle.

Therefore, in analysing the stability of the rock mass, selection of the Poisson's ratio will not result in significant errors.

***4.2.3.3 Effect of joint dip angle on rock mass displacement.*** The effect of joint dip angle to the rock mass deformation is shown in Figure 4.11. From the curves, the following conclusions can be obtained.

(i) The change of rock mass deformation is slightly affected by the joint dip angle varying between 0 and 90°. It suggests that the joint dip angle is a non-sensitive factor affecting the rock mass deformation.
(ii) The sensitivity of joint dip angle is related to the joint elastic modulus. For the joint set II with a high elastic modulus, the change in the dip angle results in little change of the rock mass deformation. On the other hand, for the joint set I with a low modulus, the change of the dip angle causes increase of the rock mass deformation as shown in Figure 4.11(c). It implies that the sensitivity of the joint dip angle increases with the decreasing joint elastic modulus. The joint dip angle is a sensitive factor when the joint elastic modulus is very low.
(iii) The vertical settlement of roof, $u_y$, is more sensitive to the change of joint dip angle, as compared with the horizontal deformation of the sidewall.
(iv) It is noted from Figure 4.11(c) that when a joint set of low modulus and dip angle is between 40 and 60°, the rock mass deformation becomes the greatest, which is most unfavourable to the rock mass stability.

***4.2.3.4 Effect of joint persistence on rock mass deformations.*** Figure 4.12 presents the relation curve of the joint persistence ($\eta_j$) and the rock mass displacement ($u$).

The displacement ($u$) shows the following character as $\eta_j$ varies:

(i) When $\eta_j \leq 0.6$, $u_x$ and $u_y$ increase more or less linearly with increasing $\eta_j$; when $\eta_j > 0.6$, the horizontal displacement, $u_x$, increases acceleratively with increasing $\eta_j$, while the vertical displacement, $u_y$, increases at a much lower rate. This indicates that the joints with a persistence greater than 0.6 have higher sensitivity to the sidewall deformation. The sensitivity of joint persistence to the roof displacement, however, is rather low.
(ii) The sensitivity of the persistence depends on the joint elastic modulus. For example, when $\eta_j^* = 0.6$ and the maximum deviation is $\pm 0.2$, for joint set I with $E_{j1}^* = 5000$ MPa, the maximum error of the sidewall displacement is 4.5%; while for joint set II with $E_{j2}^* = 14{,}000$ MPa, the error of the sidewall displacement is 1.68%. It can be seen that when the joint elastic modulus is low, the sensitivity of the joint persistence to the rock mass deformation is high.
(iii) The sensitivity of joint persistence to the sidewall displacement is higher than that to the roof displacement, as shown in Figure 4.12.

***4.2.3.5 Effect of joint aperture on rock mass deformations.*** The relation between the joint aperture and the rock mass deformation is as shown in Figure 4.13.

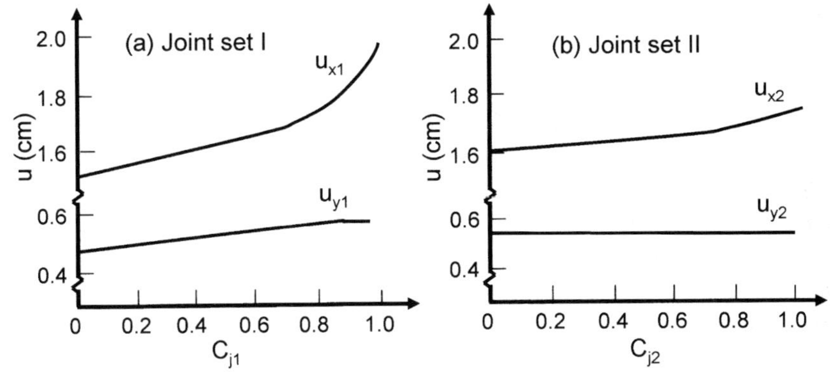


**Figure 4.12.** Effects of joint persistence on rock mass displacement.

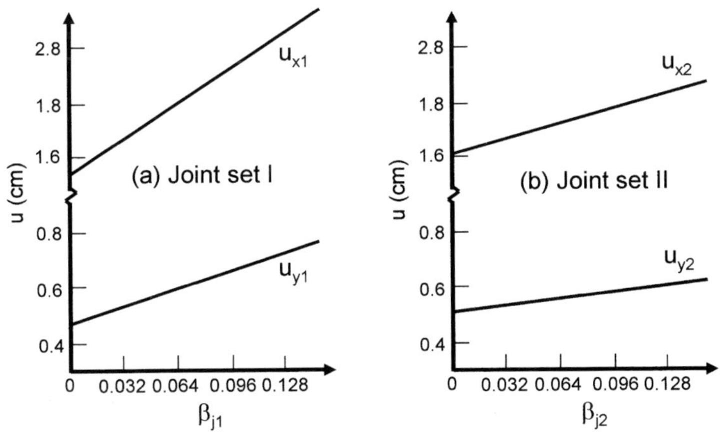


**Figure 4.13.** Effect of joint aperture on rock mass deformation.

It leads to the following conclusions:

(i) The rock mass deformation increases linearly with the increasing joint aperture.

(ii) The roof settlement is more sensitive to the change of the joint aperture than the sidewall deformation, as shown in Figure 4.13.

(iii) The sensitivity of the joint aperture to the surrounding rock mass deformation is also influenced by the joint elastic modulus. The comparison between Figures 4.13a and 4.13b shows clearly that joint set I, with lower elastic modulus has higher sensitivity factor compared with joint set II.

***4.2.3.6 Comparison of sensitivity of different parameters.*** The previous section discussed the sensitivity of various parameters to the rock mass deformation. In this section, attempt is made to compare the sensitivities of various parameters within the same domain. Since the Poisson's ratio and the joint dip angle are non-sensitive parameters, only elastic modulus, persistence and aperture of the joint sets are used in the comparison study.

The sensitivity of a parameter can be measured by the sensitivity factor $S_p$ [254], defined as the ratio of the relative deviation ($\delta p = |\Delta p/p^*|$), by which the system's character $p$ departs from a certain state of $P^*$, to the relative deviation of the parameter, hence,

$$S_p(\alpha) = \left|\Delta p/p^*\right| / \left|\Delta\alpha/\alpha^*\right| \tag{4.11}$$

A higher value of $S_p$ suggests greater effect of the parameter $\alpha$ on the system's character, $P$, i.e., the greater sensitivity of $\alpha$. The relative deviation of $\alpha$ may depend on the problem concerned.

In the comparison study below, a relative error of 50% is selected for all parameters for the same underground cavern complex analysis in Figure 4.3.

The sensitivity factors of various parameters to the sidewall displacement, the roof settlement and their relative deviations are analysed and summarised in Table 4.4.

It can be seen from Table 4.4 for that particular underground project, the sensitivity order of the joint parameters to the cavern sidewall deformation: $\eta_{j1}$ (persistence of joint set I), $E_{j1}$ (elastic modulus I), $\beta_{j1}$ (joint aperture I), $E_{j2}$ (elastic modulus II), $\eta_{j2}$ (persistence II), and $\beta_{j2}$ (joint aperture II). $\eta_{j1}$ and $E_{j1}$ are the parameters of high sensitivity and the others are of low sensitivity. Similarly, the sensitivity order of the joint parameter to the roof settlement is: $E_{j1}$, $\beta_{j1}$, $E_{j2}$, $\beta_{j2}$ and $\eta_{j2}$. $E_{j1}$ and $\beta_{j1}$ are the parameters of high sensitivity and the rest are of low sensitivity. The synthetic analysis on $u_x$ and $u_y$ shows that joint set I with a low elastic modulus is the main factor affecting the rock mass deformability. A 50% of combined error of the parameters of this joint set will result in an error of over 10% of the rock mass deformation.

**Table 4.4.** Sensitivity factors of various parameters to roof and sidewall displacement.

| Sensitivity factor | $E_{j1}$ | $\eta_{j1}$ | $\beta_{j1}$ | $E_{j2}$ | $\eta_{j2}$ | $\beta_{j2}$ |
|---|---|---|---|---|---|---|
| $Su_x$ | 0.127 | 0.203 | 0.082 | 0.058 | 0.0578 | 0.036 |
| $Su_y$ | 0.241 | 0.077 | 0.118 | 0.075 | 0.013 | 0.047 |
| $\delta Su_x$ % | 6.37 | 10.15 | 4.11 | 2.91 | 2.88 | 1.82 |
| $\delta Su_y$ % | 12.03 | 3.86 | 5.89 | 3.75 | 0.66 | 2.34 |

It should be noted that the above conclusions have been drawn on the basis that each basic parameter varies independently. The interactions between the parameters have not been considered. A more comprehensive study of sensitivity can be conducted by coupling the present method with a parameter interaction study, e.g. the Rock Engineering System approach [1–3]. Nevertheless, the present method is usually sufficient to quantify the sensitivity of various parameters.

## 4.3. SENSITIVITY ANALYSIS OF ROCK MASS PARAMETERS ON DAMAGE ZONES

This section discusses the sensitivity analysis of the rock material and joint parameters on the magnitude of the damaged zone. The rock material parameters and the rock joint parameters are regarded as dependent variables in the analysis. The stability criterion is based on the magnitude of the damaged zone. The engineering case used is the same cavern complex project as the previous section (Figure 4.3).

### *4.3.1 Failure criterion for the equivalent jointed rock mass*

The principle for analysing the deformation equivalence of the jointed rock mass is similar to that discussed in the previous section. However, each finite element includes at least one joint. The finite element may have several possible failure modes or damage patterns including: plastic flow (or shear) of the rock material, tensile or shear failure of joints. The rock mass is considered having been damaged, if any of these phenomena takes place.

Drucker–Prager criterion is used to judge if the plastic flow (or shearing) of the intact rock occurs,

$$\sqrt{3}\sin\phi I_1 + 3\sqrt{3+\sin^2\phi}\cdot\sqrt{J_2} - 3\sqrt{3}c\cos\phi \geq 0 \tag{4.12}$$

$I_1$ and $J_2$ are the first invariant of stress tensor and the second invariant of stress deviator respectively, and $c$ and $\phi$ are the cohesion and the frictional angle of the rock block.

The tensile or shear failure of the joint planes in the rock mass follows the criterion of

$$\left.\begin{aligned} -\sigma &\geq 0 \\ |\tau| &\geq \sigma \mathrm{tg}\phi'_j + C'_j \end{aligned}\right\} \tag{4.13}$$

$\sigma$ is the normal stress on the joint plane (negative sign stands for tension); $|\tau|$ is the absolute value of the shear stress on the joint plane; $c'_j$ and $\phi'_j$ are the cohesion and friction angle of the joint plane of failure. $c'_j$ and $\phi'_j$ are calculated through weighted mean method taking account of the cohesion and friction angle of the existing joint and the rock block as well as the joint persistence.

### 4.3.2 Sensitivity analysis of an underground cavern complex

The sensitivity analysis is to study the effect caused by the error in the rock mass parameters on the damaged zones of the underground cavern complex. The underground cavern complex comprises the main power cavern, transforming cavern and tailtrace surge cavern, as shown in Figure 4.3. Details of the project are outlined in the Section 4.1.2.

The single-parameter approach is adopted for the sensitivity analysis [255–261]. The method allows each parameter to vary at a time within a possible range, and then derives the corresponding variation of the damage zone in the surrounding rock mass.

In order to compare the sensitivities of all the parameters, the dimensionless sensitivity factor is defined as:

$$S(k) \triangleq \max\left\{\left(\frac{A_{k\max} - A^*}{A^*}\right)\left(\frac{A^* - A_{k\min}}{A^*}\right)\right\} \tag{4.14}$$

where $S(k)$ is the sensitivity of the parameter $k$, $A^*$ is the area of the damaged zone corresponding to the basic parameter set, $A_{k\max}$ and $A_{k\min}$ are the maximum and minimum areas of the damaged zone within the error domain of the parameter $k$. The definition of the sensitivity factor here is not the same as the one defined in the previous section.

The sensitivity analysis is aimed at the following parameters: rock material elastic modulus $E_r$, Poisson's ratio $\nu_r$, cohesion $C_r$ and internal friction angle $\phi_r$; rock joint $E_j$, $u_j$, $c_j$, $\phi_j$ and dip angle $\alpha_j$, persistence $\eta_j$ and specific width $\beta_j$. The basic values and error ranges of the parameters are summarised in Table 4.5.

### 4.3.3 Result and analysis

***4.3.3.1 Effect of parameters on damaged zones.*** The relationship between parameter's error and damaged zone area has been obtained through computation. Examples are given in Figure 4.14 showing the relationships between the main parameters ($C_r$, $\phi_r$ and $E_{j1}$) to the damaged zone area. The values of the parameters and the damaged zone area are normalised according to respective basic values.

**Table 4.5.** Basic value and error range of parameters.

| Parameter | $E_r$ ($10^3$ MPa) | $\nu_r$ | $C_r$ (MPa) | $\phi_r$ (°) | $E_{j1}$ ($10^3$ MPa) | $\nu_{j1}$ | $c_{j1}$ (MPa) | $\phi_{j1}$ (°) | $\alpha_{j1}$ (°) |
|---|---|---|---|---|---|---|---|---|---|
| Basic value | 58.0 | 0.21 | 2.5 | 48.0 | 5.0 | 0.23 | 0.1 | 24 | 70 |
| Error range | 46.4 ~ 69.6 | 0.168 ~ 0.252 | 1.5 ~ 2.5 | 28.8 ~ 67.2 | 1.0 ~ 9.0 | 0.184 ~ 0.276 | 0 ~ 0.2 | 19 ~ 29 | 50 ~ 90 |

| Parameter | $\eta_{j1}$ | $\beta_{j1}$ (%) | $E_{j2}$ ($10^3$ MPa) | $\nu_{j2}$ | $c_{j2}$ (MPa) | $\phi_{j2}$ (°) | $\alpha_{j2}$ (°) | $\eta_{j2}$ | $\beta_{j2}$ (%) |
|---|---|---|---|---|---|---|---|---|---|
| Basic value | 0.6 | 3.2 | 14.0 | 0.23 | 0.1 | 24 | 110 | 0.6 | 3.2 |
| Error range | 0.4 ~ 0.8 | 1.6 ~ 4.8 | 2.8 ~ 252 | 0.184 ~ 0.276 | 0 ~ 0.2 | 19 ~ 29 | 90 ~ 130 | 0.4 ~ 0.8 | 1.6 ~ 4.8 |

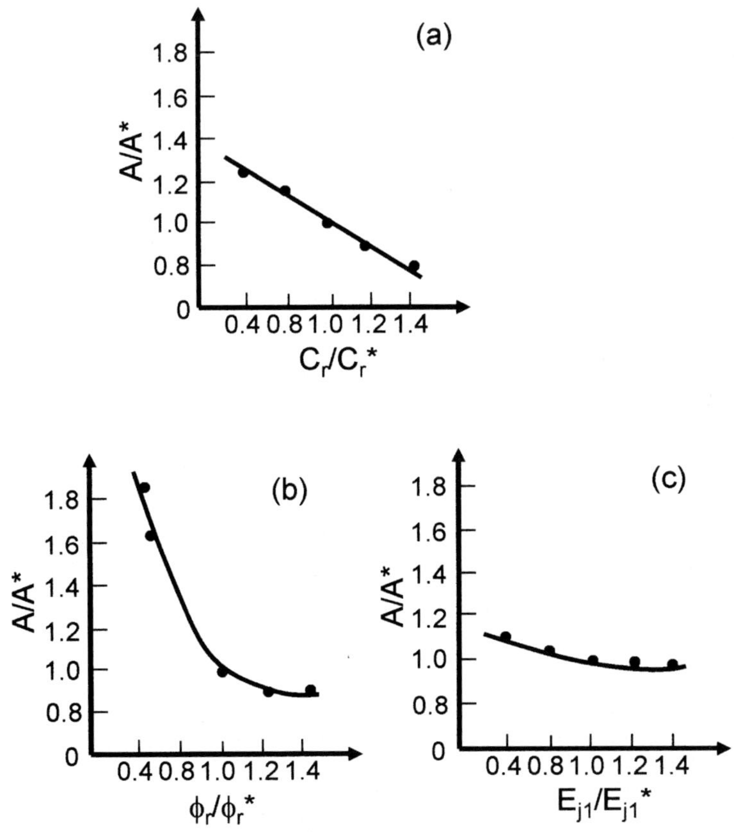


**Figure 4.14.** Effects of rock mass parameters on damaged zone area.

It can be seen from Figure 4.14 that the damaged zone area increases linearly with decreasing $c_r$, while it increases abruptly with decreasing $\phi_r$. Therefore, special attention should be paid to $\phi_r$ with respect to rock stability. The decrease of the joint elastic modulus also results in the increase of the damaged zone area. But its sensitivity is much lower than that of $\phi_r$ and $c_r$.

***4.3.3.2 Comparison between sensitivities of various parameters.*** Based upon the error ranges of parameters listed in Table 4.5, the sensitivity factors of all the parameters have been obtained as defined by equation (4.14). The sensitivity factors are summarised in Table 4.6.

A parameter with a sensitivity factor greater than 0.2 ($S \geq 0.2$) is defined as a highly sensitive parameter. $S \geq 0.2$ means that 20% of the apparent error in damage zone area is resulted from the parameter error. A parameter with $0.04 < S < 0.2$ is a moderately sensitive parameter. A parameter of $S < 0.04$ is considered non-sensitive parameter. From the results, each parameter can be arranged from high to low sensitivity as follows:

Highly sensitive parameters: $\phi_r$, $c_r$ (high to low)
Moderately sensitive parameters: $E_{j1}$, $\eta_{j1}$, $E_{j2}$, and $\alpha_{j1}$ (high to low)
Non-sensitive parameters: $\beta_{j1}$, $\nu_r$, $\eta_{j2}$, $\beta_{j2}$, $E_r$, $\alpha_{j2}$, $\nu_{j1}$, $c_{j1}$, $\phi_{j1}$, $\nu_{j2}$, $c_{j2}$, and $\phi_{j2}$.

The results show that the strength of the rock material is the most critical factor affecting the damaged zones in the surrounding rock mass. Particularly, the internal friction angle, $\phi_r$, is the most sensitive parameter. On the other hand, the deformation parameter of the rock material has little effect on the damaged zone.

The deformation and geometry properties of the joints have certain effects on the size of damaged zone in the surrounding rock mass. The comparison between various parameters of the same joint set shows that the elastic modulus is the most sensitive parameter, the persistence and dip angle have almost the same sensitivity factor. The Poisson's ratio is non-sensitive. As in the previous sections, the effect of joint parameter with a lower elastic modulus on the damaged zone area is larger than that of the joint set with a higher elastic modulus.

**Table 4.6.** Sensitivity of various parameters.

| | | | | | | | |
|---|---|---|---|---|---|---|---|
| Rock parameter | $S(E_r)=$ 0.014 | $S(\nu_r)=$ 0.035 | $S(c_r)=$ 0.264 | $S(\phi_r)=$ 0.837 | | | |
| Joint set I | $S(E_{j1})=0.101$ | $S(\nu_{j1})=0$ | $S(c_{j1})=0$ | $S(\phi_{j1})=0$ | $S(\alpha_{j1})=0.047$ | $S(\eta_{j1})=0.053$ | $S(\beta_{j1})=0.036$ |
| Joint set II | $S(E_{j2})=0.051$ | $S(\nu_{j2})=0$ | $S(c_{j2})=0$ | $S(\phi_{j2})=0$ | $S(\alpha_{j2})=0.034$ | $S(\eta_{j2})=0.035$ | $S(\beta_{j2})=0.010$ |

The computation results also show that the joint strength parameters of $c_j$ and $\phi_j$ have no effects on the damaged zone, which can be explained from the following two aspects:

(i) Under the given in situ stresses of $\sigma_x = 13.3$ MPa and $\sigma_y = 9.5$ MPa, the damage of the surrounding rock mass mainly behaves yielding flow and only slight damage takes place along the joint plane.
(ii) With the joint persistence at 0.6, rock mass strength is governed by rock material properties rather than those of the joint. Therefore, the effect of the change in joint strength parameters on the damaged zone is not being reflected.

### 4.3.4 Summary

(i) The strength parameters of rock materials are the main factors affecting the size of the damaged zone. When there is a 20% relative error in $\phi_r$ and $c_r$, they will produce about 83.7% and 26.4% increments in damaged zone areas, respectively. Therefore, careful assessment of the two parameters is important to control damage zone. Strength reinforcement such as systematic rock bolts can be applied to reduce the damaged zone.
(ii) The elastic modulus of the joint set I also affects the damaged zone area in the surrounding rock mass considerably. The existence of *weaker* joint set has greater effect on rock mass stability, and vice versa.
(iii) The effect of the geometric parameters of joint sets on the damaged zone area is relatively low and often can be neglected.

Chapter 5

# Stability Analysis of Rheologic Rock Mass

Rocks and rock masses often exhibit rheologic behaviour, especially weak and soft rocks or highly jointed rock masses. Rheologic behaviour is the time-dependent characteristics of the material deformation and strength [262–275]. For example, the deformation of a rock mass may increase under a constant loading with time elapsing, i.e., creep effect, and the strength may decrease. In the case of underground works, the phenomenon is often found that due to the rheologic behaviour, the loading on support elements gradually increases, leading to the final failure of the system [276–286]. In the vicinity of an excavated opening, weak rocks or jointed rock masses can creep and deform visco-elastically [287–301]. Often, stresses in the surrounding rock mass exceed the rock strength, causing the rock mass to become visco-plastic and increasing the visco-plastic composition in the total deformation [302–306]. The understanding of rheologic characteristics is important to the design of underground excavation and support. The reinforcement and support design for the rheologic rock mass must take into consideration the visco-elastic and visco-plastic deformation.

## 5.1. RHEOLOGICAL MECHANICAL MODELS FOR ROCKS AND ROCK MASSES

There are two basic approaches to study the rheological phenomena of a rock mass. The first approach is from the macro point of view to study synthetical and mechanical behaviour of the rock mass by taking a large volume to represent the whole rock mass containing adequate quantities of discontinuities [17–19,307]. The second is to study the intact rock material and discontinuities individually and then to assemble them together [15,20,21]. Here, emphasis is laid on the first approach.

There are three basic ideal bodies for common rheological mechanical models: Hooke's elastic solid, Newton's viscous liquid and St. Venant's plastic mass [264,291]. In the rock mechanics, these basic bodies are often used to model a variety of rock masses and to simulate different rheological characteristics of the rock masses. Table 5.1 summarises the common rheological models of rock mechanics. It should be noted that all the models are linear. The stress–strain curves of the models describe the relationship between deviator tensors. The equations are mainly applicable to the uniaxial loading state. The equations that describe multi-axial loading state can be derived through the superposition theorem.

**Table 5.1.** Typical mechanical model, formula and rock type.

| Model | Model illustration | Formula | Rheological type | Rock type |
|---|---|---|---|---|
| Hooke (H) |  | $\sigma = E\varepsilon$ | Elastic | Hard or relatively hard rock |
| Newton (N)<br>St. Venant (V) |  | $\sigma = 2\eta\dot{\varepsilon}$<br>$\sigma = \theta$ | Viscous<br>Plastic | Soft rock<br>Soft rock or rock under high confining pressure |
| Kelvin (K = H/N) |  | $\sigma = 2G\varepsilon + 2\eta\dot{\varepsilon}$ | Visco-elastic | Most medium to soft rock (elastic post effect) |
| Maxwell (M = H – N) |  | $\sigma + \frac{\eta}{G}\dot{\sigma} = 2\eta\underline{\varepsilon}$ | Visco-elastic (plastic) | Halite under long-term load or rock at depth |
| Prandtl (P = H – V) |  | $\begin{cases} \sigma = E\varepsilon \text{ when } \sigma < \theta \\ \sigma = E(\varepsilon - \varepsilon_{pl}) \text{ when } \sigma \geq \theta \end{cases}$ | Elasto-plastic | Rock under certain confining pressure |
| Poynting-Thomson (PT = H/(H – N)) |  | $\sigma + \frac{\eta m}{GM}\sigma = 2GH\varepsilon + \frac{GM + GH}{GM}\eta M\dot{\varepsilon}$ | Visco-elasto-viscous | Medium strength rock and most sedimentation rock |
| Bingham (B = H – (N/V)) |  | $\begin{cases} \sigma = E\varepsilon \text{ when } \sigma < \theta \\ \sigma + \frac{\eta}{G}p\dot{\sigma} = \theta + 2\eta p\dot{\varepsilon} \\ \text{when } \sigma \geq \theta \end{cases}$ | Elasto-visco-plastic | Soft rock and soil under high confining pressure |
| Burgers (Bu = M – K) | 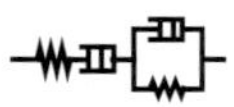 | $\sigma + \frac{\eta M(G_M + G_K) + \eta K G_M}{G_M G_K}\dot{\sigma} + \frac{\eta_M \eta_K}{G_M G_K}\ddot{\sigma} = 2\eta M\left(\dot{\varepsilon} + \frac{\eta_K}{G_K}\ddot{\varepsilon}\right)$ | Visco-elastic (plastic) | Soft rock such as clayed shale, mudstone |
| Schofield–Scott–Blair (SSB = K – (M/V)) |  | $\sigma + \frac{G_M\eta_p + G_K\eta_p + G_M\eta_K}{G_M G_K}\dot{\sigma}$<br>$\frac{\eta_p\eta_K}{G_M G_K}\ddot{\sigma} = \theta + 2\eta p\dot{\varepsilon} + \frac{2\eta_p\eta_K}{G_K}\ddot{\varepsilon}$ | Elasto-visco-plastic | Halite under long-term load |

Note: $\sigma$ – deviator component of stress tensor, $\varepsilon$ – deviator component of strain tensor, $G$ – shearing module and $\eta$ – viscosity factor.

From the testing results on the rheological characteristics of most rock materials, it can be seen that the constitutive relations of rocks and rock masses are all non-linear. This is reflected in two aspects: firstly, the rheological mechanical parameters, such as $G$, $\eta$, $E$ and $\nu$ are not constant but the functions of stress level and/or time; and secondly, strain, stress and their rates are also related non-linearly [19,194,308,309]. The universal constitutive relation of rocks and rock masses can be expressed as

$$\begin{aligned} &a(\sigma,t)(\sigma)^{n_1} + b(\sigma,t)(\dot{\sigma})^{n_2} + c(\sigma,t)(\ddot{\sigma})^{n_3} \\ &\quad = \theta + \alpha(\sigma,t)(\varepsilon)^{n_4} + \beta(\sigma,t)(\dot{\varepsilon})^{n_5} + \gamma(\sigma,t)(\ddot{\varepsilon})^{n_6} \end{aligned} \tag{5.1}$$

where $n_1, n_2, \ldots, n_n$ are all positive, $\theta$ is the threshold value of stress at which the rock material enters the plastic state.

The above non-linear constitutive relation can be used provided that all parameters concerned are available by sufficient testing of rock behaviour, i.e., the functional factors such as $a(\sigma,t)$ and power values $n_n$ of stress and strain can be obtained. However, it is difficult to determine so many parameters or variables as the rheologic tests are time consuming. In practice, simplified linear models are often adopted for engineering applications [308,309]. Equation (5.2), for example, is a common constitutive rheological model adopted in rock mechanics study to simulate a variety of rocks:

$$a\sigma + b\dot{\sigma} + c\ddot{\sigma} = \theta + \alpha\varepsilon + \beta\dot{\varepsilon} + \gamma\ddot{\sigma} \tag{5.2}$$

where $a$, $b$, $c$, $\theta$, $\alpha$, $\beta$, $\gamma$ are specially defined as constants. It can be seen that the mathematical equations of various rheological mechanical models in Table 5.1 are in fact special cases of equation (5.2) or different combinations of equation (5.2).

## 5.2. VISCO-ELASTIC SURROUNDING ROCK MASS AND SUPPORTING PROBLEM

Most rock masses involved in underground engineering exhibit characteristics of visco-elasticity to different extents except some very hard and massive rocks [194,308]. It is therefore of paramount importance to study the behaviour of the visco-elastic rock masses. Nowadays, numerical analyses including finite element method (FEM) and boundary element method (BEM) have been widely used to study the stress state of the surrounding rock masses. Nevertheless, it is still of great significance to use the analytical method as it provides theoretical solutions in limited forms. The final expression of this method contributes to direct understanding of the problems under consideration.

### *5.2.1 General solution for circular visco-elastic media*

In this section, the excavated opening, the support lining and the surrounding rock mass are treated as a generalised two-dimensional problem. They are assumed to be homogeneous, isotropic and of finite deformation.

A triple-element model of visco-elasticity (Figure 5.1) is adopted. It is equivalent to the Poynting–Thomson model (see Table 5.1). Assuming that the deformations of each element have their own independent physical equations and the Poisson's ratio is constant, then the rheological physical equations for the plane problems can be expressed as [8,52]:

$$\left.\begin{aligned}
&\tau + \Theta\dot{\tau} = G\gamma + \eta\dot{\gamma} \\
&\sigma_Z + \Theta\dot{\sigma}_Z = 2G\left(\varepsilon_Z + \frac{3\nu}{1-2\nu}\varepsilon\right) + 2\eta\left(\dot{\varepsilon}_Z + \frac{3\nu}{1-2\nu}\dot{\varepsilon}\right) \\
&\sigma_X + \Theta\dot{\sigma}_X = 2G\left(\varepsilon_X + \frac{3\nu}{1-2\nu}\varepsilon\right) + 2\eta\left(\dot{\varepsilon}_X + \frac{3\nu}{1-2\nu}\dot{\varepsilon}\right) \\
&\sigma_Y + \Theta\dot{\sigma}_Y = 2G\frac{3\nu}{1-2\nu}\varepsilon + 2\eta\frac{3\nu}{1-2\nu}\dot{\varepsilon}
\end{aligned}\right\} \tag{5.3}$$

where $\dot{\sigma}_X$, $\dot{\sigma}_Y$, $\dot{\sigma}_Z$ and $\dot{\tau}$ are the normal stress rate and shear stress rate along the directions of $X$, $Y$, $Z$; $\dot{\varepsilon}_X$, $\dot{\varepsilon}_Z$ and $\dot{\gamma}$ are the normal strain rate and the shear strain

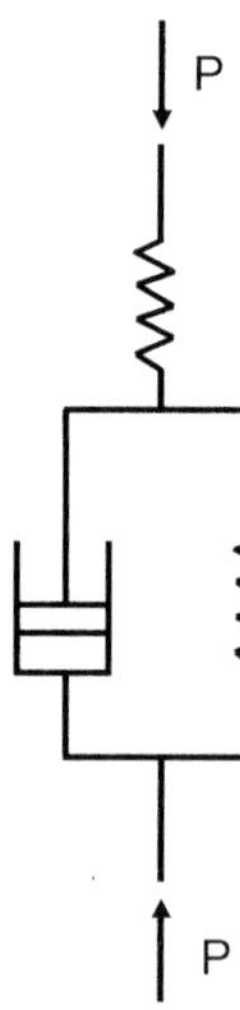


**Figure 5.1.** Rheological model of triple-element.

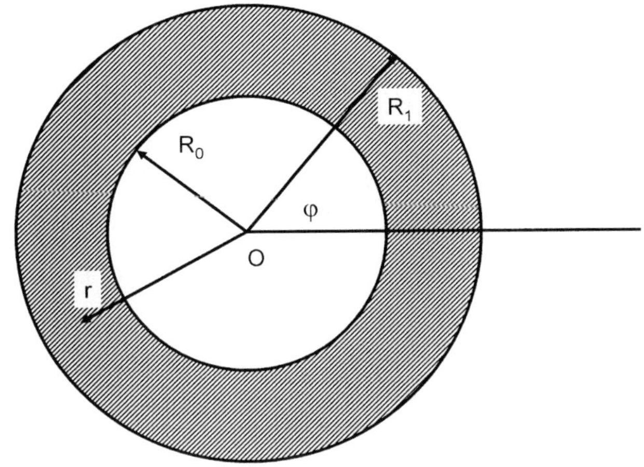


**Figure 5.2.** Coordinate system of circular medium.

rate along $X$, $Z$; $\Theta$ is the relaxation time, $G$ is the elastic shear module (infinitesimal loading rate) and $\eta$ is the viscosity of shearing deformation.

Writing equation (5.3) in the form of integration equation and substituting it into equilibrium equation and considering geometric equation, one can obtain the Volter integration equation set of the second type with the body force neglected. The equation is integral. The components of the stress in a circular medium derived from the general solution can be expressed in polar coordinates [8,52], as shown in Figure 5.2:

$$\begin{aligned} U_r(t) = \frac{r}{2G}\Bigg\{&(1-2\nu)C_0'(t) + b_0'(t)S^2 + \left[a_1'(t)(3-4\nu)S\ln\frac{1}{S}\right.\\ &\left.- b_1(t)S^3 - (1-4\nu)C_1(t)q + d_1(t)\frac{1}{q}\right]Q_1(\varphi)\\ &+ \sum_{n=2}^{\infty}\Big[A_{z,n}(\nu)a_n(t)S^n - nb_n(t)S^{n+2}\\ &+ A_{w,n}(\nu)C_n(t)q^n + nd_n(t)q^{n-2}\Big]Q_n(\varphi)\Bigg\} \end{aligned} \tag{5.4}$$

$$\begin{aligned} W_\theta(t) = \frac{r}{2G}&\left\{a_1'(t)S\left[(3-4\nu)S\ln\frac{1}{S}+1\right] + b_1(t)S^3 + (5-4\nu)C_1(t)q + d_1(t)q^{-1}\right\}\\ &\times\frac{dQ(\varphi)}{d\varphi} + \frac{r}{2G}\sum_{n=2}^{\infty}\Big[B_{z,n}(\nu)a_n(t)S^n + b_n(t)S^{n+2}\\ &\qquad + B_{w,n}(\nu)C_nq^n + d_n(t)q^{n-2}\Big]\frac{dQ_n(\varphi)}{d\varphi} \end{aligned} \tag{5.5}$$

$$\sigma_r(t) = KT\{[C_0'(t) - b_0'(t)S^2] + [(3-2\nu)Sa_1'(t) + 2b_1(t)S^3 - 2C_1(t)q]Q(\varphi)\}$$
$$+ K(1-KT)\int_{t_0}^{t}\{[C_0'(t') - b_0'(t')S^2] + [(3-2\nu)Sa_1'(t') + 2b_1(t')S^3 - 2C_1(t')q]$$
$$\times Q(\varphi)\}e^{K(t'-t)}\mathrm{d}t' - KT \times \sum_{n=2}^{\infty}[n(n-1)(n+2)a_n(t)S^n - n(n+1)b_n(t)S^{n+2}$$
$$- n(n+1)(n-2)c_n(t)q^n - n(n-1)dn(t)q^{n-2}]Q_n(\varphi) - K(1-KT)\sum_{n=2}^{\infty}\int_{t_0}^{t}[n(n-1)$$
$$\times (n+2)a_n(t')S^n - n(n+1)b_n(t')S^{n+2} - n(n+1)(n-2)c_n(t')q^n$$
$$- n(n-1)d_n(t')q^{n-2}]Q_n(\varphi)e^{K(t'-t)}\mathrm{d}t' \tag{5.6}$$

$$\sigma_\theta(t) = KT\{[C_0'(t) - b_0'(t)S^2] - [(1-2\nu)a_1'(t)S + 2b_1(t)S^3 + 6C_1(t)q]Q(\varphi)\}$$
$$+ K(1-KT)\int_{t_0}^{t}\{[C_0'(t') + b_0'(t')S^2] - [(1-2\nu)a_1'(t')S + 2b_1(t')S^3 + 6C_1(t')q]$$
$$\times Q(\varphi)\}e^{K(t'-t)}\mathrm{d}t' + KT\sum_{n=2}^{\infty}[n(n-1)(n+2)a_n(t)S^n - n(n+1)b_n(t)S^{n+2}$$
$$- n(n+1)(n+2)C_n(t)q^n - n(n-1)d_n(t)q^{n-2}]Q_n(\varphi) + K(1-KT)\sum_{n=2}^{\infty}\int_{t_0}^{t}[n(n-1)$$
$$\times (n-2)a_n(t')S^n - n(n+1)b_n(t')S^{n+2} - n(n+1)(n+2)C_n(t')q^n$$
$$- n(n-1)d_n(t')q^{n-2}]Q_n(\varphi)e^{K(t'-t)}\mathrm{d}t' \tag{5.7}$$

$$\sigma_y(t) = KT\nu\{2C_0(t) + [2a_1(t)S - 8C_1(t)q]Q(\varphi)\}$$
$$+ K(1-KT)\nu\int_{t_0}^{t}\{2C_0(t') + [2a_1(t')S - 8C_1(t')q]Q_1(\varphi)e^{K(t'-t)}\mathrm{d}t' - KT\nu$$
$$\times \sum_{n=2}^{\infty}[4n(n-1)a_n(t)S^n + 4n(n+1)C_n(t)q^n]Q_n(\varphi) - K(1-KT)\nu$$
$$\times \sum_{n=2}^{\infty}\int_{t_0}^{t}[4n(n-1)a_n(t')S^n + 4n(n+1)C_n(t')q^n]Q_n(\varphi)e^{K(t'-t)}\mathrm{d}t' \tag{5.8}$$

$$\tau(t) = KT[(1-2\nu)a_1(t)S - 2b_1(t)S^3 + 2C_1(t)q]\frac{\mathrm{d}Q(\varphi)}{\mathrm{d}(\varphi)}$$
$$+ K(1-KT)\int_{t_0}^{t}[(1-2\nu)a(t')S - 2b_1(t')S^3 + 2C_1(t')q]\frac{\mathrm{d}Q_1(\varphi)}{\mathrm{d}\varphi}e^{K(t'-t)}\mathrm{d}t'$$

$$+KT\sum_{n=2}^{\infty}[n(n-1)a_n(t)S^n-(n+1)b_n(t)S^{n+2}+n(n+1)\times C_n(t)q^n+(n-1)d_n(t)q^{n-2}]$$

$$\times\frac{\mathrm{d}Q_n(\varphi)}{\mathrm{d}\varphi}+K(1-KT)\sum_{n=2}^{\infty}\int_{t_0}^{t}[n(n-1)\times a_n(t')S^n-(n+1)b_n(t')S^{n+2}+n(n+1)C_n(t')q^n$$

$$+n(n-1)d_n(t')q^{n-2}]\frac{\mathrm{d}Q_n(\varphi)}{\mathrm{d}\varphi}e^{K(t'-t)}\mathrm{d}t' \quad (5.9)$$

$$\left.\begin{array}{ll} A_{\tau,n}(\nu)\equiv n[n+2(1-2\nu)] & A_{w,n}(\nu)=n[n-2(1-2\nu)] \\ B_{z,n}(\nu)=n+4(1-\nu) & B_{w,n}(\nu)=n+4(1-\nu) \\ a_n(t)=[a'_n(t),a''_n(t)] & b_n(t)=[b'_n(t),b''_n(t)] \\ C_n(t)=[C'_n(t),C''_n(t)] & d_n(t)=[d'_n(t),d''_n(t)] \\ Q_n(\varphi)=(\cos n\varphi,\sin n\varphi) & Q_1(\varphi)=(\cos\varphi,\sin\varphi) \end{array}\right\} \quad (5.10)$$

where

$$S=\frac{R_0}{r} \qquad q=\frac{r}{R_1} \qquad n=2,3,4\cdots$$

$$K=\frac{1}{\Theta}\ \text{(reciprocal of the relaxation time)}$$

$$T=\frac{\eta}{G}\ \text{(time of post-elastic behaviour)}$$

The functions of cos φ [with coefficients of $a'_n(t)$, $b'_n(t)$, $C'_n(t)$, $d'_n(t)$] and sin φ [with $a''_n(t)$, $b''_n(t)$, $C''_n(t)$, $d''_n(t)$] are determined from the boundary conditions.

Applying the previous equation sets describing the components of displacement and stress, one can solve all problems (which can be expressed by the Fourier series) with visco-elastic medium boundary conditions.

### 5.2.2 *Interaction of visco-elastic surrounding rock mass and elastic lining*

Long tunnels at depth can often be treated as plane problems for stability analysis. When the overlying depth ($H$) is greater than 30 times of tunnel radius, the gravitational field can be replaced by the stress state in an infinite plane approximately, as shown in Figure 5.3. The initial vertical stress is $P_z=\gamma_0 H$ and horizontal stress is $P_x=\lambda_1 P_z=\lambda_1\gamma_0 H$. $H$ is the overlying depth, $\gamma$ is the unit weight and $\lambda_1$ is the lateral pressure coefficient.

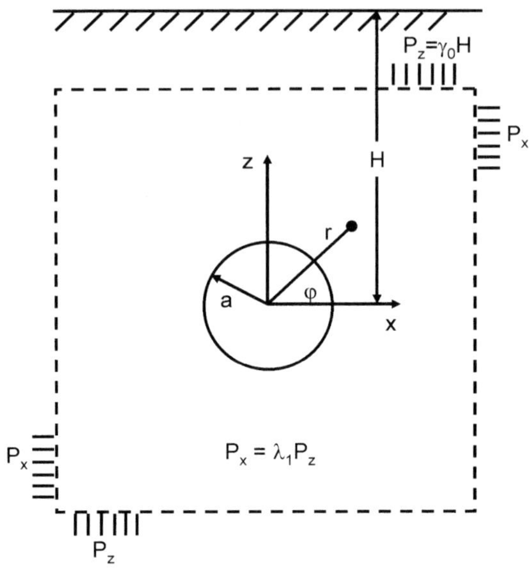


**Figure 5.3.** Plane problem of tunnel stability analysis.

Before excavation, the undisturbed rock mass is generally in an elastic state. According to the theory of elasticity, its initial stress state is

$$\left.\begin{aligned}\sigma_{r,0}^{(2)} &= -\tfrac{1}{2}p_z[(1+\lambda_1)-(1-\lambda_1)\cos 2\varphi]\\ \sigma_{\theta,0}^{(2)} &= -\tfrac{1}{2}p_z[(1+\lambda_1)+(1-\lambda_1)\cos 2\varphi]\\ \sigma_{y,0}^{(2)} &= -\lambda_1 p_z\\ \tau_0^{(2)} &= -\tfrac{1}{2}p_z(1-\lambda_1)\sin 2\varphi\end{aligned}\right\} \tag{5.11}$$

and initial displacement components are

$$\left.\begin{aligned}U_{r,0}^{(2)} &= \frac{-r}{4G_2}(1-\lambda_1)r_0H(1-\cos 2\varphi)\\ W_{\theta,0}^{(2)} &= \frac{-r}{4G_2}(1-\lambda_1)r_0H\sin 2\varphi\\ V_{y,0}^{(2)} &= 0\end{aligned}\right\} \tag{5.12}$$

If an opening is excavated, the initial stress state around the opening periphery can be described as: $(r=a)$, $S_0^{(2)} = S_0^{(2)}(\sigma_{r,0}^{(2)}, \sigma_{\theta,0}^{(2)}, \sigma_{y,0}^{(2)}, \tau_0^{(2)}, U_{r,0}^{(2)}, W_{\theta,0}^{(2)}, V_{y,0}^{(2)})$. Such a disturbed stress state does not meet the boundary conditions, and a compensating stress state, $S^{(2)}$, is produced in the same area. The new stress state, $S_c^{(2)}$, is the summation of the above two stress states, i.e.,

$$S_c^{(2)}\left(\sigma_{r,c}^{(2)},\sigma_{\theta,c}^{(2)},\sigma_{y,c}^{(2)},\tau_c^{(2)},U_{r,c}^{(2)},W_{\theta,c}^{(2)},V_{y,c}^{(2)}\right) = S_0^{(2)}\left(\sigma_{r,0}^{(2)},\sigma_{\theta,0}^{(2)},\sigma_{y,0}^{(2)},\tau_0^{(2)},U_{r,0}^{(2)},W_{\theta,0}^{(2)},V_{y,0}^{(2)}\right)$$
$$+ S^{(2)}\left(\sigma_r^{(2)},\sigma_\theta^{(2)},\sigma_y^{(2)},\tau^{(2)},U_r^{(2)},W_\theta^{(2)},V_y^{(2)}\right) \tag{5.13}$$

Compared with the stress on the lining after excavation, the stress state produced by the self weight is so small that it can be neglected. The initial stress state in the lining, $S_0^{(1)}$, can be considered as zero, so

$$S_c^{(1)}\left(\sigma_{r,c}^{(1)},\sigma_{\theta,c}^{(1)},\sigma_{y,c}^{(1)},\tau_c^{(1)},U_{r,c}^{(1)},W_{\theta,c}^{(1)},V_{y,c}^{(1)}\right) = S^{(1)}\left(\sigma_r^{(1)},\sigma_\theta^{(1)},\sigma_y^{(1)},\tau^{(1)},U_r^{(1)},W_\theta^{(1)},V_y^{(1)}\right) \tag{5.14}$$

When the overlying depth above the opening is sufficiently great and the initial in situ stress state is nearly of hydrostatic stress, $P_x = P_y = P_z = \gamma_0 H$, then from equations (5.11) and (5.12) we have

$$\left.\begin{aligned} \sigma_{r,0}^{(2)} &= \sigma_{t,0}^{(2)} = \sigma_{y,0}^{(2)} = -P \\ \tau_0^{(2)} &= 0 \\ U_{r,0}^{(2)} &= W_{t,0}^{(2)} = V_{y,0}^{(2)} = 0 \end{aligned}\right\} \tag{5.15}$$

The stress state after excavation can be derived accordingly from equations (5.4), (5.5), (5.6), (5.7), (5.8) and (5.9). In consideration of one-dimensional axial symmetry of the stress state, all terms relating to the polar angle φ are equal to zero, then

$$\begin{aligned}
\sigma_r^{(2)}(t) &= K_2T_2\left[C'_{0,2}(t) - b_{0,2}(t)S_2^2\right] + K_2(1-K_2T_2)\int_0^t \left[C'_{0,2}(t') - b'_{0,2}(t')S_2^2\right] \times e^{k_2(t'-t)}\mathrm{d}t' \\
\sigma_\theta^{(2)}(t) &= K_2T_2\left[C'_{0,2}(t) + b'_{0,2}(t)S_2^2\right] + K_2(1-K_2T_2)\int_0^t \left[C'_{0,2}(t') + b'_{0,2}(t')S_2^2\right] \times e^{k_2(t'-t)}\mathrm{d}t' \\
\sigma_y^{(2)}(t) &= K_2T_2C'_{0,2}(t) + K_2(1-K_2T_2)\int_0^t C'_{0,2}(t')e^{k_2(t'-t)}\mathrm{d}t' \\
\tau^{(2)}(t) &= 0 \\
U_r^{(2)}(t) &= \frac{1}{2G_2}r_2\left[(1-2\nu)C'_{0,2}(t) + b'_{0,2}(t)S_2^2\right] \\
W_\theta^{(2)}(t) &= 0 \\
V_y^{(2)} &= 0
\end{aligned} \tag{5.16}$$

where $K_2$, $T_2$, $G_2$ are the mechanical parameters of the rheological characteristics of the surrounding rock mass; $S_2 = a/r$, $r$ is the radial distance to any point in the rock mass from the centre of the circular tunnel.

Because the second stress state after excavation is the sum of the initial and the compensating stress state (equation (5.13)), then

$$\begin{aligned}
&\sigma_{r,c}^{(2)} = K_2T_2\left[C'_{0,2}(t) - b'_{0,2}(t)S_2\right] + K_2(1-K_2T_2)\int_0^t\left[C'_{0,2}(t') - b'_{0,2}(t')S_2^2\right]\cdot e^{k_2(t'-t)}\mathrm{d}t' - P\\
&\sigma_{\theta,c}^{(2)}(t) = K_2T_2\left[C'_{0,2}(t) + b'_{0,2}(t)S_2^2\right] + K_2(1-K_2T_2)\int_0^t\left[C'_{0,2}(t') + b'_{0,2}(t')S_2^2\right]\cdot e^{k_2(t'-t)}\mathrm{d}t' - P\\
&\sigma_{y,c}^{(2)}(t) = K_2T_2C'_{0,2}(t) + K_3(1-K_3T_3)\int_0^t C'_{0,2}(t')e^{k_2(t'-t)}\mathrm{d}t' - P\\
&\tau_c^{(2)}(t) = 0\\
&U_{r,c}^{(2)}(t) = \frac{1}{2G_2}r_2\left[(1-2\nu)C'_{0,2}(t) + b'_{0,2}(t)S_2^2\right]\\
&W_{\theta,c}^{(2)}(t) = 0
\end{aligned} \tag{5.17}$$

The stress state in the lining which is assumed to be ideally elastic can be derived from equation (5.16). Since $\theta_1$ and $T_1 = \eta_1/G_1$ have the same order of magnitude for the common solid materials, and $\theta_1 \to 0$, $T_1 \to 0$ for an elastic body, then $K_1T_1 = (T_1/\theta_1) \to 1$, i.e.,

$$\left.\begin{aligned}
K_1T_1 &= 1\\
K_1(1-K_1T_1) &= 0
\end{aligned}\right\} \tag{5.18}$$

In this case, equation (5.16) expressing the stress state in the lining is

$$\left.\begin{aligned}
&\sigma_r^{(1)}(t) = C'_{0,1}(t) - b'_{0,1}(t)S_1^2\\
&\sigma_\theta^{(1)}(t) = C'_{0,1}(t) + b'_{0,1}(t)S_1^2\\
&\sigma_y^{(1)}(t) = 2\nu_1C'_{0,1}(t)\\
&\tau^{(1)} = 0\\
&U_r^{(1)}(t) = \frac{1}{2G_1}\gamma_1\left[(1-2\nu)C'_{0,1}(t)b'_{0,1}(t)S_1^2\right]\\
&W_\theta^{(1)}(t) = V_y^{(1)}(t) = 0
\end{aligned}\right\} \tag{5.19}$$

where $\nu_1$ and $G_1$ are the mechanical parameters of the lining; $S_1 = a_0/r_1$, $r_1$ is the radial distance to a given point in the lining from the centre of the circular tunnel and $a_0$ is the inner radius of the lining.

There are four coefficients in equations (5.17) and (5.19) to be determined; $C'_{0,2}(t)$, $b'_{0,2}(t)$, $C'_{0,1}(t)$ and $b'_{0,1}(t)$, which are the functions of time. They can be obtained according to the following boundary conditions:

$$\left.\begin{array}{lll} \text{(i)} & r_2 \to \infty & \sigma^{(2)}_{r,c} = \sigma^{(2)}_{r,0} + \sigma^{(2)}_{r} = -P \\ \text{(ii)} & r_1 = a & \sigma^{(1)}_{r} = 0 \\ \text{(iii)} & r_1 = r_2 = a & \end{array}\right\} \tag{5.20}$$

$$\sigma^{(1)}_{r} = \sigma^{(2)}_{r} + \sigma^{(2)}_{r,0}$$

$$U^{(1)}_{r} = U^{(2)}_{r} - U^{(2)}_{r,(t=0)}$$

where the last term is the boundary condition of displacement, implicating that once the lining is completed, the elastic displacement at the periphery of the opening is relieved. The term of $U^{(2)}_{r(t=0)}$ can be easily obtained using the method of elasticity theory.

The above four functional coefficients can be obtained by simultaneously solving the simultaneous equation set (5.17), (5.19) and (5.20) that consists of algebraic and integration equations. And finally, the stress states in the surrounding rock mass and the lining can be derived as follows:
for the surrounding rock mass,

$$\left.\begin{array}{l} \sigma^{(2)}_{r,c} = -P\left\{1 - \dfrac{a^2}{r_2^2}[1 - \alpha'(1 - e^{-\beta_2 t})]\right\} \\ \sigma^{(2)}_{\theta,c} = -P\left\{1 + \dfrac{a^2}{r_2^2}[1 - \alpha'(1 - e^{-\beta_2 t})]\right\} \\ \sigma^{(2)}_{y,c} = -P \quad \tau^{(2)}_{c} = 0 \end{array}\right\} \tag{5.21}$$

where

$$\alpha' = \frac{\upsilon \cdot a}{\upsilon \cdot a + 2G_2}\left(1 - \frac{1}{T_2 K_2}\right)$$

$$\upsilon = 2G_1 \frac{a^2 - a_0^2}{a} \cdot \frac{1}{a_0^2 + (1 - 2\nu_1)a^2}$$

and for the lining,

$$\left.\begin{aligned}\sigma_{r,c}^{(1)} &= -\alpha_1' P(1 - e^{-\beta_2 t})\left(1 - \frac{a_0^2}{r_1^2}\right)\\ \sigma_{\theta,c}^{(1)} &= -\alpha_1' P(1 - e^{-\beta_2 t})\left(1 + \frac{a_0^2}{r_1^2}\right)\\ \sigma_{y,c}^{(1)} &= -2\nu_1\alpha_1' P(1 - e^{-\beta_2 t})\end{aligned}\right\} \tag{5.22}$$

where

$$\alpha_1' = \frac{(T_2K_2 - 1)\left[(1 - 2\nu_1) + X_0^2\right]}{T_2K_2[(1 - 2\nu_1) + X_0^2 + (1 - X_0^2)\overline{G}]}$$

$$\overline{G} = \frac{G_1}{G_2} \qquad X_0 = \frac{a_0}{a}$$

$$\beta_2 = \frac{K_2\left[(1 - 2\nu_1) + X_0^2\right] + (1 - X_0^2)\overline{G}}{K_2T_2\left[(1 - 2\nu_1) + X_0^2\right] + (1 - X_0^2)\overline{G}}$$

It clearly shows that the final stress states in the rock mass and the lining depend upon the following parameters:

(i) the initial stress state, $P$
(ii) thickness of the lining, $X_0 = a - a_0$
(iii) elastic shearing moduli of the lining and the rock mass, $\overline{G} = G_1/G_2$
(iv) relation of post-elastic behaviour time to relaxation time in the rock mass, $T_2K_2 = T_2/\theta_2$
(v) the Poisson's ratio of the lining, $\nu_1$.

The value of $\beta_2$ in the equation for stress states only affects the rate of transition from the initial state to the final state but does not affect the magnitude of the final state. The parameters that are related to $\beta_2$ include $\nu$, $X_0$, $G$, $T_2K_2$ and $T_2$.

### *5.2.3 Interaction of rock mass and lining of different visco-elastic media*

When the surrounding rock mass and the lining are composed of visco-elastic media, but different rheological parameters, the rheological mechanical stress state of the rock mass can still be expressed by equation (5.17). The stress state in lining can be expressed by the following rheological mechanical equations,

$$\left.\begin{aligned}
&\sigma_r^{(1)}(t)=K_1T_1\left[C'_{0,1}(t)-b'_{0,1}(t)S_1^2\right]+K_1(1-K_1T_1)\int_0^t\left[C'_{0,1}(t')-b'_{0,1}(t')S_1^2\right]e^{k_1(t'-t)}\mathrm{d}t'\\
&\sigma_\theta^{(1)}(t)=K_1T_1\left[C'_{0,1}(t)+b'_{0,1}(t)S_1^2\right]+K_1(1-K_1T_1)\int_0^t\left[C'_{0,1}(t')+b'_{0,1}(t')S_1^2\right]e^{k_1(t'-t)}\mathrm{d}t'\\
&\sigma_y^{(1)}(t)=2\mu_1\left[K_1T_1C'_{0,1}(t)+K_1(1-K_1T_1)\int_0^t C'_{0,1}(t')\right]e^{k_1(t'-t)}\mathrm{d}t'\\
&U_r^{(1)}(t)=\frac{r_1}{2G_1}\left[(1-2\nu_1)C'_{0,1}(t)+b'_{0,1}(t)S_1^2\right]\\
&W_\theta^{(1)}(t)=V_y^{(1)}(t)=0\quad \tau^{(1)}(t)=0
\end{aligned}\right\}\tag{5.23}$$

By assuming that the boundary condition is similar to that given in last section, several functional coefficients in the equations of the rock mass and the lining can be determined. The four boundary equations independent to each other are:

(i) when $r_2 \to \infty$

$$\left.\begin{aligned}
&\sigma_{r,0}^{(2)}+\sigma_r^{(2)}=-P\\
&-P+K_2T_2[C'_{0,2}(t)+K_2(1-K_2T_2)\times\textstyle\int_0^t C'_{0,2}(t')e^{k_2(t'-t)}\mathrm{d}t'=-P
\end{aligned}\right\}\tag{5.24a}$$

(ii) when $r_1=a_0$

$$\left.\begin{aligned}
&\sigma_r^{(1)}=0\\
&K_1T_1[C'_{0,1}(t)-b'_{0,1}(t)]+K_1(1-K_1T_1)\times\textstyle\int_0^t[C'_{0,1}(t')-b'_{0,1}(t')]\times e^{k_1(t'-t)}\mathrm{d}t'=0
\end{aligned}\right\}\tag{5.24b}$$

(iii) when $r_1=r_2=a$

$$\sigma_r^{(1)}=\sigma_r^{(2)}+\sigma_{r,0}^{(2)}$$

$$K_1T_1\left[C'_{0,1}(t)-b'_{0,1}(t)X_0^2\right]+K_1(1-K_1T_1)\times\int_0^t\left[C'_{0,1}(t')-b'_{0,1}(t')X_0^2\right]\times e^{k_1(t'-t)}\mathrm{d}t'$$

$$=K_2T_2[C'_{0,2}(t)-b'_{0,2}(t)]+K_2(1-K_2T_2)\times\int_0^t[C'_{0,2}(t')-b'_{0,2}(t')]e^{k_2(t'-t)}\mathrm{d}t'-P$$

(iv) When $r_1=r_2=a$

$$U_r^{(1)}=U_r^{(2)}-U_{r,(t=0)}^{(2)}$$

$$\frac{a}{2G_1}\left[(1-2\nu_1)C'_{0,1}(t)+b'_{0,1}(t)X_0^2\right]=\frac{a_2}{2G_2}b'_{0,2}(t)-\left(-\frac{aP}{2G_2T_2K_2}\right)$$

To obtain the solutions to these four simultaneous equations, it is necessary to solve the following second-order homogeneous differential equations,

$$\bar{a}\ddot{b}'_{0,1}(t) + \bar{b}\dot{b}'_{0,1}(t) + \bar{c}b'_{0,1}(t) = \bar{d} \tag{5.25}$$

where

$$\bar{a} = K_1 T_1 (1 - X_0^2)\overline{G} + K_2 T_2[(1 - 2\nu_1) + X_0^2]$$
$$\bar{b} = K_1 (1 - X_0^2)\overline{G}(1 + K_2 T_2) + K_2[(1 - 2\nu_1) + X_0^2](K_1 T_2 + 1)$$
$$\bar{c} = K_1 K_2[(1 - X_0^2)\overline{G} + (1 - 2\nu_1) + X_0^2]$$
$$\bar{d} = \frac{K_1 \overline{G}(1 - T_2 K_2)}{T_2} P$$

There are three possible solutions to the integration of equation (5.25), depending on the different values of the discriminant of $\Delta = b^2 - 4ac$, i.e., $\Delta > 0$, $\Delta < 0$ and $\Delta = 0$. The main results with the derivation omitted are given below:
When $\Delta \neq 0$, the stress state of the surrounding rock mass

$$\left.\begin{aligned}
\sigma_{r,c}^{(2)} &= -P\left\{1 - \frac{a^2}{r_2^2}[\alpha'_2 - \beta'_2 e^{\bar{r}_1 t} - \beta''_2 e^{\bar{r}_2 t} - r'_2 e^{-k_2 t}]\right\} \\
\sigma_{\theta,c}^{(2)} &= -P\left\{1 + \frac{a^2}{r_2^2}[\alpha'_2 - \beta'_2 e^{\bar{r}_1 t} - \beta''_2 e^{\bar{r}_2 t} - r'_2 e^{-k_2 t}]\right\} \\
\sigma_{y,c}^{(2)} &= -P \qquad \tau_c^{(2)} = 0
\end{aligned}\right\} \tag{5.26}$$

where

$$\left.\begin{aligned}
\alpha'_2 &= \frac{(1 - X_0^2)\overline{G} + [(1 - 2\nu_1) + X_0^2]T_2 K_2}{T_2 K_2[(1 - X_0^2)\overline{G} + (1 - 2\nu_1) + X_0^2]} \\
\beta'_2 &= \bar{C}_1[(1 - 2\nu_1) + X_0^2]\frac{K_2(1 + T_2\bar{\bar{r}}_1)}{K_2 + \bar{\bar{r}}_1} \\
\beta_2 &= \bar{C}_2[(1 - 2\nu_1) + X_0^2]\frac{K_2(1 + T_2\bar{\bar{r}}_2)}{K_2 + \bar{\bar{r}}_2} \\
r_2 &= (K_2 T_2 - 1)\left\{[(1 - 2\nu_1) + X_0^2]\left(\frac{\bar{C}_1}{K_2 + \bar{\bar{r}}_1} + \frac{\bar{C}_2}{K_2 + \bar{\bar{r}}_2}\right)K_2 - a'_2\right\} \\
\bar{C}_1 &= \frac{K_1(T_2 K_2 - 1)\bar{\bar{r}}_2}{T_2(\bar{\bar{r}}_2 - \bar{\bar{r}}_1)\bar{c}} + \frac{T_2 K_2 - 1}{T_2(\bar{\bar{r}}_2 - \bar{\bar{r}}_1)\{K_1 T_1 (1 - X_0^2)\overline{G} + K_2 T_2[(1 - 2\nu_1) + X_0^2]\}} \\
\bar{C}_2 &= \frac{-K_1(T_2 K_2 - 1)\bar{\bar{r}}_1}{T_2(\bar{\bar{r}}_2 - \bar{\bar{r}}_1)\bar{c}} - \frac{T_2 K_2 - 1}{T_2(\bar{\bar{r}}_2 - \bar{\bar{r}}_1)\{K_1 T_1 (1 - X_0^2)\overline{G} + K_2 T_2[(1 - 2\nu_1) + X_0^2]\}} \\
\left.\begin{matrix}\bar{\bar{r}}_1 \\ \bar{\bar{r}}_2\end{matrix}\right\} &= \frac{-b \pm \sqrt{b^2 - 4ac}}{2\bar{a}}
\end{aligned}\right\} \tag{5.27}$$

and the stress state of the lining

$$\left.\begin{aligned}
\sigma_{r,c}^{(1)} &= -\left(1 - \frac{a_0^2}{r_1^2}\right)\overline{G}P\left\{\alpha_1' - \beta_1' e^{\bar{\bar{r}}_1 t} - \beta_1'' e^{\bar{\bar{r}}_2 t} - r_1' e^{-k_1 t}\right\} \\
\sigma_{\theta,c}^{(1)} &= -\left(1 + \frac{a_0^2}{r_1^2}\right)\overline{G}P\left\{\alpha_1' - \beta_1' e^{\bar{\bar{r}}_1 t} - \beta_1'' e^{\bar{\bar{r}}_2 t} - r_1' e^{-k_1 t}\right\} \\
\sigma_{y,c}^{(1)} &= -2\nu_1 \overline{G}P\left\{\alpha_1' - \beta_1' e^{\bar{\bar{r}}_1 t} - \beta_1'' e^{\bar{\bar{r}}_2 t} - r_1' e^{-k_1 t}\right\} \\
\tau_0'' &= 0
\end{aligned}\right\} \tag{5.28}$$

where

$$\left.\begin{aligned}
\alpha_1' &= \frac{T_2 K_2 - 1}{T_2 K_2\left[\left(1 - X_0^2\right)\overline{G} + (1 - 2\nu_1) + X_0^2\right]} \\
\beta_1' &= \frac{\bar{c}_1 K_1\left(1 + T_1 \bar{\bar{r}}_1\right)}{K_1 + \bar{\bar{r}}_1} \\
\beta_1'' &= \frac{\bar{c}_2 K_1\left(1 + T_1 \bar{\bar{r}}_2\right)}{K_1 + \bar{\bar{r}}_2} \\
r_1' &= (K_1 T_1 - 1)\left[\left(\frac{\bar{c}_1}{K_1 + \bar{\bar{r}}_1} + \frac{\bar{c}_2}{K_1 + \bar{\bar{r}}_2}\right)K_1 - \alpha_1'\right]
\end{aligned}\right\} \tag{5.29}$$

When $\Delta = 0$, the derivation procedure is the same.
For the rock mass,

$$\left.\begin{aligned}
\sigma_{r,c}^{(2)} &= -P\left\{1 - \frac{a^2}{r_2^2}\left[\alpha_2' - \beta_2' e^{\bar{r}'' t} - \beta_2''(t) e^{\bar{r}' t} - r_2' e^{-k_2 t}\right]\right\} \\
\sigma_{\theta,c}^{(2)} &= -P\left\{1 + \frac{a^2}{r_2^2}\left[\alpha_2' - \beta_2' e^{\bar{r}'' t} - \beta_2''(t) e^{\bar{r}' t} - r_2' e^{-k_2 t}\right]\right\} \\
\sigma_{y,c}^{(2)} &= -P \qquad \tau_c^{(2)} = 0
\end{aligned}\right\} \tag{5.30}$$

where

$$\alpha_2' = \frac{(1 - X_0^2)\overline{G} + [(1 - 2\nu_1) + X_0^2]T_2K_2}{T_2K_2[(1 - X_0^2)\overline{G} + (1 - 2\nu_1) + X_0^2]}$$

$$\beta_2' = [(1 - 2\nu_1) + X_0^2]\frac{\bar{c}_1''K_2(1 + T_2\bar{r}) + \bar{c}_2''K_2(K_2T_2 - 1)}{(K_2 + \bar{r})^2}$$

$$\beta_2'' = [(1 - 2\nu_1) + X_0^2]\frac{\bar{c}_2''K_2(T_2\bar{r}'' + 1)}{K_2 + \bar{r}''}r_2'(K_2T_2 - 1)\left\{[(1 - 2\nu_1) + X_0^2].\right.$$

$$\left.\times \frac{K_2[\bar{c}_1''(K_2 + \bar{r}'') - \bar{c}_2'']}{(K_2 + \bar{r}'')^2} - \alpha_2'\right\} \tag{5.31}$$

$$\bar{r}'' = -\frac{b}{2a}$$

$$\bar{c}_1'' = \frac{T_2K_2 - 1}{K_2T_2[(1 - 2\nu_1) + X_0^2 + (1 - X_0^2)\overline{G}]}$$

$$\bar{c}_2'' = \frac{(T_2K_2 - 1)\bar{r}''}{K_2T_2[(1 - X_0^2)\overline{G} + (1 - 2\nu_1) + X_0^2]} - \frac{T_2K_2 - 1}{T_2\{K_2T_2[(1 - 2\nu_1) + X_0^2] + K_1T_1(1 - X_0^2)\overline{G}\}}$$

and for the lining,

$$\left.\begin{aligned}
\sigma_{r,c}^{(1)} &= -\left(1 - \frac{a_0^2}{r_1^2}\right)\overline{G}P\{\alpha_1' - \beta_1'e^{\bar{r}''t} - \beta_1''e^{\bar{r}''t} - r_1'e^{-k_1t}\} \\
\sigma_{\theta,c}^{(1)} &= -\left(1 + \frac{a_0^2}{r_1^2}\right)\overline{G}P\{\alpha_1' - \beta_1'e^{\bar{r}''t} - \beta_1''e^{\bar{r}''t} - r_1'e^{-k_1t}\} \\
\sigma_{y,c}^{(1)} &= -2\nu_1\overline{G}P\{\alpha_1' - \beta_1'e^{\bar{r}''t} - \beta_1''e^{\bar{r}''t} - r_1'e^{-k_1t}\} \\
\tau_c^{(1)} &= 0
\end{aligned}\right\} \tag{5.32}$$

where

$$\left.\begin{aligned}
\alpha_1' &= \frac{T_2K_2 - 1}{T_2K_2[(1 - X_0^2)\overline{G} + (1 - 2\nu_1) + X_0^2} \\
\beta_1' &= \frac{\bar{c}_1K_1(1 + T_1\bar{r}'')(K_1 + \bar{r}'') + \bar{c}_2K_1(K_1T_1 - 1)}{(K_1 + \bar{r}'')^2} \\
\beta_1'' &= \frac{\bar{c}_2K_1(T_1\bar{r}'' + 1)}{K_1 + \bar{r}''} \\
r_1' &= (K_1T_1 - 1)\left\{\frac{K_1[\bar{c}_1(K_1 + \bar{r}'')\bar{c}_2]}{(K_1 + \bar{r}'')^2} - \alpha_1'\right\}
\end{aligned}\right\} \tag{5.33}$$

In some limiting cases, the above equations can be simplified. When $t \to \infty$, for the rock mass ($\Delta > 0$ and $\Delta = 0$),

$$\left.\begin{aligned} \sigma_{r,c}^{(2)} &= -\alpha_2' P\left(1 - \frac{a^2}{r_2^2}\right) \\ \sigma_{\theta,c}^{(2)} &= -\alpha_2' P\left(1 + \frac{a^2}{r^2}\right) \\ \sigma_{y,c}^{(2)} &= -P \qquad \tau_c^{(2)} = 0 \end{aligned}\right\} \tag{5.34}$$

and, for the lining ($\Delta > 0$ and $\Delta = 0$),

$$\left.\begin{aligned} \sigma_{r,c}^{(1)} &= -\alpha_1' \overline{G}\left(1 - \frac{a_0^2}{r_1^2}\right) \\ \sigma_{\theta,c}^{(1)} &= -\alpha_1' \overline{G}\left(1 + \frac{a_0^2}{r_1^2}\right) \\ \sigma_{y,c}^{(1)} &= -2\nu_1 a_1' \overline{G} P \qquad \tau_c^{(1)} = 0 \end{aligned}\right\} \tag{5.35}$$

### *5.2.4 Two-dimensional stress state in surrounding visco-elastic rock mass*

In most cases, the in situ stress state of rock mass is anisotropic. The magnitude of the vertical stress generally equals to the overburden stress, i.e., directly proportional to the overlying depth. On the other hand, the horizontal stresses vary, which can approximately be derived using the elasticity theory adopting lateral pressure coefficient of $\lambda = \nu/(1-\nu)$ ($\nu$ is the Poisson's ratio of the rock mass). Generally, it is greater than the vertical stress, i.e., $\lambda > 1$ [310–313]. Therefore it is of great significance to study the stress state in surrounding visco-elastic rock masses for $\lambda \neq 1$.

From the general equations (5.4)~(5.9) for visco-elastic rock mass given in Section 5.2, it can be seen in principle that the equations can be used to solve any boundary problems of multi-layered circular media. Of course, provided that the lining is circular in shape and composed of a visco-elastic medium while the surrounding rock mass is composed of another visco-elastic medium, such two-dimensional problem can be solved, even for initial in situ stress with $\lambda \neq 1$. The solving procedure is tedious. For this reason, this section only deals with a simple case of two-dimensional problems, i.e., the variation of the stress state in the rock mass with an unlined opening.

In this case, the terms with subscript 1 in equations (5.4)–(5.9) are zero, i.e., the first circular layer does not exist. The basic equations are given as follows:
For stress components,

$$\left.\begin{aligned}
\sigma_r &= KT[c_0'(t) - b_0'(t)S^2 + K(1-KT)\int_0^t [c_0'(t') - b_0'(t')S^2]e^{k(t'-t)}\mathrm{d}t' \\
&\quad - \Big\{KT\big[8a_2(t)S^2 - 6b_2(t)S^4 - 2d_2(t)\big] + K(1-KT)\int_0^t \big[8a_2(t')S^2 - 6b_2(t')S^4 - 2d_2(t')\big] \\
&\quad \times e^{k(t'-t)}\mathrm{d}t'\Big\}\cos 2\varphi \\
\sigma_0 &= KT[c_0'(t) + b_0'(t)S^2 + K(1-KT)\int_0^t [c_0'(t') + b_0'(t')S^2]e^{k(t'-t)}\mathrm{d}t' \\
&\quad - \Big\{KT[-6b_2(t)S^4 - 24c_2(t)q^2 - 2d_2(t)] + K(1-KT) \\
&\quad \times \int_0^t [-6b_2(t')S^4 - 24c_2(t')q^2 - 2d_2(t')] \times e^{k(t'-t)}\mathrm{d}t'\Big\}\cos 2\varphi \\
\sigma_y &= \nu(\sigma_{r,c} + \sigma_{0,c}) \\
\tau &= -2\Big\{KT\big[2a_2(t)S^2 - 3b_2(t)S^4 + 6c_2(t)q^2 + d_2(t)q^0\big] \\
&\quad + K(1-KT)\int_0^t \big[2a_2(t')S^2 - 3b_2(t')S^4 + 6c_2(t')q^2 + d_2(t')q^0\big]e^{k(t'-t)}\mathrm{d}t'\Big\}\sin 2\varphi
\end{aligned}\right\} \tag{5.36}$$

For displacement components,

$$\left.\begin{aligned}
U_r &= \frac{r}{2G}\{(1-2\nu')c_0'(t) + b_0'(t)S^2 + [8(1-\nu')a_2(t)S^2 - 2b_2(t)S^4 \\
&\quad + 8\nu' c_2(t)q^2 + d_2(t)q]\}\cos 2\varphi \\
W_\theta &= -\frac{r}{2G}\{4(1-2\nu')a_2(t)S^2 + 2b_2(t)S^4 + 4(3-2\nu')c_2(t)q^2 + 2d_2(t)q^0\}\sin 2\varphi \\
V_y &= -\frac{\nu' - \nu}{2G\nu}y\{2c_0'(t) - \big[8a_2(t)S^2 + 24c_2(t)q^2\big]\cos 2\varphi\}
\end{aligned}\right\} \tag{5.37}$$

For strain components,

$$
\left.
\begin{aligned}
\varepsilon_r &= \frac{\partial U_r}{\partial r} \\
&= \frac{1}{2G}\{(1-2v')c_0'(t) - b_0'(t)S^2 - \left[8(1-v')a_2(t)S^2 - 6b_2(t)S^4 \right.\\
&\qquad \left. -24v'c_2(t)q^2 - 2d_2(t)q^0\right]\cos 2\varphi\} \\
\varepsilon_\theta &= \frac{1}{r}\frac{\partial W_t}{\partial \varphi} + \frac{U_r}{r} \\
&= \frac{1}{2G}\{(1-2v')c_0'(t) + b_0'(t)S^2 + \left[8v'a_2(t)S^2 - 6b_2(t)S^4 \right.\\
&\qquad \left. -24(1-v')c_2(t)q^2 - 2d_2(t)q^0\right]\cos 2\varphi\} \\
\bar{\varepsilon} &= \varepsilon_r + \varepsilon_\theta \\
&= \frac{1}{2G}\{2(1-2v')c_0'(t) - \left[8(1-2v')a_2(t)S^2 - 24(1-2v')c_2(t)q^2\right]\cos 2\varphi\} \\
2r &= \frac{\partial W_t}{\partial r} - \frac{W_t}{r} + \frac{1}{r}\frac{\partial U_r}{\partial \varphi} \\
&= \frac{-1}{2G}\{8a_2(t)S^2 - 12b_2(t)S^4 + 24c_2(t)q^2 + 4d_2(t)\}\sin 2\varphi
\end{aligned}
\right\} \tag{5.38}
$$

The above equations indicate that the undetermined coefficients of $c_0'(t)$, $b_0'(t)$ only exist in the terms with $n=0$ (without trigonometric functional terms), whereas those of $a_2(t)$, $b_2(t)$, $c_2(t)$, $d_2(t)$ only exist in the terms with $n=2$. Therefore, these undetermined coefficients can be divided into two groups with respect to boundary equations, i.e., a group of $n=0$ and a group of $n=2$, which can be conveniently solved. The equations for solving these coefficients for the following boundary conditions are given below:

(a) When $r=a$,

$$
\begin{aligned}
&\sigma_{r,c} = \sigma_r + \sigma_{r,0} = 0 \qquad \begin{cases} \sigma_{r,c}(n=0) = 0 \\ \sigma_{r,c}(n=2) = 0 \end{cases} \\
&\tau = \tau + \tau_0 = 0 \qquad (n=2)
\end{aligned}
\tag{5.39}
$$

(b) When $r \to \infty$,

$$
\begin{aligned}
&\sigma_{r,c} = \sigma_r + \sigma_{r,0} = \sigma_{r,0} \qquad \begin{cases} \sigma_{r,c}(n=0) = \sigma_{r,0} & (n=0) \\ \sigma_{r,c}(n=2) = \sigma_{r,0} & (n=2) \end{cases} \\
&\tau_c = \tau + \tau_0 \qquad (n=2)
\end{aligned}
$$

Hence, six simultaneous equations can be given according to equation (5.36) and the above boundary conditions:

$$\left.\begin{aligned}
&\text{when } r=R_0=a,\\
&S=\frac{R_0}{r}=\frac{a}{a}=1 \qquad q=\frac{r}{R_1}=\frac{a}{R_1\to\infty}\to 0\\
&\quad(\sigma_{r,c}, n=0)\\
&KT\left[c_0'(t)-b_0'(t)\right]+K(1-KT)\int_0^t\left[c_0'(t')-b_0'(t')\right]e^{k(t'-t)}\mathrm{d}t'-\frac{1}{2}p_z(1+\lambda_1)=0\\
&(\sigma_{r,c}, n=2)\\
&-KT[8a_2(t)-6b_2(t)-2d_2(t)]-K(1-KT)\int_0^t[8a_2(t')-6b_2(t')-2d_2(t')]\\
&\quad\times e^{k(t'-t)}\mathrm{d}t'+\frac{1}{2}p_z(1-\lambda_1)=0\\
&\quad(\tau_c, n=2)\\
&-2KT[2a_2(t)-3b_2(t)]-2K(1-KT)\int_0^t[2a_2(t')-3b_2(t')]e^{k(t'-t)}\mathrm{d}t'-\frac{1}{2}p_z(1-\lambda_1)=0\\
&\text{when } r=R_1\to\infty\\
&S=\frac{R_0}{r}=\frac{a}{R_1\to\infty}\to 0 \qquad q=\frac{r}{R_1}=\frac{R_1}{R_1}=1\\
&\quad(\sigma_{r,c}, n=0)\\
&KTc_0'(t)+K(1-KT)\int_0^t c_0'(t')e^{k(t'-t)}\mathrm{d}t'-\frac{1}{2}p_z(1+\lambda_1)=-\frac{1}{2}p_z(H\lambda_1)\\
&(\sigma_{r,c}, n=2)\\
&2KTd_2(t)+2K(1-KT)\int_0^t d_2(t')e^{k(t'-t)}\mathrm{d}t'+\frac{1}{2}p_z(1-\lambda_1)=\frac{1}{2}p_z(1-\lambda_1)\\
&(\tau_c, n=2)\\
&-2KT[6c_2(t)+d_2(t)]-2K(1-KT)\int_0^t[6c_2(t')+d_2(t')]e^{k(t'-t)}\mathrm{d}t'-\frac{1}{2}p_z(1-\lambda_1)\\
&\quad=-\frac{1}{2}p_z(1-\lambda_1)
\end{aligned}\right\}\tag{5.40}$$

The coefficients in the form of a function can be completely determined from these six simultaneous equations. The explicit formulae of the coefficients can be obtained by converting the above integral equations into differential equations in accordance with the initial conditions. Then each component of the stress state is divided into

two parts: static and rheologic. The latter is related to deformation rate. The two expressions for each stress component are shown below, the superscripts of *s* and *r* stand for static and rheologic respectively:
for radial stress,

$$\left.\begin{aligned}\sigma_{r,c}&=\sigma_{r,c}^{s}+\sigma_{r,c}^{r}\\ \sigma_{r,c}^{s}&=-\frac{1}{2}p_z(1+\lambda_1)\left[1-(1-e^{-t/T})\frac{a^2}{r^2}\right]+\frac{1}{2}p_z(1-\lambda_1)\left\{1-(1-e^{-t/T})\left(4\frac{a^2}{r^2}-3\frac{a^4}{r^4}\right)\right\}\cos 2\varphi\\ \sigma_{r,c}^{r}&=\frac{1}{2}p_z(1+\lambda_1)e^{-t/T}\frac{a^2}{r^2}-\frac{1}{2}p_z(1-\lambda_1)e^{-t/T}\left(4\frac{a^2}{r^2}-3\frac{a^4}{r^4}\right)\cos 2\varphi\end{aligned}\right\} \tag{5.41}$$

for tangential stress,

$$\left.\begin{aligned}\sigma_{\theta,c}^{s}&=-\frac{1}{2}p_z(1+\lambda_1)\left[1+(1-e^{-t/T})\frac{a^2}{r^2}\right]-\frac{1}{2}p_z(1-\lambda_1)\left[1+3(1-e^{-t/T})\frac{a^4}{r^4}\right]\cos 2\varphi\\ \sigma_{\theta,c}^{r}&=-\frac{1}{2}p_z(1+\lambda_1)e^{-t/T}\frac{a^2}{r^2}-\frac{3}{2}p_z(1-\lambda_1)e^{-t/T}\frac{a^4}{r^4}\cos 2\varphi\end{aligned}\right\} \tag{5.42}$$

for shearing stress

$$\left.\begin{aligned}\tau_{c}^{s}&=-\frac{1}{2}p_z(1-\lambda_1)\left[1+(1-e^{-t/T})\left(2\frac{a^2}{r^2}-3\frac{a^4}{r^4}\right)\right]\sin 2\varphi\\ \tau_{c}^{r}&=-\frac{1}{2}p_z(1-\lambda_1)e^{-t/T}\left(2\frac{a^2}{r^2}-3\frac{a^4}{r^4}\right)\sin 2\varphi\end{aligned}\right\} \tag{5.43}$$

The displacement components after excavation surrounding the opening are:

$$\left.\begin{aligned}U_r&=\frac{P_z r}{4G}(1+\lambda_1)\left(1-\frac{KT-1}{KT}e^{-t/T}\right)\frac{a^2}{r^2}+\frac{P_z r}{4G}(1-\lambda_1)\\ &\quad\times\left(1-\frac{KT-1}{KT}e^{-t/T}\right)\left[4(1-\nu')\frac{a^2}{r^2}-\frac{a^4}{r^4}\right]\cos 2\varphi\\ W_\vartheta&=-\frac{P_z r}{4G}(1-\lambda_1)\left(1-\frac{KT-1}{KT}e^{-t/T}\right)\left[2(1-2\nu')\frac{a^2}{r^2}+\frac{a^4}{r^4}\right]\sin 2\varphi\end{aligned}\right\} \tag{5.44}$$

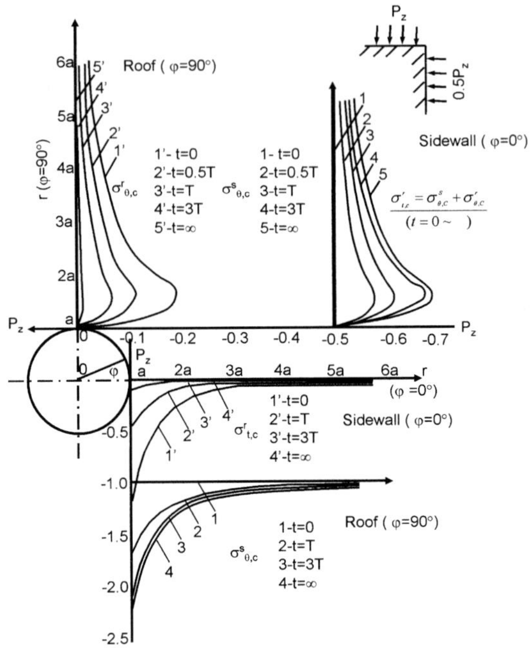


**Figure 5.4.** Variation of stress components in surrounding rock mass with time at roof (φ = 90°) and sidewall (φ = 0°).

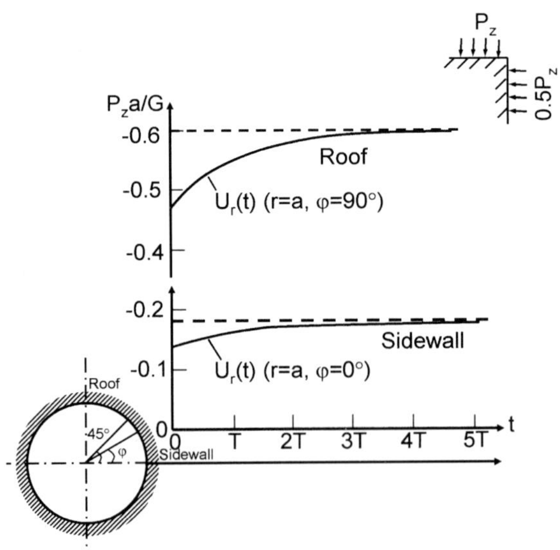


**Figure 5.5.** Variation of radial displacement of the opening periphery with time at roof and sidewall.

Figure 5.4 plots the variations of the tangential stress with time at two points of the circular opening ($\varphi = 0°$ and $90°$), given $\lambda_1 = 0.5$. The curves for the static and the dynamic components are plotted separately.

Figure 5.5 shows the variation of radial displacement with time at the same points under the same condition as in Figure 5.4, given $KT = 1.25$.

## 5.3. INTERACTION BETWEEN THE VISCO-ELASTIC–PLASTIC SURROUNDING ROCK AND LINING

This section studies the interaction between the visco-elastic rock mass within plastic zones and lining.

### 5.3.1 *Stress state in plastic zones of rock mass*

The occurrence and configuration of plastic zones in the surrounding rock mass are often non-symmetrical and they depend on the rock mass characteristics and in situ stress. However, in order to give an explicit equation using analytical method, some assumptions are made to simplify the solutions. As a typical condition, in situ stress field is isometric, and the surrounding rock mass is homogeneous, isotropic, visco-elastic and continuous.

Such a problem can be treated as a symmetrical one. If the plastic zone is assumed incompressible, then the expression for strain state, when using polar coordinates, becomes very simple:

$$\left.\begin{aligned} &\varepsilon_\theta + \varepsilon_r = 0 \\ &\varepsilon_r = \frac{du}{dr} \qquad \varepsilon_\theta = \frac{u}{r} \end{aligned}\right\} \tag{5.45}$$

Rearrangement and integration of the above two equations yield:

$$u = \frac{A(t)}{r} \qquad \varepsilon_\theta = \frac{A(t)}{r^2} \qquad \varepsilon_r = \frac{-A(t)}{r^2} \tag{5.46}$$

Given that the physical equations of the medium within the plastic zone are

$$\left.\begin{aligned} \sigma_\theta - \sigma &= 2M\varepsilon_\theta \\ \sigma_r - \sigma &= 2M\varepsilon_r \end{aligned}\right\} \tag{5.47}$$

where $\sigma$ is the average stress.

Based upon experimental data available and Mogi's criterion [128], the plastic condition of surrounding rock mass can be expressed as

$$\tau_{\mathrm{OCT}} = f(\sigma_1 + \sigma_3 + \alpha\sigma_2)$$

Because of symmetry,

$$\sigma_m = \frac{1}{3}(\sigma_1 + \sigma_3 + \sigma_2) = P = \text{const}$$

then

$$\left.\begin{aligned} &\sigma_1 + \sigma_3 + \alpha\sigma_2 = \text{const} \\ &(\sigma_1 - \sigma_2)^2 + (\sigma_2 - \sigma_3)^2 + (\sigma_3 - \sigma_1)^2 = K_p^2 = \text{const} \end{aligned}\right\} \tag{5.48}$$

Also for a symmetrical problem

$$\sigma_1 = \sigma_\theta \qquad \sigma_3 = \sigma_r \qquad \sigma_2 = \sigma_y$$

By substituting equation (5.47) into equation (5.48), we have

$$M = \frac{-K_p}{2\sqrt{6}\varepsilon_0} \tag{5.49}$$

By substituting equation (5.49) into equation (5.47), it gives

$$\left.\begin{aligned} \sigma_\theta &= \sigma - \frac{K_p}{\sqrt{6}} \\ \sigma_r &= \sigma + \frac{K_p}{\sqrt{6}} \end{aligned}\right\} \tag{5.50}$$

Therefore,

$$\sigma_\theta - \sigma_r = \frac{-2K_p}{\sqrt{6}} \tag{5.51}$$

The equilibrium equation of the plastic zone is

$$r\frac{d\sigma_r}{dr} + \sigma_r - \sigma_\theta = 0 \tag{5.52}$$

The load applied to the lining at the opening periphery ($r=a$) is $\sigma^{(1)}_{r=a}(t)$. By substituting equation (5.51) into (5.52) and through rearrangement and integration, the stress components in this area are:

$$\left.\begin{aligned}\sigma_r^{(2)}(t) &= \frac{-2K_p}{\sqrt{6}}\ln\frac{r}{a} + \sigma^{(1)}_{r=a}(t)\\ \sigma_\theta^{(2)}(t) &= \frac{-2K_p}{\sqrt{6}}\left(1+\ln\frac{r}{a}\right) + \sigma^{(1)}_{r=a}(t)\end{aligned}\right\} \tag{5.53}$$

### 5.3.2 *Interaction between surrounding rock mass and lining*

The area beyond the plastic zone in the surrounding rock mass is still in visco-elastic state. The stress field can be determined using the method stated in previous sections, i.e., the new stress field ($S_c^{(3)}$) after excavation is the summation of the initial stress field ($S_0^{(3)}$) and the compensating stress ($S^{(3)}$) field:

$$S_c^{(3)} = S_0^{(3)} + S^{(3)}$$

where the initial stress field is the same as in equation (5.23). The compensating stress field can be simplified because of axial symmetry. The global components of the stress state are

$$\left.\begin{aligned}&\sigma^{(3)}_{r,c}(t)=K_3T_3\left[C'_{0,3}(t)-b'_{0,3}(t)S_3^2\right]+K_3(1-K_3T_3)\int_{t_0}^{t}\left[C'_{0,3}(t')-b'_{0,3}(t')S_3^2\right]e^{k_3(t'-t)}\mathrm{d}t'-P\\ &\sigma^{(3)}_{0,c}(t)=K_3T_3\left[C'_{0,3}(t)+b'_{0,3}(t)S_3^2\right]+K_3(1-K_3T_3)\int_{t_0}^{t}\left[C'_{0,3}(t')+b'_{0,3}(t')S_3^2\right]e^{k_3(t'-t)}\mathrm{d}t'-P\\ &\sigma^{(3)}_{y,c}(t)=K_3T_3C'_{0,3}+K_3(1-K_3T_3)\int_{t_0}^{t}C'_{0,3}(t')e^{k_3(t'-t)}\mathrm{d}t'-P\\ &\tau^{(3)}_c(t)=0\\ &U^{(3)}_{r,c}(t)=\frac{1}{2G_3}r_3\left[(1-2\nu)C'_{0,3}(t)+b'_{0,3}(t)S_3^2\right]\\ &W^{(3)}_{0,c}(t)=0\end{aligned}\right\} \tag{5.54}$$

where $K_3$, $T_3$, $G_3$ are the mechanical parameters in the visco-elastic area of the surrounding rocks (as illustrated in Figure 5.6); $S_3 = R/r_3$, $R$ is the radius of the plastic zone and $r_3$ is the radial coordinate of the point under consideration.

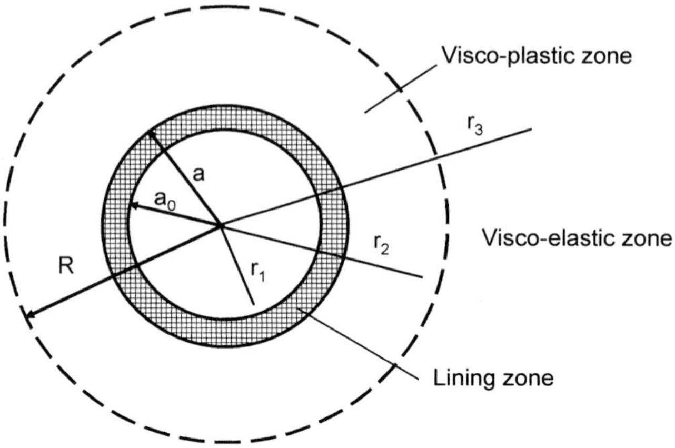


**Figure 5.6.** Division of different damage zones around an opening.

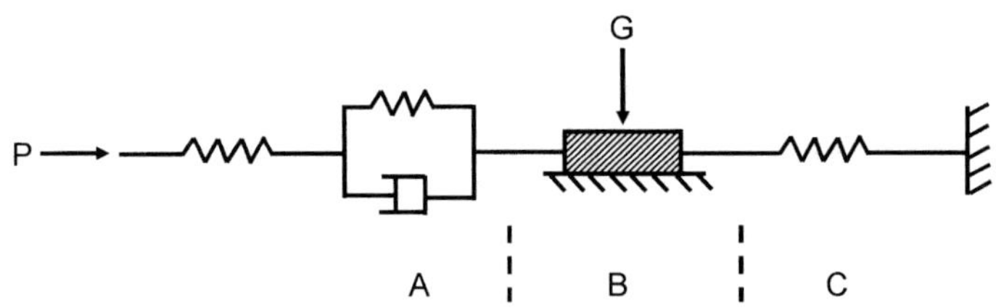


**Figure 5.7.** Synthetic rheological mechanical model for plastic zone, visco-elastic area and elastic lining, where $G - \beta K_p$; $\beta$ – constant related to the frictional coefficient of rock block; A – visco-elastic area; B – plastic zone; C – lining.

The lining material is assumed to be a Hooke's elastic body, its basic equation is the same as equation (5.23).

The synthetic mechanical model for the plastic zone, visco-elastic area and elastic lining can be described approximately using the diagram shown in Figure 5.7.

Each coefficient in equation (5.54) can be determined using the following boundary and physical conditions.

(a) At an infinite point, i.e., $r_3 \to \infty$

$$\sigma_{r,c}^{(3)} = -P$$

(b) On the inner boundary of the lining, i.e., $r_1 = a_0$,

$$\sigma_{r=a_0}^{(1)} = 0$$

(c) Along the interface between the lining and the rock mass, i.e., $r_1 = r_2 = a$, and provided that the lining is applied with the moment of $t = t_1$, then

$$U_r^{(1)} = U_r^{(2)} - U_{r(t=t_1)}^{(2)} \tag{5.55}$$

(d) Along the interface of visco-elastic and plastic zones, i.e., $r_2 = r_3 = R$,

For plastic condition

$$\sigma_\theta^{(3)} - \sigma_r^{(3)} = \frac{-2}{\sqrt{6}} K_p$$

For condition of continuous displacements

$$U_r^{(2)} = U_r^{(3)}$$

Based upon the above boundary conditions, the corresponding equations can be given. By assuming that the time for applying lining is $t = t_1$, the excavating time is $t_0 = 0$ and solving integral and algebraic equations in connection with equations (5.54), (5.23), (5.45) and (5.55), functions of $C'_{0,3}(t)$, $C'_{0,1}(t)$, $b'_{0,1}(t)$, $b'_{0,3}(t)$ and $A(t)$ can be determined. They finally give the stress components:
For the lining

$$\left.\begin{aligned}
\sigma_r^{(1)}(t) &= -\frac{\overline{G}R^2}{a^2[(1-2\nu)+X_0^2]} \frac{K_p(\delta K_3 T_3 - 1)}{\sqrt{6}K_3 T_3}(e^{-t_1/T_3} - e^{-t/T_3})(1 - S_1^2) \\
\sigma_\theta^{(1)}(t) &= -\frac{\overline{G}R^2}{a^2[(1-2\nu)+X_0^2]} \frac{K_p(\delta K_3 T_3 - 1)}{\sqrt{6}K_3 T_3}(e^{-t_1/T_3} - e^{-t/T_3})(1 + S_1^2) \\
\sigma_y^{(1)}(t) &= -\frac{2\nu_1\overline{G}R^2}{a^2[(1-2\nu)+X_0^2]} \frac{K_p(\delta K_3 T_3 - 1)}{\sqrt{6}K_3 T_3}(e^{-t_1/T_3} - e^{-t/T_3})
\end{aligned}\right\} \tag{5.56}$$

where $G = G_1/G_3$ is the stress state in the visco-elastic zone, when $r \geqslant R$,

$$\left.\begin{aligned}
&\left.\begin{aligned}\sigma_{r,c}^{(3)}(t)\\ \sigma_{\theta,c}^{(3)}(t)\end{aligned}\right\} = \pm\frac{KTK_p}{\sqrt{6}}\left[1 + \frac{1}{KT}(1-KT)e^{-t/T_3}\right]\frac{R^2}{r^2} \pm K(1-KT)\int_0^t \frac{K_p}{\sqrt{6}} \\
&\qquad \times \left[1 + \frac{1}{KT}(1-KT)e^{-t'/T_3}\right]\frac{R^2}{r^2}e^{K(t'-t)}\mathrm{d}t' - P \\
&\sigma_{y,c}^{(3)}(t) = -P
\end{aligned}\right\} \tag{5.57}$$

For the plastic zone

$$\left.\begin{aligned}\sigma_r^{(2)}(t) &= -\frac{2K_p}{\sqrt{6}}\ln\frac{r}{a} - \frac{\overline{G}R^2(1-X_0^2)}{a^2[(1-2\nu)+X_0^2]} \times \frac{K_p(K_3T_3-1)}{\sqrt{6}K_3T_3}(e^{-t_1/T_3} - e^{-t/T_3})\\ \sigma_\theta^{(2)}(t) &= -\frac{2K_p}{\sqrt{6}}(1+\ln\frac{r}{a}) - \frac{\overline{G}R^2(1-X_0^2)}{a^2[(1-2\nu)+X_0^2]} \times \frac{K_p(K_3T_3-1)}{\sqrt{6}K_3T_3}(e^{-t_1/T_3} - e^{-t/T_3})\end{aligned}\right\} \tag{5.58}$$

Given $\sigma_{r,c}^{(3)}(t) = \sigma_{r,c}^{(2)}(t)$, then the boundary, $R$, between the visco-elastic and plastic zones can be obtained from equations (5.53) and (5.57),

$$R = a\exp\left[\frac{\sqrt{6}}{2K_p}\left(\sigma_{r,c}^{(3)}(t) - \sigma_{r,c}^{(2)}(t)\right)\right] \tag{5.59}$$

It can be seen from the above equations that the loading on the lining is a function of (a) the plastic zone size ($R$) at the time when the lining is applied ($t_1$), (b) the ratio of the shearing moduli of the lining and the rock mass ($\overline{G}$), and, (c) other related physical, mechanical and geometrical parameters of the lining and the surrounding rock mass.

## 5.4. STRESS STATE IN VISCO-ELASTIC–VISCO-PLASTIC SURROUNDING ROCK MASSES

The preceding section discussed the stress states in the surrounding rock mass and the lining when there exist plastic zones. The plastic zone is time-dependent, in other words, the plastic zone is visco-plastic, whereas the rock masses at further distance still behaves visco-elastically [320–323]. This section discusses the interaction between the surrounding rock mass and the lining under such condition. Figure 5.8 shows the mechanical model of the whole system. The mechanical equations for

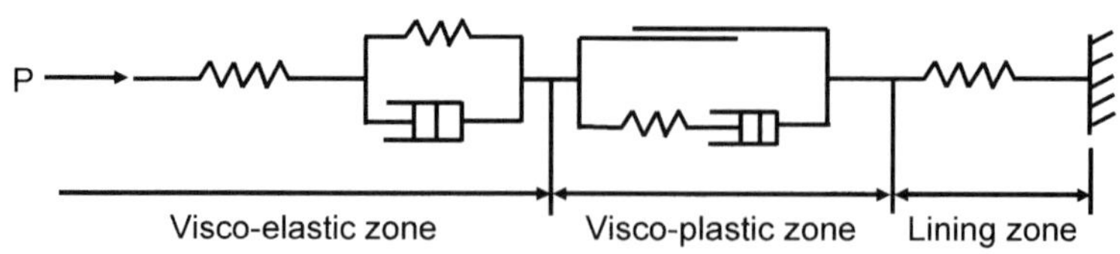


**Figure 5.8.** Mechanical model for the surrounding rock and the lining.

the visco-elastic rock mass and the lining zone are the same as that given in the preceding section; and the stress in the visco-plastic zone consists of two parts, i.e.,

$$\sigma^{(2)} = \sigma^{(2)'} + \sigma^{(2)''} \tag{5.60}$$

where the first part, $\sigma^{(2)'}$ is the stress component of the Maxwell medium, and the second part, $\sigma^{(2)''}$ is the stress component of the St. Venant medium. The physical equation for the Maxwell medium is

$$\left.\begin{aligned} \dot{\varepsilon}_\theta - \dot{\varepsilon} &= \frac{\left(\sigma_\theta^{(2)} - \sigma_\theta^{(2)}\right) - \sigma}{2\eta_2} + \frac{\left(\dot{\sigma}_\theta^{(2)} - \dot{\sigma}_\theta^{(2)}\right) - \dot{\sigma}}{2G_2} \\ \dot{\varepsilon}_r - \dot{\varepsilon} &= \frac{\left(\sigma_r^{(2)} - \sigma_r^{(2)}\right) - \sigma}{2\eta_2} + \frac{\left(\dot{\sigma}_r^{(2)} - \dot{\sigma}_r^{(2)}\right) - \dot{\sigma}}{2G_2} \end{aligned}\right\} \tag{5.61}$$

in which $\sigma$ is the average stress, $\dot{\sigma}$ is the average stress rate, $\dot{\varepsilon}$ is the average strain rate.

The general solutions to the differential equation of (5.61) can be obtained. The assumptions similar to the last section are made and the visco-plastic zone is approximately regarded as an incompressible medium. By applying the solutions of equations (5.45) and (5.46), equation (5.61) becomes:

$$u = \frac{A(t)}{r} \qquad \varepsilon_\theta = \frac{A(t)}{r^2} \qquad \varepsilon_r = -\frac{A(t)}{r^2} \tag{5.62a}$$

hence

$$\dot{u} = \frac{\dot{A}(t)}{r} \qquad \dot{\varepsilon}_\theta = \frac{\dot{A}(t)}{r^2} \qquad \dot{\varepsilon}_r = -\frac{\dot{A}(t)}{r^2} \tag{5.62b}$$

From equation (5.61),

$$\dot{\varepsilon}_\theta - \dot{\varepsilon}_r = \frac{1}{2\eta_2}\left(\sigma_\theta^{(2)'} - \sigma_r^{(2)'}\right) + \frac{1}{2G_2}\left(\dot{\sigma}_\theta^{(2)'} - \dot{\sigma}_r^{(2)'}\right) \tag{5.63a}$$

Substituting (5.62a) into (5.63a), then

$$\frac{1}{2G_2}\frac{\partial(\sigma_\theta - \sigma_r)}{\partial t} + \frac{\sigma_\theta - \sigma_r}{2\eta_2} = \frac{2}{r^2}\frac{\partial A(t)}{\partial t} \tag{5.63b}$$

Examining the visco-elastic zone and its boundary with the visco-plastic zone, the condition of continuous displacements in the boundary leads to

$$U^{(2)}_{r=R} = U^{(3)}_{r=R} \tag{5.64}$$

and the plastic condition (see equation (5.51)) leads to

$$\sigma_\theta^{(3)} - \sigma_r^{(3)} = -\delta K_p \tag{5.65}$$

A solution can be obtained by using equations (5.16), (5.62), (5.63) and (5.65) for stress and displacement in the visco-elastic zone, i.e.,

$$A(t) = \frac{R^2 \delta K_p}{4G_3}\left[\left(1 - \frac{1}{K_3 T_3}\right)e^{-t/T_3} - 1\right] \tag{5.66}$$

By substituting equation (5.66) into equation (5.63a) and solving the corresponding differential equation, we have

$$\sigma_\theta - \sigma_r = \frac{M}{r^2} e^{-t/T_3} + c_3 e^{-(G_2/\eta_2)t} \tag{5.67}$$

where

$$M = \frac{R^2 G_2 \delta K_p}{G_3 T_3}\left(\frac{1}{K_3 T_3} - 1\right)\frac{1}{((G_2/\eta_2) - (1/T_3))} \tag{5.68}$$

In equation (5.67), $c_3$ is an undetermined constant. In consideration of $t=0$, the stress state in the surrounding rock mass after excavation starts transiting from elastic state to visco-plastic state, i.e.,

$$\sigma_\theta - \sigma_{r|t=0} = -2P\frac{a^2}{r^2} \tag{5.69a}$$

In substitution of this equation into equation (5.67), we have $c_3 = -1/r^2(2Pa^2 + M)$. Again, upon substitution of the so-obtained $c_3$ into equation (5.67), we have

$$\sigma_\theta - \sigma_r = \frac{1}{r^2}\left[M(e^{-t/T_3} - e^{-t/T_2}) - 2Pa^2 e^{-t/T_2}\right] \tag{5.69b}$$

It can be obtained according to the equilibrium condition of the medium,

$$\sigma_\theta - \sigma_r = r\frac{\partial \sigma_r}{\partial r} \tag{5.69c}$$

Solving equations (5.69b) and (5.69c), by (5.69c) separating variables and integrating, we have the solution in the form of

$$\left.\begin{aligned} \sigma_r &= B(t) + \left(c - \frac{d}{r^2}\right)T(t) \\ \sigma_\theta &= B(t) + \left(c + \frac{d}{r^2}\right)T(t) \end{aligned}\right\} \tag{5.70}$$

Subtraction of these two equations gives

$$\sigma_\theta - \sigma_r = \frac{2dT(t)}{r^2}$$

By substituting this equation into equation (5.69), we obtain

$$T(t) = \frac{1}{2d}\left[M\left(e^{-t/T_3} - e^{-t/T_2}\right) - 2Pa^2 e^{-t/T_2}\right] \tag{5.71}$$

According to the condition of continuous displacements at the boundary between the lining and the rock mass, we have

$$U^{(1)}_{r=a}(t) = U^{(2)}_{r=a}(t) - U^{(2)}_{r=a}(t_1) \tag{5.72}$$

where $t_1$ is the time of applying the lining.

Making use of equations (5.62), (5.66) and (5.64) and substituting them into equation (5.72), we have

$$\sigma^{(1)}_{r=a}(t) = \frac{\delta K_p}{2}\cdot\frac{G_2}{G_3}\cdot\frac{R^2}{a^2}\cdot\frac{(1 - X_0^2)}{(1 - 2\nu_1 + X_0^2)}\cdot\left(1 - \frac{1}{K_3 T_3}\right)\left(e^{-t/T_3} - e^{-t_1/T_3}\right) \tag{5.73}$$

In the case of unlined opening, i.e., $t \leq t_1$,

$$\sigma^{(2)}_{r=a}(t) = \beta(t) + \left(c - \frac{d}{a^2}\right)T(t) = 0$$

$$B(t) = \left(\frac{1}{2a^2} - \frac{c}{2d}\right)\left[M\left(e^{-t/T_3} - e^{-t/T_2}\right) - 2Pa^2 e^{-t/T_2}\right]$$

Substituting the expressions for $B(t)$ and $T(t)$ gives

$$\left.\begin{array}{l}\sigma_r^{(2)}(t)\\ \sigma_\theta^{(2)}(t)\end{array}\right\} = \left(\frac{1}{2a^2} \mp \frac{c}{2r^2}\right)\left[M\left(e^{-t/T_3} - e^{-t/T_2}\right) - 2Pa^2e^{-t/T_2}\right] \quad (t \le t_1) \tag{5.74}$$

while in the case of lined opening, i.e., $t \ge t_1$,

$$\sigma_{r=a}^{(2)}(t) = \beta(t) + \left(c - \frac{d}{a^2}\right)T(t) = \sigma_{r=a}^{(1)}(t)$$

Substitution of the results from equations (5.73) and (5.71) into the above equation gives

$$B(t) = \frac{\delta K_p}{2} \cdot \frac{G_2}{G_3} \cdot \frac{R^2}{a^2} \cdot \frac{(1 - X_0^2)}{(1 - 2\nu_1 + X_0^2)} \cdot \left(1 - \frac{1}{K_3 T_3}\right)\left(e^{-t/T_3} - e^{-t_1/T_3}\right)$$
$$+ \left(\frac{1}{2a^2} - \frac{c}{2d}\right)\left[M\left(e^{-t/T_3} - e^{-t/T_2}\right) - 2Pa^2e^{-t/T_2}\right]$$

Substituting this equation and equation (5.71) into equation (5.70) gives

$$\left.\begin{array}{l}\sigma_r^{(2)}(t)\\ \sigma_\theta^{(2)}(t)\end{array}\right\} = \frac{\delta K_p}{2} \cdot \frac{G_2}{G_3} \cdot \frac{R^2}{a^2} \cdot \frac{(1 - X_0^2)}{(1 - 2\nu_1 + X_0^2)} \cdot \left(1 - \frac{1}{K_3 T_3}\right)\left(e^{-t/T_3} - e^{-t_1/T_3}\right)$$
$$+ \left(\frac{1}{2a^2} \mp \frac{1}{2r^2}\right)\left[M\left(e^{-t/T_3} - e^{-t/T_2}\right) - 2Pa^2e^{-t/T_2}\right] \quad \text{for } t \ge t_1 \tag{5.75}$$

According to the plastic condition (5.65) and the boundary condition, $\sigma_r^{(3)} \to -P$ for $r \to \infty$, we have at the boundary of $r_2 = r_3 = R$

$$\left.\begin{array}{l}\sigma_r^{(3)}(0) = -P + \dfrac{\delta K_p}{2}\\ \sigma_\theta^{(3)}(0) = -P - \dfrac{\delta K_p}{2}\end{array}\right\}$$

Let the stress $\sigma_r^{(t)}$ remain continuous on the boundary of $r = R$, then when $t \le t_1$,

$$\left(\frac{1}{2a^2} - \frac{1}{2R^2}\right)\left[M\left(e^{-t/T_3} - e^{-t/T_2}\right) - 2Pa^2e^{-t/T_2}\right] = -P + \frac{\delta K_p}{2} \tag{5.76}$$

and when $t \geq t_1$,

$$\left(\frac{1}{2a^2} - \frac{1}{2R^2}\right)\left[M\left(e^{-t/T_3} - e^{-t/T_2}\right) - 2Pa^2e^{-t/T_2}\right] + \frac{\delta K_p}{2} \cdot \frac{G_2}{G_3} \cdot \frac{R^2}{a^2} \cdot \frac{(1 - X_0^2)}{(1 - 2\nu_1 + X_0^2)}$$
$$\times \left(1 - \frac{1}{K_3 T_3}\right)\left(e^{-t/T_3} - e^{-t_1/T_3}\right) = -P + \frac{\delta K_p}{2}. \quad (5.77)$$

Equations (5.76) and (5.77) describe the change of the radius ($R$) of the visco-plastic zone, with time elapsing in the form of an explicit function. And equations (5.74), (5.75), (5.76) and (5.77) comprise the complete solution to the stress state of the visco-plastic zone.

For example, given that $\delta = 0.8$, $K_p = 1.5\,P$, $a = 2\,m$, lining thickness $= 40\,cm$, $\nu = 0.25$, $T_2 = T_3 = 12$ days, $K_2 = K_3 = 1/8$ days and $G_2/G_3 = 0.5$, then for $t = 0$, $R = 2.31\,m$ and for $t \to \infty$, $R = 2.623\,m$ (as shown in Figure 5.9).

It is easy to prove that the boundary conditions are met for $t = 0$ and $t \to \infty$ under limit condition. Given $t \to \infty$ in equation (5.75), we have the stress state in the visco-plastic zone in infinite time after applying lining:

$$\left.\begin{array}{l}\sigma_r^{(2)}(t)\\ \sigma_\theta^{(2)}(t)\end{array}\right\} = \frac{\delta K_p}{2} \cdot \frac{G_2}{G_3} \cdot \frac{R^2}{a^2} \cdot \frac{(1 - X_0^2)}{1 - 2\nu_1 + X_0^2} \cdot \left(1 - \frac{1}{K_3 T_3}\right)e^{-t_1/T_3} \quad (5.78)$$

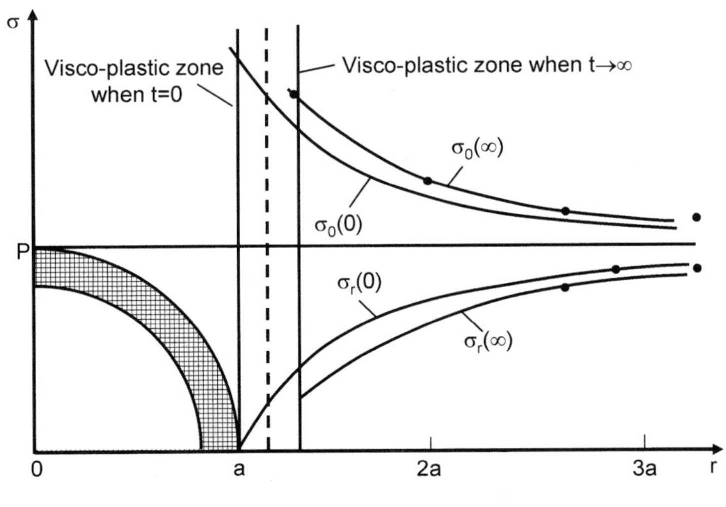


**Figure 5.9.** Stress distribution in various zones in the surrounding rock mass under limit condition.

It can be seen from equation (5.78) that the rock mass in the visco-plastic zone is unable to bear long-term shearing loads because its physical property is close to fluids. Finally, the two main stress components in the whole region tend to balance and the final loading on the lining is the same as that given in the above equation, i.e.,

$$\sigma_{r=a}^{(1)}(t \to \infty) = \sigma_r^{(2)}(t \to \infty)$$

It can be seen from above that for this type of visco-elastic and visco-plastic rock mass, the final loading on the lining is related not only to the physical and geometrical parameters of the visco-elastic zone in the rock mass and the lining but also to the time when the lining is applied ($t_1$) and the magnitude of the radius of the visco-plastic zone. It is different from the previously stated viscoelastic-plastic media. In the present case the final loading on the lining is related to the ratio of the shearing deformation moduli $G_2/G_3$ of the visco-plastic and visco-elastic zones, but not related to the shearing modulus $G_1$ of the lining itself. Additionally, it also can be seen that to reach a mechanical equilibrium state, the visco-plastic zone extends beyond the lining into the rock mass with the radius reaching $R$.

## 5.5. RHEOLOGICAL ANALYSIS WITH DILATION AND SOFTENING OF THE ROCK MASS

The previous sections assume that the plastic medium is incompressible. However, in most cases there exists a broken-swelling effect after plastic zone produced by excavation. The purpose of this section is to study the stress state accompanied by dilation effect and meanwhile to consider the softening phenomenon and its effect in the post-failure region in the rock mass [324–329]. For this reason, a more complete rheological mechanical model is proposed.

### *5.5.1 Mechanical model of surrounding rock mass*

The assumption of continuum mechanics is still adopted. The media involved is considered to be isotropic, and the initial stress field is assumed to be uniform, so axisymmetrical analysis is allowable.

Assume that the rock mass in the far region after excavation is in elastic state while the rock mass in the near region is in the visco-plastic state in which certain residual strength still exists. The lining is assumed to be an elastic medium, as shown in Figure 5.10.

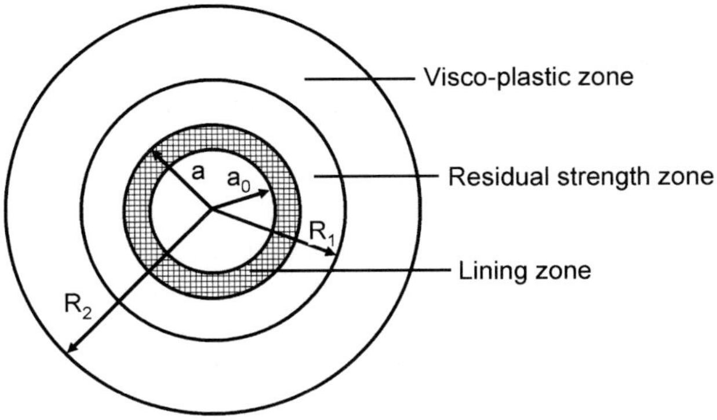


**Figure 5.10.** Schematic division of zones.

On the boundary between visco-elastic and visco-plastic zones, the stress state is assumed to meet the Mohr–Coulumb criterion, i.e.,

$$\sigma_\theta - \sigma_r = (\sigma_\theta + \sigma_r)\sin\varphi + 2c\cos\varphi \tag{5.79}$$

The stress components in the visco-elastic zone are the same as in equation (5.54).

It is known from equation (5.54) that when $r \to \infty$, $U_r(t) = 0$; then $c_0(t) = 0$ and the following equation is valid:

$$\frac{1}{2}(\sigma_\theta + \sigma_r) = -P$$

By substitution of this equation into Mohr–Coulumb criterion, gives

$$\sigma_\theta - \sigma_r = -2P\sin\varphi + 2c\cos\varphi = K_p \tag{5.80}$$

Supposing the minimum strength in the residual strength region is $\delta$ times ($\delta < 1$) $K_p$, i.e., equal to $\delta K_p$, then

$$\delta K_p = -2\delta(P\sin\varphi - c\cos\varphi) \tag{5.81}$$

By substituting equation (5.80) into equation (5.54), we can obtain the undetermined functional coefficient and then obtain the radial and tangential stresses

$$\left.\begin{aligned} \sigma_r &= -P + \frac{R_2^2}{r^2}(P\sin\varphi - c\cos\varphi) \\ \sigma_\theta &= -P - \frac{R_2^2}{r^2}(P\sin\varphi - c\cos\varphi) \end{aligned}\right\} \tag{5.82}$$

and the radial displacement

$$u_r = \frac{R_2^2}{4Gr}(P\sin\varphi - c\cos\varphi)\left[\left(1 - \frac{1}{K_3T_3}\right)e^{-t/T_3} - 1\right] \tag{5.83}$$

### 5.5.2 *Visco-plastic model considering dilation and softening*

***5.5.2.1 Physical model.*** The rock mass in the post-failure region displays visco-plastic softening behaviour when the peak strength is reached and the dilation phenomenon takes place simultaneously [325–327]. For this reason, the differential equation for the Maxwell medium that describes the visco-plastic behaviour is adopted and applied to describe the softening and dilation characteristics, it is expressed as follows:

$$(\sigma_\theta - \sigma_r) + \frac{\eta}{G}(\dot{\sigma}_\theta - \dot{\sigma}_r) = 2\eta(\dot{\varepsilon}_\theta - \dot{\varepsilon}_r) + \frac{2\nu G}{1-2\nu}(\varepsilon_\theta + \varepsilon_r) - 2G(\varepsilon_\theta - \varepsilon_r) + D \tag{5.84}$$

This equation not only reflects the softening and relaxation characteristics with visco-plastic property of the rock mass (the third term on the right-hand side of the equation), but also guarantees the stress difference in the rock mass.

***5.5.2.2 Geometric equation.*** Assuming that the problem under consideration meets the small deformation concept, we have, for an axisymmetrical problem,

$$\varepsilon_r = \frac{\partial u}{\partial r} \qquad \varepsilon_\theta = \frac{u}{r}$$

If the dilation phenomenon takes place in the visco-plastic zone and the radial strain is $\lambda/r$ times of the tangential strain, i.e., dilation occurs near to the opening periphery, then

$$\varepsilon_r = -\frac{\lambda}{r}\varepsilon_\theta$$

$$\frac{\partial u}{\partial r} = -\frac{\lambda}{r}\cdot\frac{u}{r} = -\lambda\frac{u}{r^2}$$

Integrate $u$ with respect to $r_1$, then

$$\left.\begin{aligned} u &= A(t)e^{\lambda/r} \\ \varepsilon_r &= -\frac{\lambda}{r^2}e^{\lambda/r}A(t) \\ \varepsilon_\theta &= \frac{1}{r}e^{\lambda/r}A(t) \end{aligned}\right\} \tag{5.85}$$

and strain rate

$$\left.\begin{aligned} \dot{\varepsilon}_r &= -\frac{\lambda}{r^2} e^{\lambda/r} \dot{A}(t) \\ \dot{\varepsilon}_\theta &= \frac{1}{r} e^{\lambda/r} \dot{A}(t) \end{aligned}\right\} \tag{5.86}$$

**5.5.2.3 *Equilibrium equation.*** For an axisymmetrical problem, we have

$$\left.\begin{aligned} r\frac{\partial \sigma_r}{\partial_r} &= \sigma_\theta - \sigma_r \\ \sigma_r &= \frac{1}{r}\int (\sigma_\theta - \sigma_r)\mathrm{d}r + f(t) \end{aligned}\right\} \tag{5.87}$$

### 5.5.3 *Stress components in each zone*

**5.5.3.1 *Visco-plastic zone.*** By substitution of equations (5.85) and (5.86) into equation (5.84) and further interpretation, we have

$$\dot{\sigma}_\theta - \dot{\sigma}_r + \frac{G}{\eta}(\sigma_\theta - \sigma_r) = \overline{A} \cdot \dot{A}(t) - \overline{B} \cdot A(t) + \overline{C}$$

where

$$\left.\begin{aligned} \bar{A} &= 2G\frac{1}{r}\left(1 + \frac{\lambda}{r}\right)e^{\lambda/r} \\ \bar{B} &= \frac{2G^2}{\eta} \cdot \frac{1}{r} e^{\lambda/r}\left[\frac{2\nu}{1-2\nu}\left(1 - \frac{\lambda}{r}\right) + \left(1 + \frac{\lambda}{r}\right)\right] \\ \bar{C} &= \frac{G}{\eta} D \end{aligned}\right\} \tag{5.88}$$

Equation (5.88) is the inhomogeneous partial differential equation of $(\sigma_\theta - \sigma_r)$, with the following solution

$$\sigma_\theta - \sigma_r = e^{-(G/\eta)t}\left\{-\left(\bar{A}\frac{G}{\eta} + \bar{B}\right)\int A(t)e^{(G/\eta)t}\mathrm{d}t + \bar{A} \cdot e^{(G/\eta)t}A(t) + \bar{C} \cdot \frac{\eta}{G}e^{(G/\eta)t} + \bar{D}(r)\right\} \tag{5.89}$$

By substituting equation (5.89) into equilibrium equation (5.87) and integrating it with respect to $r$, we have

$$\sigma_r = \frac{4G^2}{\eta(1-2\nu)}\left(\frac{4\nu-2}{\lambda}+\frac{3\nu-1}{r}\right)e^{\lambda/r}\cdot e^{-}(G/\eta)t$$
$$\times\int A(t)e^{(G/\eta)t}\,\mathrm{d}t - 2G\left(\frac{2}{\lambda}+\frac{1}{r}\right)e^{\lambda/r}A(t)+\overline{C}\frac{\eta}{G}\ln r+\int\frac{1}{r}\overline{D}(r)\,\mathrm{d}r+f(t) \quad (5.90)$$

It is easy to determine $\bar{D}(r)$ in equation (5.90), if analysing the transient response. The equation of $\bar{D}(r) = -D$ is given by Zhu *et al.* (1988). Considering this result in equation (5.90), we obtain

$$\sigma_\theta - \sigma_r = e^{-(G/\eta)t}\left\{-\left(\bar{A}\frac{G}{\eta}+\bar{B}\right)\int_0^t A(t')e^{-(G/\eta)t'}\mathrm{d}t' + \bar{A}e^{(G/\eta)t}A(t)+D\left(e^{-(G/\eta)t}-1\right)\right\} \quad (5.91)$$

Substituting equation (5.91) into equilibrium equation (5.87) and integrating the result with respect to $r$ gives

$$\sigma_r = e^{-(G/\eta)t}\left\{-\int_0^t A(t')e^{(G/\eta)t'}\mathrm{d}t'\cdot\frac{4G^2}{\eta(1-2\nu)}\left(\frac{4\nu-2}{\lambda}+\frac{3\nu-1}{r}\right)e^{\lambda/r}+e^{(G/\eta)t}A(t)\cdot 2G\left(\frac{2}{\lambda}+\frac{1}{r}\right)e^{\lambda/r}\right.$$
$$\left.+D\left(e^{(G/\eta)t}-1\right)\ln r\right\}+f(t) \quad (5.92)$$

Applying the condition $\sigma_{r,\nu-t} = \sigma_{r,\nu-p}$ on the boundary between $R_2$ visco-elastic and visco-plastic zones and making equality between equations (5.82) and (5.92) results in

$$f(t) = -p+(p\sin\varphi - c\cos\varphi)+e^{-(G/\eta)t}\left\{\frac{4G^2}{\eta(1-2\nu)}\left(\frac{4\nu-2}{\lambda}+\frac{3\nu-1}{R_2}\right)e^{\lambda/R_2}\right.$$
$$\left.\times\int_0^t A(t')e^{(G/\eta)t'}\mathrm{d}t' - 2G\left(\frac{2}{\lambda}+\frac{1}{R_2}\right)e^{\lambda/R_2}\cdot e^{(G/\eta)t}A(t) - D(e^{(G/\eta)t}-1)\ln R_2\right\} \quad (5.93)$$

By substituting equation (5.93) into equation (5.92), we can derive the explicit formula of $\sigma_r$. Applying equations (5.91), (5.92) and (5.93) gives

$$\sigma_\theta = e^{-(G/\eta)t}\left\{-\left(\bar{A}\frac{G}{\eta}+\bar{B}\right)\int_0^t A(t')e^{(G/\eta)t'}\mathrm{d}t' + \bar{A}e^{(G/\eta)t}A(t)+D\left(e^{(G/\eta)t}-1\right)\right\}+\sigma_r \quad (5.94)$$

***5.5.3.2 Residual strength zone.*** Because the residual strength is related to the confining pressure, the strength is higher at greater depth [328–334]. Owing to the small size of this zone, it can be assumed that the residual strength and $r$ have a linear relationship. By examining equations (5.80) and (5.81), the physical equation to describe this zone can be approximately expressed as

$$\sigma_\theta - \sigma_r = c'r + \delta K_p \tag{5.95}$$

Substituting equation (5.95) into equation (5.87) and integrating the result, we have

$$\left.\begin{aligned} \sigma_r &= c'r + \delta K_p \ln r + \varphi(t) \\ \sigma_\theta &= 2c'r + (1 + \ln r)\delta K_p + \varphi(t) \end{aligned}\right\} \tag{5.96}$$

In the case of unlined opening, $r = a$, $\sigma_r = 0$, then, from equation (5.96),

$$c' = -\frac{1}{a}[\delta K_p \ln r + \varphi(t)] \tag{5.97}$$

The term of $\varphi(t)$ in the above equation can be determined from the condition that the radial stresses $\sigma_r$ along the boundary $R_1$ between the visco-plastic zone and the residual strength zone are equal, i.e., the equation (5.92) is equal to equation (5.96).

The stress components in the visco-elastic zone are expressed by equations (5.82) and (5.83) respectively.

### 5.5.4 *Stress state without lining*

***5.5.4.1 Visco-plastic zone, $R_1 < r < R_2$.*** In the previous equations, there is a function $A(t)$ that is undetermined. It can be derived from the condition of continuous displacements on the boundary between the visco-elastic and the visco-plastic zones ($R_2$). By letting $u_{v-p} = u_{v-e}$, then we have its expression from equations (5.83) and (5.89),

$$A(t) = \frac{R_2}{4G} e^{-\lambda/R_2} (P \sin\varphi - C \cos\varphi)\left[\left(1 - \frac{1}{K_3 T_3}\right) e^{-t/T_3} - 1\right] \tag{5.98}$$

By substituting equation (5.98) into equations (5.92) and (5.94) and integrating the result, we have the expression of the stress state in the visco-plastic zone,

$$\sigma_r = -P + (p\sin\varphi - c\cos\varphi) + e^{-(G/\eta)t}\left\{\frac{GR_2}{\eta(1-2\nu)}e^{-\lambda/R_2}\cdot(p\sin\varphi - c\cos\varphi)\cdot\left[\frac{1-(1/K_3T_3)}{G/(\eta-1/T_3)}\right.\right.$$
$$\left.\times\left(e^{((G/\eta)-(1/T_3))t}-1\right)-\frac{\eta}{G}(e^{(G/\eta)t}-1)\right]\cdot\left[\left(\frac{4\nu-2}{\lambda}+\frac{3\nu-1}{R_2}\right)e^{\lambda/R_2}-\left(\frac{4\nu-2}{\lambda}+\frac{3\nu-1}{r}\right)e^{\lambda/r}\right]$$
$$-e^{(G/\eta)t}\frac{R_2}{2}\cdot e^{-\lambda/R_2}\times(p\sin\varphi - c\cos\varphi)\left[\left(1-\frac{1}{K_3T_3}\right)e^{-t/T_3}-1\right]\left(\frac{2}{\lambda}+\frac{1}{R_2}\right)e^{\lambda/R_2}-\left(\frac{2}{\lambda}+\frac{1}{r}\right)e^{\lambda/r}$$
$$\left.-D(e^{(G/\eta)t}-1)(\ln R_2 - \ln r)\right\} \quad (5.99)$$

$$\sigma_\theta = e^{-(G/\eta)t}\left\{-\left(\bar{A}\frac{G}{\eta}+\bar{B}\right)\frac{R_2}{4G}e^{-\lambda/R_2}\cdot(p\sin\varphi - c\cos\varphi)\right.$$
$$\times\left[\frac{1-(1/K_3T_3)}{(G/\eta)-(1/T_3)}\left(e^{((G/\eta)-(1/T_3))t}-1\right)-\frac{\eta}{G}\left(e^{(G/\eta)t}-1\right)\right]$$
$$+\bar{A}\cdot e^{(G/\eta)t}\cdot\frac{R_2}{4G}\cdot e^{-\lambda/R_2}\times(p\sin\varphi - c\cos\varphi)\left[\left(1-\frac{1}{K_3T_3}\right)e^{-t/T_3}-1\right] \quad (5.100)$$
$$\left.+D\left(e^{(g/\eta)t}-1\right)+\sigma_r\right\}$$

***5.5.4.2 Residual strength zone, $a < r < R_1$.*** As stated previously, the explicit formula $\varphi(t)$ can be derived from equations (5.96), (5.97) and (5.99) by applying the condition of equal $\sigma_r$ on the boundary between two adjacent zones ($R_1$):

$$\varphi(t) = -P + (p\sin\varphi - c\cos\varphi) - e^{-(G/\eta)t}\left\{\frac{GR_2}{\eta(1-2\nu)}e^{-\lambda/R_2}\cdot(p\sin\varphi - c\cos\varphi)\right.$$
$$\times\left[\frac{1-(1/K_3T_3)}{(G/\eta)-(1/T_3)}\times\left(e^{((G/\eta)-(1/T_3))t}-1\right)-\frac{\eta}{G}(e^{(G/\eta)t}-1)\right]$$
$$\times\left[\left(\frac{4\nu-2}{\lambda}+\frac{3\nu-1}{R_2}\right)e^{\lambda/R_2}-\left(\frac{4\nu-2}{\lambda}+\frac{3\nu-1}{R_1}\right)e^{\lambda/R_1}-\frac{1}{2}R_2\right.$$
$$\left.\times e^{(G/\eta)t}\cdot e^{-(\lambda/R_2)}\times(p\sin\varphi - c\cos\varphi)\right]\left[\left(1-\frac{1}{K_3T_3}\right)e^{-(t/T_3)}-1\right]\left[\left(\frac{2}{\lambda}+\frac{1}{R_2}\right)e^{\lambda/R_2}\right.$$
$$\left.\left.-\left(\frac{2}{\lambda}+\frac{1}{R_1}\right)e^{\lambda/R_1}-D(e^{G/\eta t}-1)(\ln R_2-\ln R_1)\right]-\delta K_p\left(\ln R_1-\frac{R_1}{a}\ln a\right)\right\}\Big/\left(1-\frac{R_1}{a}\right)$$
$$(5.101)$$

The stress components in this zone are,

$$\left.\begin{aligned}\sigma_r &= \delta\cdot K_p\left(\ln r - \frac{r}{a}\ln a\right) + \varphi(t)\left(1 - \frac{r}{a}\right)\\ \sigma_\theta &= \left(1 + \ln r - \frac{2r}{a}\ln a\right)\cdot\delta K_p + \varphi(t)\cdot\left(1 - \frac{2r}{a}\right)\end{aligned}\right\}\tag{5.102}$$

***5.5.4.3 Determination of boundary $R_2$.*** Since the rates of displacement components on the boundary between the visco-elastic and visco-plastic zones are equal, we can determine the value of $R_2$ by letting

$$\dot{u}_r^{(2)} = \dot{u}_r^{(3)},$$

and derive the corresponding displacement components in equations (5.83) and (5.85) with respect to $t$ and let the results be equal, then

$$R_2 = \lambda \tag{5.103}$$

Although this equation is extremely simple, its physical meaning is rationally clear. The assumption of $\varepsilon_r = -\lambda/r\varepsilon_\theta$ in the geometric equation in the previous sections, when $r = R_2$, $\varepsilon_r = -\varepsilon_\theta$, which is the starting point where the volumetric deformation in the visco-plastic zone begins to dilate in volume. Therefore, the result of $R_2 = \lambda$ coincides with the assumption made initially.

***5.5.4.4 Determination of boundary $R_1$.*** Similarly, the stress differences between two sides of the boundary are equal when $r = R_1$, $R_1$ can be determined.

For the residual strength zone, we have, from equation (5.102)

$$\sigma_\theta - \sigma_r = \delta K_p\left(1 - \frac{r}{a}\ln a\right) - \frac{r}{a}\varphi(t) \tag{5.104}$$

By comparing the stress difference of $(\sigma_\theta - \sigma_r)_{v-p}$ in the visco-plastic zone shown in equation (5.104) to equations (5.91), (5.93) and (5.98), we have

$$\begin{aligned}&\delta K_p\left(1 - \frac{R_1}{a}\ln a\right) - \varphi(t)\frac{R_1}{a}\\ &= -e^{-(G/\eta)t}\left\{\frac{4G}{\eta}\cdot\frac{1}{R_1}e^{\lambda/R_1}\left[\frac{\nu}{1-2\nu}\left(1 - \frac{\lambda}{R_1}\right) + \left(1 + \frac{\lambda}{R_1}\right)\right]\right.\\ &\times\left\{\frac{R_2}{4}e^{-\lambda/R_2}\cdot(p\sin\varphi - c\cos\varphi)\cdot\left[\frac{1-(1/K_3T_3)}{(G/\eta)-(1/T_3)}\left(e^{((G/\eta)-(1/T_3))t} - 1\right) - \frac{\eta}{G}\left(e^{(G/\eta)t} - 1\right)\right]\right\}\\ &+ \frac{1}{2}\cdot\frac{R_2}{R_1}\left(1 + \frac{\lambda}{R_1}\right)\cdot e^{\lambda(R_2 - R_1)/R_1R_2}\cdot(p\sin\varphi - c\cos\varphi)\cdot\left[\left(1 - \frac{1}{K_3T_3}\right)e^{-t/T_3} - 1\right]\\ &+ De^{-(G/\eta)t}\left(e^{(G/\eta)t} - 1\right)\end{aligned}\tag{5.105}$$

By substituting the values of $R_2$ and $\varphi(t)$ into equation (5.105), we can obtain the value of $R_1$ using trial and error method. In the limit case, we can derive the following equation for $t=0$

$$\delta K_p\left(1-\frac{R_1}{a}\ln a\right)-\frac{R_1}{a}\left\{-p+(p\sin\varphi-c\cos\varphi)-\frac{1}{2}R_2\cdot e^{-\lambda/R_2}\cdot(p\sin\varphi-c\cos\varphi)\right.$$
$$\left.\times\frac{-1}{K_3T_3}\left[\left(\frac{2}{\lambda}+\frac{1}{R_2}\right)e^{\lambda/R_2}\cdot\left(\frac{2}{\lambda}+\frac{1}{R_1}\right)e^{\lambda/R_1}\right]-\delta K_p\left(\ln R_1-\frac{R_1}{a}\ln a\right)\right\}\left(1-\frac{R_1}{a}\right)$$
$$\times\frac{R_2}{2R_1}\left(1+\frac{\lambda}{R_1}\right)\cdot e^{\lambda(R_2-R_1)/R_1R_2}\cdot(p\sin\varphi-c\cos\varphi)\left(\frac{-1}{K_3T_3}\right) \tag{5.106}$$

when $t\to\infty$

$$\delta K_p\left(1-\frac{R_1}{a}\ln a\right)-\frac{R_1}{a}\left\{-p+(p\sin\varphi-c\cos\varphi)+\frac{R_2}{1-2\nu}\cdot e^{-\lambda/R_2}\cdot(p\sin\varphi-c\cos\varphi)\right.$$
$$\times\left[\left(\frac{4\nu-2}{\lambda}+\frac{3\nu-1}{R_2}\right)e^{\lambda/R_2}-\left(\frac{4\nu-2}{\lambda}+\frac{3\nu-1}{R_1}\right)e^{\lambda/R_1}\right]$$
$$-\frac{R_2}{2}\cdot e^{\lambda/R_2}(p\sin\varphi-c\cos\varphi)$$
$$\times\left[\left(\frac{2}{\lambda}+\frac{1}{R_2}\right)\cdot e^{\lambda/R_2}-\left(\frac{2}{\lambda}+\frac{1}{R_1}\right)e^{\lambda/R_1}\right]+D(\ln R_2-\ln R_1)$$
$$\left.-\delta K_p\left(\ln R_1-\frac{R_1}{a}\ln a\right)\right\}\Big/\left(1-\frac{R_1}{a}\right)$$
$$=\frac{R_2}{R_1}\cdot e^{\lambda(R_2-R_1)/R_1R_2}\cdot(p\sin\varphi-c\cos\varphi)\left[\frac{\nu}{1-2\nu}\left(1-\frac{\lambda}{R_1}\right)+\left(1+\frac{\lambda}{R_1}\right)\right]$$
$$-\frac{R_2}{2R_1}e^{\lambda(R_2-R_1)/R_1R_2}\cdot\left(1+\frac{\lambda}{R_1}\right)(p\sin\varphi-\cos\varphi)+D \tag{5.107}$$

### 5.5.5 *Stress state with lining*

Assume that circular lining of elastic media is applied soon after the excavation, i.e., $t=0$, the stress state in this case can be expressed as [8,52]:

$$\left.\begin{aligned}\sigma_r(t)&=c(t)-b(t)\frac{a_0^2}{r^2}\\ \sigma_\theta(t)&=c(t)+b(t)\frac{a_0^2}{r^2}\end{aligned}\right\} \tag{5.108}$$

If the inner surface of the lining is free from loading, i.e., $\sigma_r = 0$ when $r = a_0$, then,

$$c(t) = b(t)$$

So equation (5.108) becomes,

$$\left.\begin{aligned}\sigma_r(t) &= c(t)\left(1 - \frac{a_0^2}{r^2}\right)\\ \sigma_\theta(t) &= c(t)\left(1 + \frac{a_0^2}{r^2}\right)\end{aligned}\right\} \tag{5.109}$$

Again, $\sigma_r$ and $\sigma_\theta$ at the boundary ($r = a$) of two adjacent zones are equal, and the values of stress differences ($\sigma_\theta - \sigma_r$) are also equal at $r = R_1$. The following undetermined parameter or function can be determined when solving the simultaneous equations (5.102), (5.109), (5.91), (5.95) and (5.98),

$$\begin{aligned}c' = \Bigg\{ &-e^{-(G/\eta)t} \cdot \frac{2G}{\eta} \cdot \frac{1}{R_1} e^{\lambda/R_1}\left[\frac{\nu_1}{1-2\nu}\left(1 - \frac{\lambda}{R_1}\right) + \left(1 + \frac{\lambda}{R_1}\right)\right]\\ &\times \left\{\frac{R_2}{2} e^{-\lambda/R_2} \cdot (p\sin\varphi - c\cos\varphi) \cdot \left[\frac{1-(1/K_3T_3)}{(G/\eta)-(1/T_3)}\left(e^{((G/\eta)-(1/T_3))t} - 1\right) - \frac{\eta}{G}\left(e^{(G/\eta)t} - 1\right)\right]\right\}\\ &+ \frac{R_2}{2R_1}\left(1 + \frac{\lambda}{R_1}\right) \cdot e^{\lambda(R_2-R_1)/R_1R_2}\\ &\times (p\sin\varphi - c\cos\varphi) \cdot \left[(1-(1/K_3T_3))e^{-t/T_3} - 1\right] + De^{-(G/\eta)t}\left(e^{(G/\eta)t} - 1\right) - \delta K_p \Bigg\} \Big/ R_1\end{aligned} \tag{5.110}$$

$$\left.\begin{aligned}c(t) &= \frac{a^2}{2a_0^2}(c'a + \delta K_p)\\ \varphi(t) &= \left(\frac{a^2}{2a_0^2} - \frac{3}{2}\right)ac' + \delta K_p\left[\left(\frac{a^2}{2a_0^2} - \frac{1}{2}\right) - \ln a\right]\end{aligned}\right\} \tag{5.111}$$

The term of $f(t)$ in the stress component in the visco-plastic zone can be determined using the condition of equal $\sigma_r$ at $r = R_1$ of the two adjacent zones,

$$\begin{aligned}f(t) = {} & c'R_1 + \delta K_p \ln R_1 + \varphi(t) + \frac{R_2 G}{\eta(1-2\nu)}(p\sin\varphi - c\cos\varphi)\left(\frac{4\nu-2}{\lambda} + \frac{3\nu-1}{R_1}\right) \cdot e^{\lambda(R_2-R_1)/R_1R_2}\\ &\times \left\{\frac{1-(1/K_3T_3)}{(G/\eta)-(1/T_3)}\left[\left(e^{((G/\eta)-(1/T_3))t} - 1\right] - \frac{\eta}{G}\left(e^{(G/\eta)t} - 1\right)e^{-(G/\eta)t}\right\} - \frac{R_2}{2} e^{\lambda(R_2-R_1)/R_1R_2}\\ &\times (p\sin\varphi - c\cos\varphi)\left(\frac{2}{\lambda} + \frac{1}{R_1}\right) \cdot \left[\left(1 - \frac{1}{K_3T_3}\right)e^{-t/T_3} - 1\right]\end{aligned} \tag{5.112}$$

The stress components in the visco-plastic zone, residual strength zone and the lining can be obtained from the function or parameter determined using equations (5.110), (5.111) and (5.112). The locations of $R_1$ and $R_2$ are then determined according to the boundary condition that the stress or displacement components are equal at $r = R_2$ of two adjacent zones.

For simplification, only the stress component in the lining for $r \to \infty$ is given by the equation below,

$$\left.\begin{array}{l}\sigma_r\\ \sigma_\theta\end{array}\right\} = \left(\mp\frac{a_0^2}{r^2}\right)\left\{\frac{a^3}{2a_0^2}\cdot\frac{1}{R_1}\left\{\frac{R_2}{R_1}(p\sin\varphi - c\cos\varphi)e^{\lambda(R_2-R_1)/R_1R_2}\left[\frac{\nu}{1-2\nu}\left(1-\frac{\lambda}{R_1}\right)+\left(1+\frac{\lambda}{R_1}\right)\right]\right.\right.$$
$$\left.\left. -\frac{R_2}{2}e^{\lambda(R_2-R_1)/R_1R_2}(p\sin\varphi - c\cos\varphi)\left(1+\frac{\lambda}{R_1}\right)+D-\delta K_p\right\}+\frac{a^2}{2a_0^2}-\delta K_p\right\} \tag{5.113}$$

## 5.6. EFFECT OF BOLT REINFORCEMENT IN VISCO-ELASTIC ROCK MASS

Bolt and shotcrete support is to reinforce surrounding rock masses, and to control deformation. The technique effectively controls the rock deformation, block loosening and cave-in. It reinforces the complete rock mass system by adjusting the stress distribution in the rock mass, and by mobilising the self-supporting ability of the surrounding rock mass [66,194–196,330–339].

The mechanism of bolting technique has not yet been thoroughly understood, although the technology has been well developed and advanced in recent years [180,218,340–381]. The purpose of this section is to examine the interaction between the rock mass and the bolts, to study the use of bolt reinforcement and its effect on the stress state of the rock mass.

### *5.6.1 Stress state in different zones*

Similar to the previous cases, the rock mass surrounding an opening in far region is regarded as a visco-elastic medium, whereas the rock mass in the bolted region is considered to be another visco-elastic medium having different parameters, as shown in Figure 5.11. The assumptions on the continuum mechanics adopted in the previous sections are still used, and the initial stress state is assumed to be uniform and the surrounding rock mass is assumed to be isotropic and homogeneous. The visco-elastic stress state in the far region is the same as in equation (5.54).

Assume that bolts are applied when $t = t_1$, the bolts are $(R - a)$ in length and create a bolted circular ring. The parameters of $T$, $K$, $G$ in the bolted region have basically

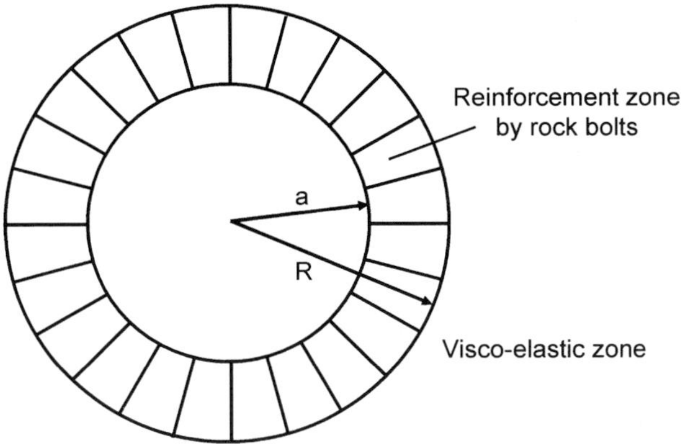


**Figure 5.11.** Schematic representation of surrounding rock mass zones.

the same value as those in the visco-elastic zone, the bolts only restrain the displacement of the bolted rock mass.

According to the principle of superposition, the components of stress and displacements in the bolted region should meet the following conditions:

$$\sigma_r^{(2)} = K_3T_3[c_2(t) - b_2(t)S_2^2] + K_3(1 - K_3T_3)\int_{t_1}^{t}[c_2(t') - b_2(t')S_2^2]e^{K_3(t'-t)}\mathrm{d}t' - P(1 - S_2^2) \tag{5.114}$$

$$\sigma_\theta^{(2)} = K_3T_3[c_2(t) + b_2(t)S_2^2] + K_3(1 - K_3T_3)\int_{t_1}^{t}[c_2(t') + b_2(t')S_2^2]e^{K_3(t'-t)}\mathrm{d}t' - P(1 + S_2^2) \tag{5.115}$$

$$U_r^{(2)} = \frac{1}{2G_3}r[(1 - 2\nu)c_2(t) + b_2(t)S_2^2] + \frac{1}{2G_3}PrS_2^2\left[\left(1 - \frac{1}{K_3T_3}\right)e^{-t/T_3} - 1\right] \tag{5.116}$$

where $S_2 = a/r$ and $S_3 = R/r$.

It should also satisfy the boundary condition below.

(i) When $r \to \infty$, because $U_r^{(3)} = 0$, we have, from equation (5.54)

$$c_3(t) = 0 \tag{5.117}$$

(ii) When $r = a$, the stress condition is

$$\sigma_r^{(2)} = \alpha\Delta U + \sigma_0 = \alpha\left|U_{r=a}^{(2)} - U_{r=R}^{(2)}\right| + \sigma_0 \tag{5.118}$$

where $\alpha$ is a constant related with the length, cross-sectional area of rock bolt, and bolting density and mechanical property of the bolts, i.e.,

$$\alpha = \frac{E_1 B_1}{(R - a)B_2} \tag{5.119}$$

In the above equation, $E_1$ is the elastic modulus of the bolts; $B_1$ is the sum of the cross-sectional areas of all bolts in unit length of the opening; $B_2$ is the total area of the opening face in unit length; $\sigma_0$ is the equivalent normal stress converted from the axial stress in the bolt; $|U^{(2)}_{r=a} - U^{(2)}_{r=R}|$ is the displacement difference in the bolted circular ring after bolts installed at $t \geq t_1$.

This boundary condition reflects the action of the bolt, i.e., if relative displacements take place in the bolted region after bolting, they will be restrained by bolts. In addition, the bolt's bearing force in the surrounding rock mass is proportional to the displacement.

Substituting equation (5.116) into equation (5.118), we have

$$\begin{aligned}\sigma^{(2)}_{r=a} = \frac{\alpha}{2G_3}&[(1 - 2\nu)(a - R)c_2(t) + a(1 - X_2)b_2(t)\\ &+ Pa\left(1 - \frac{1}{K_3 T_3}\right)(1 - X_2)(e^{-t/T_3} - e^{-t_1/T_3})] + \sigma_0\end{aligned} \tag{5.120}$$

where $X_2 = a/R$

Substituting $r = a$ into equation (5.114) gives

$$\sigma^{(2)}_{r=a} = K_3 T_3[c_2(t) - b_2(t)] + K_3(1 - K_3 T_3)\int_{t_1}^{t}[c_2(t') - b_2(t')]e^{K_3(t'-t)}\mathrm{d}t' \tag{5.121}$$

From equations (5.120) and (5.121), we have

$$\begin{aligned}&K_3 T_3[c_2(t) - b_2(t)] + K_3(1 - K_3 T_3)\int_{t_1}^{t}[c_2(t') - b_2(t')]e^{K_3(t'-t)}\mathrm{d}t'\\ &= \frac{\alpha}{2G_3}[(1 - 2\nu) \times (a - R)c_2(t) + a(1 - X_2)b_2(t)\\ &\quad + Pa\left(1 - \frac{1}{K_3 T_3}\right)(1 - X_2)(e^{-t/T_3} - e^{-t_1/T_3})] + \sigma_0\end{aligned}$$

Multiplying the two sides of the above equation by $e^{K_3 t}$, deriving the result with respect to $t$ and eliminating the term of $e^{K_3 t}$, we have

$$\left[\frac{\alpha}{2G_3}(1-2\nu)(a-R)-K_3T_3\right]\dot{c}_2(t)+\left[\frac{\alpha}{2G_3}(1-X_2)+K_3T_3)\dot{b}_2(t)\right]$$
$$+K_3\left[\frac{\alpha}{2G_3}(1-2\nu)(a-R)-1\right]c_2(t)+K_3\left[\frac{\alpha}{2G_3}a(1-X_2)+1\right]b_2(t)$$
$$=\frac{\alpha Pa}{2G_3}\left(1-\frac{1}{K_3T_3}\right)(1-X_2)\left[K_3e^{-t_1/T_3}-\left(K_3-\frac{1}{T_3}\right)e^{-t/T_3}\right]-K_3\sigma_0 \tag{5.122}$$

(iii) When $r=R$, the displacement condition is

$$U_r^{(2)}=U_r^{(3)}$$

According to equations (5.54), (5.116) and (5.117), we have

$$b_3(t)=(1-2\nu)c_2(t)+X_2^2b_2(t) \tag{5.123}$$

(iv) When $r=R$, the stress condition is

$$\sigma_{r=a}^{(3)}=\sigma_{r=R}^{(2)}-X_2\sigma_{r=R}^{(2)}$$

The last term in this equation reflects the action of the bolts. Substituting equations (5.54), (5.114) and (5.117) into the above equation and interpreting the results, we have

$$K_3T_3[(X_2-1)\dot{c}_2(t)+X_2(X_2-1)\dot{b}_2(t)-\dot{b}_3(t)]$$
$$+K_3T_3[(X_2-1)c_2(t)+X_2(X_2-1)b_2(t)-b_3(t)]=0$$

By solving this differential equation and substituting the result into equation (5.123), we have

$$b_2(t)=\left(1+\frac{2R(\nu-1)}{a}\right)c_2(t)+C_1e^{-t/T_3} \tag{5.124}$$

where $C_1$ is an undetermined constant.

Substituting equation (5.124) back into equation (5.122) gives

$$\left[\frac{\alpha}{2G_3}(1-X_2)(a+4\nu R-3R)+\frac{2R}{a}K_3T_3(\nu-1)\right]\dot{c}_2(t)$$
$$+K_3\left[\frac{\alpha}{2G_3}(1-X_2)(a+4\nu R-3R)+\frac{2R}{a}(\nu-1)\right]c_2(t)$$
$$=\frac{\alpha Pa}{2G_3}\left(1-\frac{1}{K_3T_3}\right)(1-X_2)\left[K_3e^{-t/T_3}-\left(K_3-\frac{1}{T_3}\right)e^{-t/T_3}\right]$$
$$-K_3\sigma_0+\frac{\alpha a}{2G_3}(1-X_2)\left(\frac{1}{T_3}-K_3\right)C_1e^{-t/T_3}$$

then solving this equation gives

$$c_2(t)=C_2e^{-(N_2/N_1)t}+\frac{a}{3R-a-4\nu R}C_1e^{-t/T_3}-\frac{Pa(K_3T_3-1)}{K_3T_3(a+4\nu R-3R)}e^{-t/T_3}$$
$$+\frac{(1-X_2)\alpha PK_3a^2}{N_2}\left(1-\frac{1}{K_3T_3}\right)e^{-t_1/T_3}-\frac{2K_3G_3a}{N_2}\sigma_0 \tag{5.125}$$

where

$$\left.\begin{aligned}N_1&=\alpha a(1-X_2)(a+4\nu R-3R)+4G_3K_3T_3(\nu-1)R\\N_2&=K_3\alpha a(1-X_2)(a+4\nu R-3R)+4G_3RK_3(\nu-1)\end{aligned}\right\} \tag{5.126}$$

and $C_1$, $C_2$ are undetermined constants.

For equation (5.114), when $t=t_1$, $r=a$ and $r=R$, we have respectively

$$c_2(t_1)-b_2(t_1)=0 \quad \text{and} \quad c_2(t_1)-X_2^2b_2(t_1)=0$$

From this relation, we have

$$c_2(t_1)=b_2(t_1)=0$$

If we substitute this result back into equations (5.124) and (5.125), then the undetermined constants of $C_1$ and $C_2$ can be obtained:

$$\left.\begin{aligned}C_1&=0\\C_2&=\left[\frac{Pa(K_3T_3-1)}{K_3T_3(a+4\nu R-3R)}-\frac{\alpha PK_3a^2(1-X_2)}{N_2}\left(1-\frac{1}{K_3T_3}\right)\right]\\&\quad\times e^{((N_2/N_1)-(1/T_3))t}+\frac{2K_3G_3a\sigma_0}{N_2}e^{(N_2/N_1)t_1}\end{aligned}\right\} \tag{5.127}$$

Substituting equation (5.127) back into equations (5.124) and (5.125) gives

$$c_2(t) = C_2 e^{-(N_2/N_1)t} - \left[\frac{Pa(K_3T_3 - 1)}{K_3T_3(a + 4\nu R - 3R)} e^{-t/T_3} + \frac{\alpha P K_3 a^2(1 - X_2)}{N_2}\left(1 - \frac{1}{K_3T_3}\right)\right] e^{-t/T_3} - \frac{2K_3G_3 a}{N_2}\sigma_0 \tag{5.128}$$

$$b_2(t) = \left[1 + \frac{2R(\nu - 1)}{a}\right] C_2 e^{-(N_2/N_1)t} - \frac{P(K_3T_3 - 1)(a + 2\nu R - 2R)}{K_3T_3(a + \nu R - 3R)} e^{-t/T_3} + \left[\frac{\alpha P K_3 a(1 - X_2)(a + 2\nu R - 2R)}{N_2}\left(1 - \frac{1}{K_3T_3}\right)\right] e^{-t_1/T_3} - \frac{2K_3G_3(a + 2\nu R - 2R)}{N_2}\sigma_0 \tag{5.129}$$

Furthermore, by substituting equations (5.128) and (5.129) back into equations (5.114), (5.115) and (5.116) and integrating the results, we have the expressions for the stress and strain components in the bolted region:

$$\begin{aligned}\sigma_r^{(2)} = {} & \frac{N_1 - T_3N_2}{K_3N_1 - N_2} K_3\left[1 - S_2^2 - \frac{2S_2^2R(\nu - 1)}{a}\right] C_2 e^{-(N_2/N_1)t} - \frac{N_1}{K_3N_1 - N_2} K_3(1 - K_3T_3) \\ & \times \left[1 - S_2^2 - \frac{2S_2^2R(\nu - 1)}{a}\right] c_2 \exp\left[\left(K_3 - \frac{N_2}{N_1}\right)t_1 - K_3 t\right] - \frac{P(K_3T_3 - 1)}{a + 4\nu R - 3R} \\ & \times [a - S_2^2(a + 2\nu R - 2R)] \exp\left[\left(K_3 - \frac{1}{T_3}\right)t_1 - K_3 t\right] + \frac{\alpha P K_3 a(1 - X_2)}{N_2} \\ & \times [a - S_2^2(a + 2\nu R - 2R)]\left(1 - \frac{1}{K_3T_3}\right) e^{-t_1/T_3} - \frac{\alpha P K_3 a(1 - X_2)(1 - K_3T_3)}{N_2} \\ & \times [a - S_2^2(a + 2\nu R - 2R)]\left(1 - \frac{1}{K_3T_3}\right) \exp\left[-K_3 t + \left(K_3 - \frac{1}{T_3}\right)t_1\right] - \frac{2K_3G_3\sigma_0}{N_2} \\ & \times [a - S_2^2(a + 2\nu R - 2R)][1 - (1 - K_3T_3)] \exp[K_3(t_1 - t)] - P(1 - S_2^2)\end{aligned} \tag{5.130}$$

$$\begin{aligned}\sigma_\theta^{(2)} = {} & \frac{N_1 - T_3N_2}{K_3N_1 - N_2} K_3\left[1 + S_2^2 + \frac{2S_2^2R(\nu - 1)}{a}\right] C_2 e^{-(N_2/N_1)t} - \frac{N_1}{K_3N_1 - N_2} K_3(1 - K_3T_3) \\ & \times \left[1 + S_2^2 + \frac{2S_2^2R(\nu - 1)}{a}\right] C_2 \exp\left[\left(K_3 - \frac{N_2}{N_1}\right)t_1 - K_3 t\right] - \frac{P(K_3T_3 - 1)}{a + 4\nu R - 3R} \\ & \times [a + S_2^2(a + 2\nu R - 2R)] \exp\left[\left(K_3 - \frac{1}{T_3}\right)t_1 - K_3 t\right] + \frac{\alpha P K_3 a(1 - X_2)}{N_2} \\ & \times [a + S_2^2(a + 2\nu R - 2R)]\left(1 - \frac{1}{K_3T_3}\right) e^{-t_1/T_3} - \frac{\alpha P K_3 a(1 - X_2)(1 - K_3T_3)}{N_2} \\ & \times [a + S_2^2(a + 2\nu R - 2R)]\left(1 - \frac{1}{K_3T_3}\right) \exp\left[\left(K_3 - \frac{1}{T_3}\right)t_1 - K_3 t\right] - \frac{2K_3G_3\sigma_0}{N_2} \\ & \times [a + S_2^2(a + 2\nu R - 2R)]\{1 - (1 - K_3T_3) \exp[K_3(t_1 - t)]\} - P(1 + S_2^2)\end{aligned} \tag{5.131}$$

$$U_r^{(2)} = \frac{r}{2G_3}\left\{\left[1 - 2\nu + S_2^2 + \frac{2S_2^2R(\nu-1)}{a}\right]C_2e^{-(N_2/N_1)t} - \frac{P(K_3T_3-1)}{K_3T_3(a+4\nu R-3R)}\right.$$
$$\times [a(1-2\nu) + S_2^2(a+2\nu R-2R)]e^{-t/T_3} + \frac{\alpha PK_3a(1-X_2)}{N_2}$$
$$\times [a(1-2\nu) + S_2^2(a+2\nu R-2R)]\left(1 - \frac{1}{K_3T_3}\right)e^{-t_1/T_3} - \frac{2K_3G_3\sigma_0}{N_2}$$
$$\left.\times[a(1-2\nu) + S_2^2(a+2\nu R-2R)]\right\} + \frac{1}{2G_3}PrS_2^2\left[\left(1 - \frac{1}{K_3T_3}\right)e^{-t/T_3} - 1\right] \tag{5.132}$$

By substituting equations (5.128), (5.129) back into equation (5.123), we can obtain the explicit formula of $b_3(t)$. By further substituting $b_3(t)$ and equation (5.117) back into equation (5.54) and integrating the results, we have the expression for the stress and strain components in the visco-elastic zone as following:

$$\sigma_r^{(3)} = -\frac{N_1 - T_3N_2}{K_3N_1 - N_2}K_3S_3^2\left[1-2\nu+X_2^2+\frac{2X_2^2R(\nu-1)}{a}\right]C_2e^{-(N_2/N_1)t}$$
$$+\frac{N_1}{N_1K_3-N_2}K_3S_3^2(1-K_3T_3)\left[1-2\nu+X_2^2+\frac{2X_2^2R(\nu-1)}{a}\right]C_2$$
$$\times\exp\left[\left(K_3-\frac{N_2}{N_1}\right)t_1-K_3t\right]+\frac{P(K_3T_3-1)}{a+4\nu R-3R}S_3^2[a(1-2\nu)+X_2^2(a+2\nu R-2R)]$$
$$\times\exp\left[\left(K_3-\frac{1}{T_3}\right)t_1-K_3t\right]-\frac{\alpha PK_3a(1-X_2)}{N_2}S_3^2[a(1-2\nu)+X_2^2(a+2\nu R-2R)]$$
$$\times\left(1-\frac{1}{K_3T_3}\right)e^{t_1/T_3}+\frac{\alpha PK_3a(1-X_2)}{N_2}S_3^2[a(1-2\nu)+X_2^2(a+2\nu R-2R)]$$
$$\times\left(1-\frac{1}{K_3T_3}\right)\exp\left[\left(K_3-\frac{1}{T_3}\right)t_1-K_3t\right]+\frac{2K_3G_3\sigma_0}{N_2}S_3^2[a(1-2\nu)+X_2^2(a+2\nu R-2R)]$$
$$\times\{1-(1-K_3T_3)\exp[K_3(t_1-t)]\}-P(1-S_2^2) \tag{5.133}$$

$$\sigma_\theta^{(3)} = \frac{N_1 - T_3N_2}{K_3N_1 - N_2}K_3S_3^2\left[1-2\nu+X_2^2+\frac{2X_2^2R(\nu-1)}{a}\right]C_2e^{-(N_2/N_1)t}$$
$$-\frac{N_1}{K_3N_1-N_2}K_3S_3^2(1-K_3T_3)\left[1-2\nu+X_2^2+\frac{2X_2^2R(\nu-1)}{a}\right]C_2$$
$$\times\exp\left[\left(K_3-\frac{N_2}{N_1}\right)t_1-K_3t\right]-\frac{P(K_3T_3-1)}{a+4\nu R-3R}S_3^2[a(1-2\nu)+X_2(a+2\nu R-2R)]$$
$$\times\exp\left[\left(K_3-\frac{1}{T_3}\right)t_1-K_3t\right]+\frac{\alpha PK_3a(1-X_2)}{N_2}S_3^2[a(1-2\nu)+X_2^2(a+2\nu R-2R)]$$

$$\times\left(1-\frac{1}{K_3T_3}\right)e^{-t_1/T_3}-\frac{\alpha PK_3a(1-X_2)}{N_2}S_3^2[a(1-2\nu)+X_2^2(a+2\nu R-2R)]$$
$$\times\left(1-\frac{1}{K_3T_3}\right)\exp\left[\left(K_3-\frac{1}{T_3}\right)t_1-K_3t\right]$$
$$-\frac{2K_3G_3\sigma_0}{N_2}S_3^2[a(1-2\nu)+X_2^2(a+2\nu R-2R)]$$
$$\times\left[1-(1-K_3T_3)e^{K_3(t_1-t)}\right]-P(1+S_2^2) \tag{5.134}$$

$$U_r^{(3)}=\frac{rS_3^2}{2G_3}\left\{\left[1-2\nu+X_2^2+\frac{2X_2^2R(\nu-1)}{a}\right]C_2e^{-(N_2/N_1)t}-\frac{P(K_3T_3-1)}{K_3T_3(a+4\nu R-3R)}\right.$$
$$\times[a(1-2\nu)+X_2^2(a+2\nu R-2R)]e^{-t/T_3}+\frac{\alpha PK_3a(1-X_2)}{N_2}$$
$$\times[a(1-2\nu)+X_2^2(a+2\nu R-2R)]\left(1-\frac{1}{K_3T_3}\right)e^{-t_1/T_3}-\frac{2K_3G_3\sigma_0}{N_2}$$
$$\left.\times[a(1-2\nu)+X_2^2(a+2\nu R-2R)]\right\}+\frac{1}{2G_3}PrS_2^2\left[\left(1-\frac{1}{K_3T_3}\right)e^{-t/T_3}-1\right] \tag{5.135}$$

By allowing $t\to\infty$, the expression of stress and strain components can be obtained for the different zones.

(i) In the bolted region

$$\sigma_r^{(2)}=\frac{\alpha PK_3a(1-X_2)}{N_2}[a-S_2^2(a+2\nu R-2R)]\left(1-\frac{1}{K_3T_3}\right)e^{-t_1/T_3}$$
$$-\frac{2K_3G_3\sigma_0}{N_2}[a-S_2^2(a+2\nu R-2R)]-P\left(1-S_2^2\right)$$

$$\sigma_\theta^{(2)}=\frac{\alpha PK_3a(1-X_2)}{N_2}[a+S_2^2(a+2\nu R-2R)]\left(1-\frac{1}{K_3T_3}\right)e^{-t_1/T_3}$$
$$-\frac{2K_3G_3\sigma_0}{N_2}[a+S_2^2(a+2\nu R-2R)]-P(1+S_2^2) \tag{5.136}$$

$$U_r^{(2)}=\frac{\alpha PK_3a(1-X_2)}{2G_3N_2}r[a(1-2\nu)+S_2^2(a+2\nu R-2R)]\left(1-\frac{1}{K_3T_3}\right)e^{-t_1/T_3}$$
$$-\frac{K_3\sigma_0}{N_2}r[a(1-2\nu)+S_2^2(a+2\nu R-2R)]-\frac{1}{2G_3}PrS_2^2 \tag{5.137}$$

(ii) In the visco-elastic zone

$$\sigma_r^{(3)} = -\frac{\alpha P K_3 a(1-X_2)}{N_2} S_3^2[a(1-2\nu)+X_2^2(a+2\nu R-2R)]\left(1-\frac{1}{K_3T_3}\right)e^{-t_1/T_3} + \frac{2K_3G_3\sigma_0}{N_2} S_3^2[a(1-2\nu)+X_2^2(a+2\nu R-2R)] - P(1-S_2^2)$$

$$\sigma_\theta^{(3)} = \frac{\alpha P K_3 a(1-X_2)}{N_2} S_3^2[a(1-2\nu)+X_2^2(a+2\nu R-2R)]\left(1-\frac{1}{K_3T_3}\right)e^{-t_1/T_3} - \frac{2K_3G_3\sigma_0}{N_2} S_3^2[a(1-2\nu)+X_2^2(a+2\nu R-2R)] - P(1+S_2^2) \quad (5.138)$$

$$U_r^{(3)} = \frac{\alpha P K_3 a(1-X_2)}{2G_3N_2} rS_3^2[a(1-2\nu)+X_2^2(a+2\nu R-2R)]\left(1-\frac{1}{K_3T_3}\right)e^{-t_1/T_3} - \frac{K_3\sigma_0}{N_2} rS_3^2[a(1-2\nu)+X_2^2(a+2\nu R-2R)] - \frac{1}{2G_3}PrS_2^2 \quad (5.139)$$

It can be seen clearly from equation (5.139) that the expression of stress or displacement comprises three terms, each having distinct physical meaning. The first term is the effect of the time ($t_1$) of bolt installation on the surrounding rock mass. It indicates that the earlier the installation the better to improve the stress state in the rock mass, i.e., reducing $t_1$ as fully as possible. The second term reflects the effect of prestress of the bolts ($\sigma_0$). The higher $\sigma_0$ gives the more effective improvement of the rock mass. The third term reflects the effect of the initial in situ stress, $P$.

### *5.6.2 Discussion and application*

Figure 5.12 shows stress distribution in the surrounding rock mass reinforced with bolts.

According to equations (5.126) and (5.119), we have

$$N_2 = K_3\alpha a(1-X_2)(a+4\nu R-3R)+4G_3RK_3(\nu-1) = -\frac{E_1B_1}{B_2}K_3a\left(3-4\nu-\frac{a}{R}\right)-4G_3RK_3(1-\nu)$$

Because $B_2 \gg B_1$, $4G_3RK_3(1-\nu) >> (E_1B_1/B_2)K_3a(3-4\nu-(a/R))$. The above equation can be approximately reduced to

$$N_2 = -4G_3RK_3(1-\nu) \quad (5.140)$$

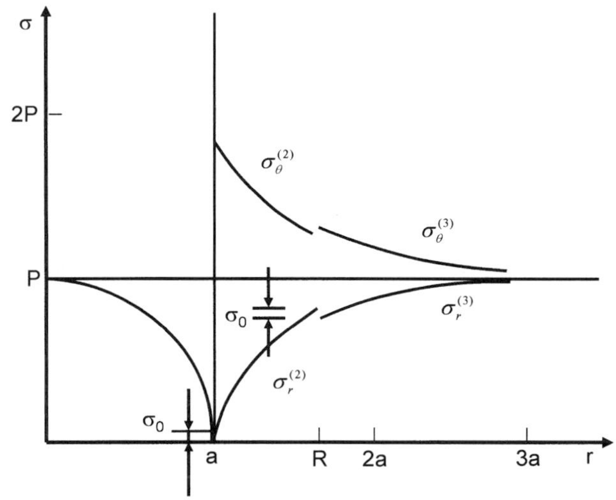


**Figure 5.12.** Stress distribution in the surrounding rock mass after bolting.

Substituting equations (5.119) and (5.140) into equations (5.136) and (5.137) gives the approximate solution to the stress and strain in the bolted zones:

$$\sigma_r^{(2)} = -\frac{E_1 B_1 P}{4B_2 G_3(1-\nu)}\frac{a}{R}\left[\frac{a}{R} + S_{\frac{2}{2}}^{2}\left(2 - 2\nu - \frac{a}{R}\right)\right]\left(1 - \frac{1}{K_3 T_3}\right)e^{-t_1/T_3} + \frac{\sigma_0}{2(1-\nu)}\left[\frac{a}{R} + S_{\frac{2}{2}}^{2}\left(2 - 2\nu - \frac{a}{R}\right)\right] - P\left(1 - S_{\frac{2}{2}}^{2}\right) \tag{5.141}$$

$$\sigma_\theta^{(2)} = \frac{E_1 B_1 P}{4B_2 G_3(1-\nu)}\frac{a}{R}\left[S_{\frac{2}{2}}^{2}\left(2 - 2\nu - \frac{a}{R}\right) - \frac{a}{R}\right]\left(1 - \frac{1}{K_3 T_3}\right)e^{-t_1/T_3} + \frac{\sigma_0}{2(1-\nu)}\left[\frac{a}{R} - S_{\frac{2}{2}}^{2}\left(2 - 2\nu - \frac{a}{R}\right)\right] - P\left(1 - S_{\frac{2}{2}}^{2}\right) \tag{5.142}$$

$$U_r^{(2)} = \frac{r}{2G_3}\left\{-\frac{E_1 B_1 P}{4B_2 G_3(1-\nu)}\frac{a}{R}\left[\frac{a}{R}(1-2\nu) - S_{\frac{2}{2}}^{2}\left(2 - 2\nu - \frac{a}{R}\right)\right]\left(1 - \frac{1}{K_3 T_3}\right)e^{-t_1/T_3} + \frac{\sigma_0}{2(1-\nu)}\left[\frac{a}{R}(1-2\nu) - S_{\frac{2}{2}}^{2}\left(2 - 2\nu - \frac{a}{R}\right)\right] - PS_{\frac{2}{2}}^{2}\right\} \tag{5.143}$$

From the plastic condition of equation (5.51) given in Section 5.3, we have

$$\sigma_\theta - \sigma_r = -\frac{2}{\sqrt{6}}\delta K_p \tag{5.144}$$

Considering equations (5.141) and (5.142), we have

$$\sigma_\theta^{(2)} - \sigma_r^{(2)} = \frac{E_1 B_1 P S_2^2}{2B_2 G_3(1-\nu)}\frac{a}{R}\left(2 - 2\nu - \frac{a}{R}\right)\left(1 - \frac{1}{K_3 T_3}\right)e^{-t_1/T_3} + \frac{-\sigma_0}{(1-\nu)} S_2^2\left(2 - 2\nu - \frac{a}{R}\right) - 2PS_2^2 \tag{5.145}$$

Therefore, in order to prevent the opening wall ($S_2 = 1$) from becoming plastic state after an infinite long time, we may apply the following criterion according to equations (5.144) and (5.145):

$$\frac{E_1 B_1 P}{2B_2 G_3(1-\nu)}\frac{a}{R}\left(2 - 2\nu - \frac{a}{R}\right)\left(1 - \frac{1}{K_3 T_3}\right)e^{-t_1/T_3} + \frac{-\sigma_0}{(1-\nu)}\left(2 - 2\nu - \frac{a}{R}\right) - 2P \le -\frac{2}{\sqrt{6}}\delta K_p \tag{5.146}$$

To meet the condition in equation (5.146), it is necessary to enlarge the first and the second terms. To enlarge the second term, however, it needs to increase both prestress and length of the bolt, i.e., $\sigma_0$ and $R$. However, there is a limit to increase the prestress of bolts, because a bolt of a certain cross-sectional area can bear only a limited prestress corresponding to the tensile strength. Therefore, efforts should be made to increase the first term as far as possible. There are several ways to increase this term, for instance, install bolts as early as possible (decreasing $t_1$), or enlarge the bolt cross-section area (i.e. increase $B_1$) and increase the bolting density (i.e. decrease $B_2$), or change the bolt length. The first two measures are more effective.

When the prestress ($\sigma_0$) is not taken into consideration or the bolt is not prestressed, equation (5.145) becomes

$$\sigma_\theta^{(2)} - \sigma_r^{(2)} = \frac{E_1 B_1 P}{2B_2 G_3(1-\nu)}\frac{a}{R}\left(2 - 2\nu - \frac{a}{R}\right)\left(1 - \frac{1}{K_3 T_3}\right)e^{-t_1/T_3} - 2P \tag{5.147}$$

This equation takes the status of sidewalls of the opening into account, i.e., $S_2 = 1$.

The relationship between $\sigma_\theta^{(2)} - \sigma_r^{(2)}$and $a/R$ is expressed in diagram shown in Figure 5.13.

Given that $E_1 = 2.1 \times 10^5$ MPa, $B_1 = \pi \times 1\ \text{cm}^2$, $B_2 = \pi \times 50\ \text{cm}^2$, $G_3 = 4 \times 10^3$ MPa, $\mu = 0.25$, $T_3 = 12d$, $K_2 = 1/8d$, $t_1 = 0$, then from the figure, $D_1 = (0.01 - 2)P$, $D_2 = (0.006 - 2)P$, $D_3 = (0.004 - 2)P$.

It can be seen from equation (5.147) or from Figure 5.13 that $\sigma_\theta^{(2)} - \sigma_r^{(2)}$ has the minimum value when $a/R = 1 - \nu$.

The above analysis leads to the conclusion that in order to avoid the surrounding rock mass damage due to stress concentration or to make the stress difference in

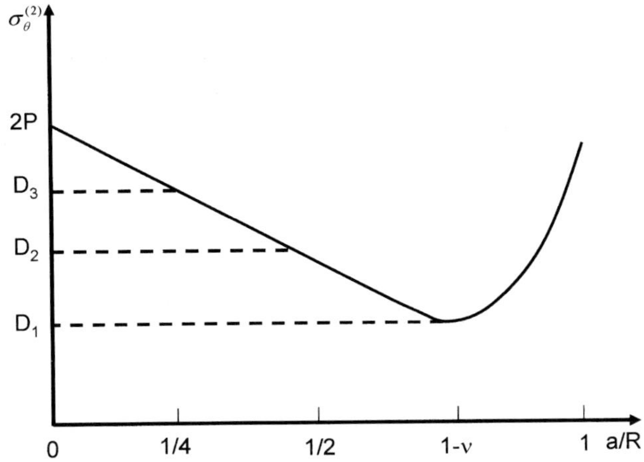


**Figure 5.13.** Effect of relative bolt length on stress difference of surrounding rock mass.

opening walls reach the lowest level: (a) the bolts must be installed as early as possible, (b) the prestress in the bolts must be sufficiently large, (c) the bolts must have sufficient length and cross-sectional area, (d) bolting should be installed with sufficient density. The optimal bolt length is $\nu/(1-\nu)$ times opening radius, i.e., $a/R = 1-\nu$, when non-prestressed bolts are used. If bolts are installed late, the first term in equation (5.145) becomes basically non-functional. The second term in the equation shows that the higher $R$, the better, i.e., long bolts should be selected, up to the optimal length.

If failure is determined by the magnitude of displacements, the conditions of which the displacement of the opening wall becomes in minimum should be examined. From equation (5.143) and substituting $S_2=1$, $r=a$ into equation (5.143), we can obtain

$$U^{(2)}_{r=a} = \frac{a}{2G_3}\left[\frac{E_1B_1P}{2B_2G_3}\frac{a}{R}\left(1-\frac{a}{R}\right)\left(1-\frac{1}{K_3T_3}\right)e^{-t_1/T_3} + (-\sigma_0)\left(1-\frac{a}{R}\right) - P\right]$$

If the prestress $\sigma_0 = 0$, then the above equation becomes,

$$U^{(2)}_{r=a} = \frac{a}{2G_3}\left[\frac{E_1B_1P}{2B_2G_3}\frac{a}{R}\left(1-\frac{a}{R}\right)\left(1-\frac{1}{K_3T_3}\right)e^{-t_1/T_3} - P\right]$$

Under the conditions given in this section, $[\overline{D}]_n$ is proven to be symmetrical and to have the expression similar to that of matrix $[D]$. In the case of $\omega = 0$, $[\overline{D}]_n$ becomes

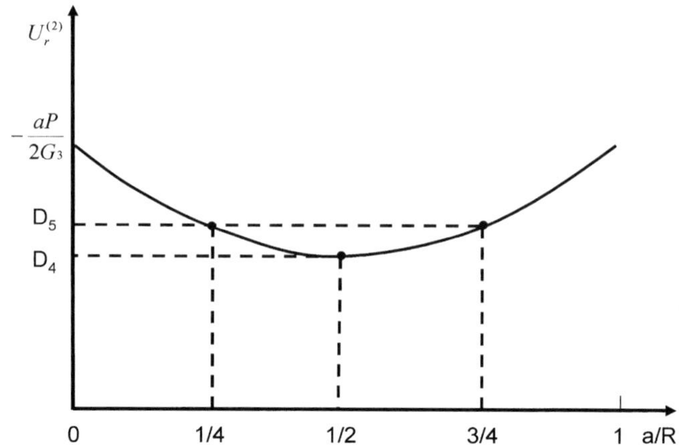


**Figure 5.14.** Effect of relative bolt length on displacement of surrounding rock mass.

an elastic matrix and for $\omega$ = constant with no damage development occurring, the above proposed algorithm constitutes the constant stiffness algorithm.

In Figure 5.14, $D_4 = (0.0025 - 1) \times (aP/2G_3)$ and $D_5 = (0.0018 - 1) \times (aP/2G_3)$. Figure 5.14 indicates that $U_r^{(2)}$ reaches the minimum value when $a/R = 1/2$, i.e., when bolt length equals to the opening radius, opening wall has minimum displacement.

Similarly, in order to minimise the displacement of opening walls, bolts should be installed as early as possible and the prestress of bolts should be increased, in addition to enlarge bolt cross-sectional area and to increase bolting density. In the case of non-prestressed bolts, the optimum length of bolt should have the ratio of $a/R = 1/2$. If bolts are installed late, the same conclusion can be drawn from equation (5.147) that long bolts should be used.

In summary, the optimum time for installing bolts and the magnitude of prestress applied in the bolts can be determined quantitatively using the method presented in this section. A clear conclusion can be drawn that the most rational bolt length $(R - a)$ for visco-elastic surrounding rock mass should be $a/(1 - \nu)$, where $a$ is the opening radius, $R$ is the radius of the visco-elastic zone, $\nu$ is the Possion's ratio of the rock mass.

## 5.7. RHEOLOGICAL DAMAGE ANALYSIS OF THE ROCK MASS STABILITY

Previous sections have studied the problems of the stress state in surrounding rock masses and the interaction of the rock mass and support using various rheological models. The present section combines damage mechanics and rheological mechanics to study the stability of rock mass. When excavation is in soft rock or at great depth, the rock mass deforms. A great portion of the deformation is caused by the loosening due to relatively high stresses as compared to the strength of rock around the opening. It produces cracks in the surrounding rock mass, and results in remarkable dilatancy phenomena reflected by obvious volume increment in the rock mass [39,206,382–389]. As a consequence, the conventional analysis methods are not adequate to study the rock mass exhibiting the above behaviour. The concept and method of damage mechanics combining with rheological mechanics are applied to analyse the rock mass in this section.

### 5.7.1 *Damage evolution equation*

The basic concepts and assumptions of damage mechanics have been discussed in Chapter 3. This section adopts the damage equation by Desai [291,292,382] based upon the isotropic damage concept to express the effective stress and apparent stress, i.e.,

$$\sigma_{ij} = (1 - \omega)\sigma_{ij}^* + \frac{\omega}{3}\delta_{ij}\sigma_{kk}^* \tag{5.148}$$

where $\omega$ is the damage variable.

Considering that the rupture failure planes have residual strength, a specific maximum damage value $\omega_p$ less than 1 of the rock mass can be assumed. The value of $\omega_p$ depends upon the type of the rock.

Most rocks have brittle characteristics. Under loading, microcracks initiate and propagate and a failure plane forms. The failure can be caused by tensile, shear or combined tensile and shear. The rock dilates before the final failure as cracks are growing in number and extending gradually. The dilatancy phenomenon is the most important warning sign of rock failure.

There are three main types of damage evolutions for brittle material such as rocks:

(i) Evolution and development of damages are governed mainly by strains, especially by the tensile strain.
(ii) Damage evolution and development are governed mainly by stresses, especially by the deviator stress or the major principal stress.
(iii) Such evolution is governed by energy or relief rate of energy.

Tan [39] studied the mechanism and criterion of rock dilatancy phenomenon and suggested the condition of dilatancy generation,

$$\left(\frac{\sigma_{\text{oct}}}{f_3}\right)^{k_2} - 1 \tag{5.149}$$

where $f_3 = k - mp$ is the strength envelope curve (or surface) for a critical surface, which is roughly corresponding to the long-term envelope (surface); $p$ is the mean spherical stress; $k$ and $m$ are the strength index; $k_2$ is a real number greater than unity that is determined from test, and $\sigma_{oct}$ is the second variant of deviator stress tensor.

The damage evolution equation adopted in the present section that follows the assumption of isotropic damage expressed by equation (5.149) is,

$$\omega = K_1(1 - \omega)\left[F\left(\frac{\sigma_{oct}}{f_3}\right) - 1\right] \tag{5.150}$$

and,

$$\left[F\left(\frac{\sigma_{oct}}{f_3}\right)\right] = \begin{cases} \left(\frac{\sigma_{oct}}{f_3}\right)^{k_2} - 1 & \text{when } \left(\frac{\sigma_{oct}}{f_3}\right) > 1 \\ 0 & \text{when } \left(\frac{\sigma_{oct}}{f_3}\right) \leq 1 \end{cases}$$

in which tensile stress is taken as positive.

The above damage evolution equation is associated to both increments in number and extension of the cracks and the dilatancy of the material.

### 5.7.2 *Viscoelastic–viscoplastic-damage constitutive equation and FEM method*

***5.7.2.1 Constitutive equation.*** The common constitutive equation of non-damaged viscous materials has the following increment expression,

$$\{d\sigma\} = [D](\{d\varepsilon\} - \{d\varepsilon_V\}) \tag{5.151}$$

where $[D]$ is the elastic matrix; $\{d\sigma\}$,$\{d\varepsilon\}$ and $\{d\varepsilon_V\}$ are the apparent stress increment, overall strain increment and viscous strain increment in the form of column matrix respectively.

Based on the concept of equivalent strain, i.e., the effect of damage on the strain behaviour is reflected only by the equivalent stress. This means that the constitutive relation of a damaged material can use equation (5.151) by replacing the stress term with the equivalent stress. Hence, we have the following constitutive equation for viscoelastic–viscoplastic-damage rock masses:

$$\{d\sigma^*\} = [D](\{d\varepsilon\} - \{d\varepsilon_V\}) \tag{5.152}$$

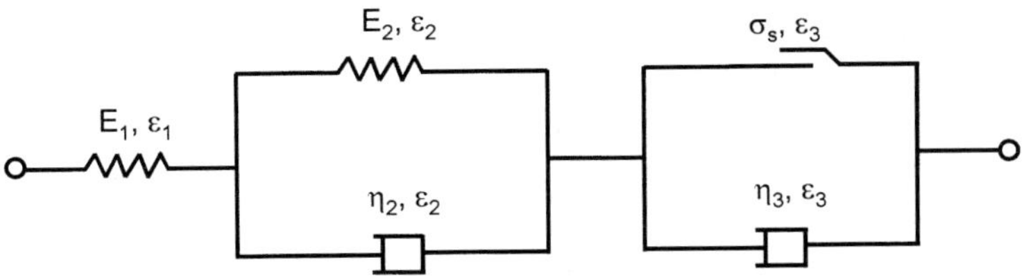


**Figure 5.15.** A rheological model ($\sigma_s$ is the plastic limit).

A typical rheological model with visco-elastic and visco-plastic characteristic is shown in Figure 5.15.

In the case of one-dimensional problem, the constitutive equation of the model can be divided into two stages according to the stress level,

$$\sigma + \frac{\eta_2}{E_1 + E_2}\dot{\sigma} = \frac{E_1 E_2}{E_1 + E_2}\varepsilon + \frac{E_1 \eta_2}{E_1 + E_2}\dot{\varepsilon} \quad \text{for } \sigma \leq \sigma_s \tag{5.153}$$

and

$$\sigma - \sigma_s + \left(\frac{\eta_3}{E_1} + \frac{\eta_2 + \eta_3}{E_2}\right)\dot{\sigma} + \frac{\eta_2 \eta_3}{E_1 E_2}\ddot{\sigma} = \eta_2 \dot{\varepsilon} \frac{\eta_2 \eta_3}{E_2}\ddot{\varepsilon} \quad \text{for } \sigma > \sigma_s \tag{5.154}$$

in the case of creep, the equation becomes

$$\varepsilon = \frac{\sigma}{E_1} + \frac{\sigma}{E_2}\left(1 - e^{-(E_2/\eta_2)t}\right) \quad \text{for } \sigma \leq \sigma_s \tag{5.155}$$

and

$$\varepsilon = \frac{\sigma}{E_1} + \frac{\sigma}{E_2}\left(1 - e^{-(E_2/\eta_2)t}\right) + \frac{\sigma - \sigma_s}{\eta_3}t \quad \text{for } \sigma > \sigma_s \tag{5.156}$$

and in the case of creep damage,

$$\varepsilon = \frac{\sigma^*}{E_1} + \frac{\sigma^*}{E_2}\left(1 - e^{-(E_2/\eta_2)t}\right) \quad \text{for } \sigma \leq \sigma_s \tag{5.157}$$

and

$$\varepsilon = \frac{\sigma^*}{E_1} + \frac{\sigma^*}{E_2}\left(1 - e^{-(E_2/\eta_2)t}\right) + \frac{\sigma^* - \sigma_s}{\eta_3}t \quad \text{for } \sigma > \sigma_s \tag{5.158}$$

***5.7.2.2 FEM method.*** According to the virtual work principle, we have the following equilibrium equation,

$$\int_V [B]^T \{\Delta\sigma\}_n \mathrm{d}V = \{\Delta f\}_n \tag{5.159}$$

where $[B]$ is the strain matrix; $\{\Delta\sigma\}_n$, $\{\Delta f\}_n$ are the stress increment and the exterior loading with the time interval of $\Delta t_n$; $\{\Delta f\}_n = \int_{S_t} [N]^T \{\Delta t\}_n \mathrm{d}S + \int_v [N]^T \{\Delta p\}_n \mathrm{d}V$; $[N]$ is the matrix of shape functions; $\{\Delta t\}_n$ is the surface force exerted on the boundary of $S_t$ within $\Delta t_n$; and $\{\Delta p\}_n$ is the body force within $\Delta t_n$.

Integration on equation (5.148) gives

$$\mathrm{d}\sigma_{ij} = (1-\omega)\mathrm{d}\sigma_{ij}^* - \frac{1}{3}\delta_i\omega_j \mathrm{d}\sigma_{kk}^* - S_{ij}^*\, \mathrm{d}\omega \tag{5.160}$$

where $S_{ij}^*$ is the effective deviator stress.

Equation (5.160) can be written in the matrix form,

$$\{\mathrm{d}\sigma\} = [C_\omega]\{\mathrm{d}\sigma^*\} - \mathrm{d}\omega\{S^*\} \tag{5.161}$$

where $[C_\omega]$ is the matrix related to the damage variable $\omega$, called damage matrix, for a three-dimensional problem, it is expressed as

$$[C_\omega] = \begin{vmatrix} \left(1-\omega+\frac{\omega}{3}\right) & \frac{\omega}{3} & \frac{\omega}{3} & 0 & 0 & 0 \\ & \left(1-\omega+\frac{\omega}{3}\right) & \frac{\omega}{3} & 0 & 0 & 0 \\ & & \left(1-\omega+\frac{\omega}{3}\right) & 0 & 0 & 0 \\ & \text{symmetry} & & (1-\omega) & 0 & 0 \\ & & & & (1-\omega) & \\ & & & & & (1-\omega) \end{vmatrix} \tag{5.162}$$

Writing the differential expression of equation (5.186) in the increment form gives

$$\{\Delta\sigma\}_n = [C_\omega]_n \{\Delta\sigma^*\}_n - \Delta\omega_n \{S^*\}_n \tag{5.163}$$

Equation (5.163) shows that within the time interval of $\Delta t_n$, the apparent stress increment is correlated to the effective stress state, the damage state at the current moment, as well as the increment of damage and the increment of effective stress within $\Delta t_n$.

Within $\Delta t_n$, from equation (5.152), we have

$$\{\Delta\sigma^*\}_n = [D](\{\Delta\varepsilon\}_n - \{\Delta\varepsilon_V\}_n) \tag{5.164}$$

Substituting equations (5.163) and (5.164) into the equilibrium equation (5.159), we have the equilibrium equation that describes the damage effect,

$$\int_V [B]^T[C_\omega]_n[D][B]\mathrm{d}V\{\Delta u\}_n = \{\Delta f\}_n + \int_V \Delta\omega_n[B]^T\{S^*\}_n\mathrm{d}V + \int_V [B]^T[C_\omega]_n[D]\{\Delta\omega_V\}_n\mathrm{d}V \tag{5.165}$$

By letting

$$[\bar{D}]_n = [C_\omega]_n[D] \tag{5.166}$$

$$\{S^*_{\Delta\omega}\}_n = \Delta\omega_n\{S^*\}_n \tag{5.167}$$

we can rewrite equation (5.165) as

$$[k]\{\Delta u\}_n = \{\Delta f\}_n + \{\Delta F^*\}_n + \{\Delta F'\}_n \tag{5.168}$$

where

$$[k] = \int_V [B]^T[\overline{D}]_n[B]\,\mathrm{d}V \tag{5.169}$$

is the stiffness matrix.

$$\{\Delta F^*\}_n = \int_V [B]^T\{S^*_{\Delta\omega}\}_n\mathrm{d}V \tag{5.170}$$

is the additional force caused by damage evolution.

The following equation

$$\{\Delta F'\}_n = \int_V [B]^T[\overline{D}]_n\{\Delta\varepsilon_V\}_n\mathrm{d}V \tag{5.171}$$

is the additional force caused by the viscous stress increment.

Under the conditions given in this section, $[\overline{D}]_n$ is proven to be symmetrical and to have the expression similar to that of matrix $[D]$. In the case of $\omega = 0$, $[\overline{D}]_n$ becomes an elastic matrix and for $\omega$ = constant with no damage development occurring, the above proposed algorithm constitutes the constant stiffness algorithm.

### *5.7.3 Application to stability analysis of an underground opening*

Analysis of underground excavation project using damaged rheological model is conducted and compared with non-damage analysis. The underground project model described is a metal mine located in a very complicated geological environment where the old metamorphic rock mass has experienced many times of tectonic movement. In addition to the faults of various sizes, the joints and fractures are densely distributed in the mining area. The surrounding rock mass displays remarkable rheological characteristics. In order to study the surrounding rock mass stability and to rationalise the supporting scheme, a special testing gallery has been excavated, with a variety of monitoring instrumentation. Large-size in situ triaxial rheological tests have been conducted in the gallery to study the rheological behaviour. The gallery has overlying depth of 500 m and a span of 3 m with arch roof. The testing gallery has been supported using initial shotcrete-bolting that follow the advance of working face closely and secondary shotcrete-steel net supporting. The grouted bolts are 1.5–1.8 m in length and the total thickness of the shotcreting layer is about 25 cm. Two sections for multi-point borehole extensometer were installed in the testing segment, each measuring 10 m in depth with six monitoring points. The observed deformation curve shows obvious 3-stage rheological deformability. Owing to the uniformly cracked or fractured structure of the surrounding rock mass, it can be considered approximately as a quasi-continuous and homogeneous rheological medium.

The rock mass stability around the opening is studied using the viscoelastic–viscoplastic-damage method discussed in Section 5.7.2, incorporated in a FEM program. The results from the analysis are compared with the in situ measurements to check the effectiveness of the model and the method. To make comparison between the methods, calculations have also been carried out using the rheological model with no damage evolution.

***5.7.3.1 Decomposition of the rheological deformation.*** According to the model (Figure 5.15), the overall deformation comprises three parts.

(a) *Transient elastic deformation* It can be obtained according to Hooke's law

$$\{\Delta\varepsilon_e\} = \frac{1}{E_1}[A]\{\Delta\sigma\} \tag{5.172}$$

where $E_1$ is the transient elastic module; $[A]$ is the constant matrix related to Poisson's ratio. In the case of plane strain,

$$[A] = \begin{bmatrix} 1-\nu^2 & -\nu(1+\nu) & 0 \\ & 1-\nu^2 & 0 \\ \text{symmetry} & & 2(1+\nu) \end{bmatrix} \tag{5.173}$$

(b) *Visco-elastic deformation.* It is calculated from the following equation, for a time interval $\Delta t_n$,

$$\{\Delta\varepsilon_{ve}\}_n = \frac{1}{E_2}[A]\{\sigma\}_n(1 - e^{-(E_2/\eta_2)\Delta t_n}) - (1 - e^{-(E_2/\eta_2)\Delta t_n})\{\varepsilon_{ve}\}_n \tag{5.174}$$

where $E_2$ is the delay elastic modulus; $\eta_2$ is the visco-elastic coefficient; $[A]$ is the constant matrix related to Poisson's ratio.

(c) *Visco-plastic deformation.* When the material is proven having entered the yield state according to the criterion of viscoplastic yield, this part of deformation can be calculated by the following methods.

*(i) Visco-plastic yield criterion.* It adopts Mohr–Coulomb yield criterion, whose expression under a complex stress state is

$$F = \frac{1}{3}\sin\varphi I_1 + \left(\cos\theta - \frac{1}{\sqrt{3}}\sin\theta\sin\varphi\right)(J_2)^{1/2} - C\cos\varphi = 0 \tag{5.175}$$

where $I_1 = \sigma_{ij}$, is the first variant of stress tensor; $J_2 = \frac{1}{2}\sigma_{ij}\sigma_{ij}$, is the second variant of deviator stress tensor; $\theta$ is Lode parameter,

$$\sin 3\theta = -\frac{3\sqrt{3}}{2}\frac{J_3}{(J_2)^{3/2}}; \quad J_3 = \frac{1}{3}\sigma_{ij}\sigma_{jk}\sigma,$$

is the third variant of deviator stress tensor.

(ii) *Visco-plastic stress increment.* It can be expressed by:

$$\{\dot{\varepsilon}_{vp}\} = \frac{1}{\eta_3}\langle\phi(F)\rangle\frac{\partial Q}{\partial\{\sigma\}} \tag{5.176}$$

where $\eta_3$ is the visco-plastic flowing coefficient; and

$$\langle \phi(F) \rangle = \begin{cases} \phi(F) & (F > 0) \\ 0 & (F \leq 0) \end{cases}$$

$\phi(F)$ has the form of $\phi(F) = F/F_0$ where $F_0$ is a reference value. The yield function value of $F$ is dimensionless, and $F_0 = C \cos \phi$. $Q$ is the plastic potential function. $Q \neq F$ stands for irrelevant flowing, and $Q = F$ stands for relevant flowing; the relevant flowing law is used here.

From Equation (5.20), the visco-plastic strain increment that occurs in the time interval of $\Delta t_n = t_{n+1} - t_n$ can be calculated using the following equation,

$$\{\Delta\varepsilon_{vp}\}_n = \Delta t_n[(1 - \Theta)\{\dot{\varepsilon}_{vp}\}_n + \Theta\{\dot{\varepsilon}_{vp}\}_{n+1}] \quad (5.177)$$

When $\Theta = 0$, we have Euler's time integration method and the strain increment is determined by the strain rate at the current moment, $t_n$; when $\Theta = 1$, we have the complete implicit integration method and the strain increment is determined by the strain rate at the end of the time interval; when $\Theta = 1/2$, The implicit trapezoid method can be adopted.

The above equation is for strain increment of non-damaged material. For damaged material, the equation is similar, except that the effective stress term is used instead of the apparent stress term, based on the effective strain concept.

***5.7.3.2 Determination of Model's parameters.*** The Model's parameters used for analysis are obtained through field triaxial compressive rheological tests on the rock mass. Each parameter used in the damage evolution equation is obtained by fitting the elastic module of specimens from multi-loading-stage creep tests, i.e.,

$$\omega = 1 - \frac{E}{E_0} \quad (5.178)$$

where $E_0$ is the elastic modulus of the tested body with no damage evolution; the failure damage value is calculated using equation (5.178) from the elastic modulus.

The values of cohesion ($C$) and friction angle ($\varphi$) are determined from the stress-volumetric strain curve at maximum volumetric strain. At this stress level, the swelling deformation initiates. It can be seen from equation (5.150) that the criterion for damage development is to judge if dilatancy of the rock mass occurs. Therefore, the method also reveals that the yield condition of the rock mass is the damage evolution condition. The property parameters are listed in Table 5.2 and the initial in situ stress field has the vertical and horizontal components of $\sigma_x = 20.0$ and

Table 5.2. Material's properties.

| Material | Parameters | | | | | | | | | |
|---|---|---|---|---|---|---|---|---|---|---|
| | $E_1$ (MPa) | $E_2$ (MPa) | $\mu$ | $\eta_2$ (MPa·h) | $\gamma = 1/\eta_3$ [1/(MPa·h)] | $C$ (MPa) | $\phi(^\circ)$ | $K_1$ | $K_2$ | $\omega_p$ |
| Rock mass | 7500 | 6500 | 0.3 | 174400 | $0.45 \times 10^{-4}$ | 1.0 | 33.0 | $5.87 \times 10^{-3}$ | 4.82 | 0.40 |
| Shotcrete layer | 22000 | | 0.18 | | | | | | | |

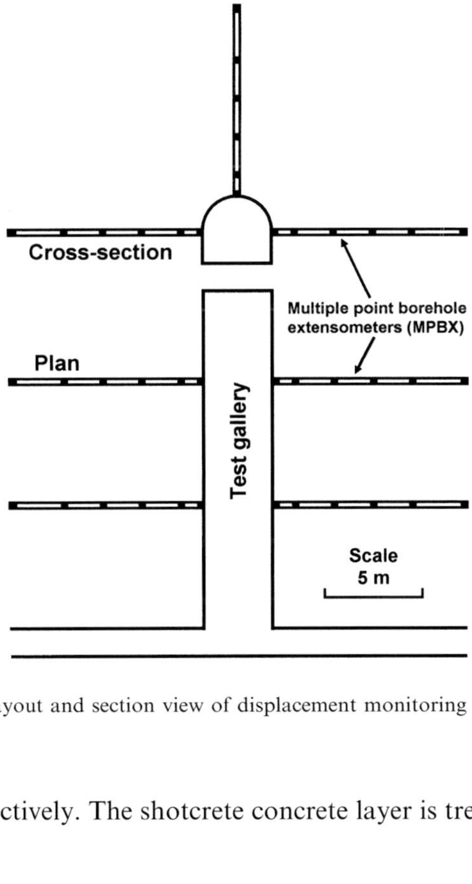


**Figure 5.16.** Layout and section view of displacement monitoring in the test gallery.

$\sigma_y = 14.0$ MPa, respectively. The shotcrete concrete layer is treated as a linear elastic medium.

***5.7.3.3 Comparison between calculated and in situ measured results.*** The layout of monitoring points in the testing gallery in the mining under consideration is shown as in Figure 5.16. The measured results obtained from extensometer observation have been compared with the calculated results.

Figures 5.17–5.20 compare the calculated results and the measured results. From these results, we can see that the displacement of the gallery periphery calculated by taking the damage evolution of the rock mass into account is somewhat higher than the result from conventional viscous calculation. With respect to the attenuation law of the displacement rate at the periphery and to the distribution of displacements in the depth of the surrounding rock mass, the computed results by damage model are better coincident with the measured one. On the distribution of displacements along depths, the difference between the conventional calculation and the damage computation lies in the plastic zone. In the plastic zone around the gallery, the

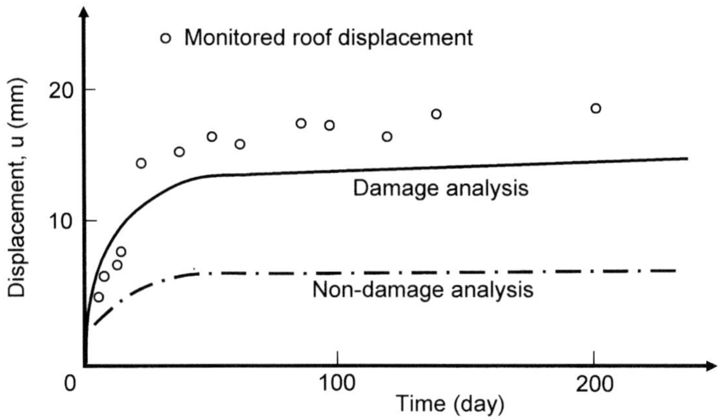


**Figure 5.17.** Displacement–time curve of the gallery roof.

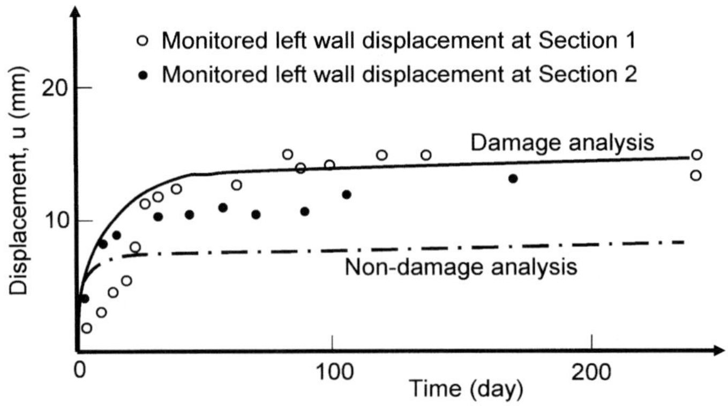


**Figure 5.18.** Displacement–time curve of the sidewall.

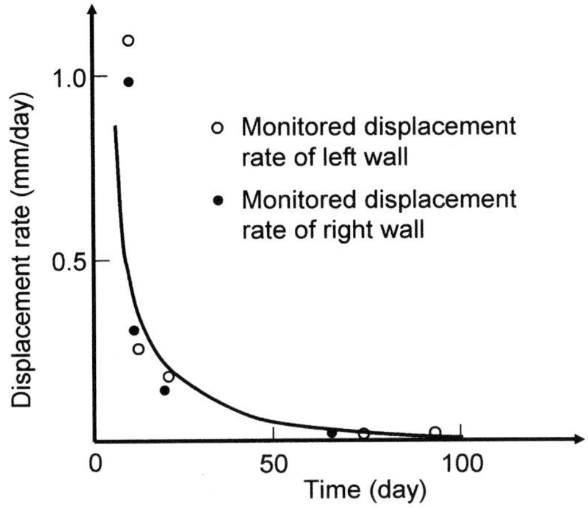


**Figure 5.19.** Calculated displacement–time curve of the sidewall with damage evolution of rock mass.

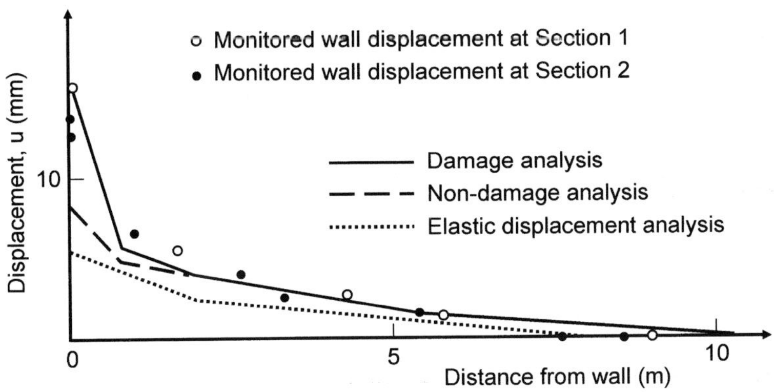


**Figure 5.20.** Displacement distribution with depth on the sidewalls.

displacement gradient obtained from the damage analysis is greater than that from the conventional analysis. The conventional value from the damage analysis is closer to the field measurement. It shows that the visco-elasto-plastic calculation with rock mass damage evolution can describe more realistically the loosening failure characteristics in the rheological process of the surrounding rock mass than the conventional analytical method.

Chapter 6

# Back Analysis and Observational Methods

In recent years, the back analysis method has been widely applied in geotechnical engineering especially in the underground works. Since Peck [390] proposed this method in various related analytical methods and numerical methods have been developed [53–58,391–400].

As a relatively new branch of mechanics, rock mechanics has been developed by adopting traditional analytical methods of mechanics based on continuum mechanics. The typical mathematical and mechanical analytical model is somewhat to find the stress field and deformation field in material by applying the known external loads (or internal loads) with the known material constitutive relationship, geometrical shape and mechanical parameters [53–56]. However, back analysis is a reverse procedure, which is to solve the external load or partial material parameters, based on the known deformation and stresses at limited points and the partially known material parameters. In the stability analysis of the geotechnical engineering problems, it is often necessary to know the in situ stress field, material mechanical parameters and even the mechanical model, by utilising the monitored physical information such as deformation, strain, stress and pressure during constructions. Such a methodology is called back analysis method. The back analysis method, based on the required input physical information, can be divided into deformation back analysis method, stress back analysis method and coupled back analysis method [53]. The physical information in the coupled back analysis method requires both deformation and stress.

The back analysis has been applied to various geotechnical engineering projects, particularly to the rock engineering projects. The reason is due to the fact that the rock masses are very complicated and inhomogeneous. Excavation may disturb the rock mass to different extents. Therefore, the construction process is not a close system behaviour, and is affected by the environment and at the same time affects the environment [63–70,401,402]. In underground excavation and stability analysis, we face two problems. Firstly, the information available is generally 'grey' and uncertain. It is therefore very difficult to obtain explicitly the exact solution by using the conventional mathematical and mechanical analytical methods. Secondly, since the excavation is a process of forming an underground opening, the exchange of energy would occur between the rock mass and its surrounding environment.

In other words, the rock mass would absorb some external energy and at the same time, release some energy to the external environment. This would result in some energy concentration and energy relaxation zones.

Generally, the energy concentration and relaxation zones are obtained by using conventional continuum-based mechanical methods and they do not account for the energy dissipation. In fact, actual rock masses do not obey the continuity assumption, because some cracks and fractures are induced due to stress concentration and blast effect during the construction. The inhomogeneity and relative slips at fracture interfaces cause energy dissipation, e.g., thermal energy. It is often difficult to obtain exact results by using the continuum-based analytical method to such a discontinuous medium. Therefore, the analysis of rock excavation must combine both structure analysis and behaviour analysis, since the information monitored on-site is related to the rock behaviour during the excavation. The information can supplement the results obtained by using the conventional analytical method. The deformation monitoring and back analysis is to identify the 'grey' problems by using the information of rock mass behaviour. The rock engineering problems usually involve several to several thousand meters. It is very difficult to determine representative or average external loads (in situ stresses in most cases) and material parameters (such as the Young's modulus, the Poisson's ratio, cohesion and friction coefficient), by testing rock samples. Large-scale in situ tests are generally very expensive, and often conducted at very few numbers. Hence, it is difficult to ensure the result as a representative one reflecting the whole project based on in situ tests. On the other hand, the information used for back analysis is monitored directly from the actual site. As compared with the in situ large-scale mechanical tests, the deformation monitoring and back analysis have following features and advantages:

(a) Monitored deformation is the average response of a rock mass in a large scale, from several to several tens of meters, sometimes several hundreds of meters.
(b) It corresponds the actual engineering response to the excavation.
(c) It gives large quantity of information, as each monitoring location produces different information.
(d) It can be correlated with laboratory tests to generate the correlation relationship.
(e) It can monitor the response of rock outside the blast influence zone.
(f) The deformation monitoring is easy to perform.
(g) Monitoring is cost-effective and can be widely used in projects of various sizes.

With reliable monitoring data obtained from construction site, it is possible to back analyse the external load and mechanical parameters, by using proper physical material model and the analytical method.

## 6.1. ELASTIC BACK ANALYSIS AND STRESS DISTRIBUTION ANALYSIS

### 6.1.1 Elastic back analysis

Elastic deformation back analysis method can be classified into three types: force method, mathematical regression method and graphic method [53–56]. The force back analysis method can be analytical or a numerical method. The analytical method is usually used when the excavation shape is simple and analytical solution can be found. The numerical method, firstly proposed by Sakurai *et al.* [53,54], is to assume the deformation at a point in a direction being sum of the individual deformations produced by individual loads at the same point in the same direction. Corresponding equation set can then be created. The number of equations or the number of monitored deformations should be equal to or more than the number of variables. The variables can then be solved by using damping least square method. For homogeneous media, the deformation modulus can also be solved. This is termed the inverse approach of the back analysis method.

***6.1.1.1. Basic formulation.*** The basic formula can be expressed as

$$\{P_0\} = \int_V [B]^T \{\sigma^0\} \mathrm{d}V \tag{6.1}$$

where $\{\sigma^0\}$ is the in situ stress; $\{P_0\}$ is the nodal load acting on the excavation face; $V$ is the excavated volume; $[B]$ is a matrix relating to the element geometry.

The relationship of nodal load $\{P\}$ and nodal deformation $\{U\}$ is

$$\{P\} = [K]\{U\} \tag{6.2}$$

where $[K]$ is the stiffness matrix. Using $E^R$ and $E^L$ to represent the modulus of rock and lining, respectively, we have

$$[K] = E^R\left([K^R] + \frac{E^L}{E^R}[K^L]\right) = E^R[K^*] \tag{6.3}$$

where $[K^R]$ and $[K^L]$ denote the stiffness matrixes when $E^R = 1$ and $E^L = 1$, respectively.

The nodal displacement $\{U\}$ in equation (6.2) can be divided into two parts: displacements at monitoring points and displacements at other points. They are:

$$\{U^m\} = \frac{1}{E^R}[A]\{\sigma^0\} = [A]\{\bar{\sigma}_0\} \tag{6.4}$$

where $[A]$ is the compliance matrix which is a function of the Poisson's ratio and the co-ordinates of monitoring points, $\bar{\sigma}_0$ is the normalised in situ stress expressed as

$$\bar{\sigma}_0 = \left\{\sigma_x^0/E \ \sigma_y^0/E \ \tau_{xy}^0/e\right\}^T \tag{6.5}$$

Replacing the absolute displacements in equation (6.4) with relative monitored displacements, then

$$\{\Delta U^m\} = [T][A]\{\bar{\sigma}_0\} = [A^*]\{\bar{\sigma}_0\}$$

or

$$\{\bar{\sigma}_0\} = \left[[A^*]^T[A^*]\right]^{-1}[A^*]^T\{\Delta U^m\} \tag{6.6}$$

***6.1.1.2. Deformation monitoring and back analysis.*** An example is given in the following sections to illustrate deformation monitoring and back analysis technique.

(i) *Deformation monitoring of surrounding rock mass.* A large underground hydropower complex is located in a high in situ stress zone in southwest China. Before the construction, a rectangular test tunnel of 2.5 m wide, 5 m high and 30 m long was excavated for deformation monitoring and stability analysis. The layout of testing tunnel is shown in Figure 6.1. The tunnel is parallel to the exploratory tunnel. The distance between the two tunnels is 15 m. Three monitoring sections are set in the test tunnel. Smooth blasting is used around the monitoring sections. Partial monitoring results obtained from a section are illustrated and discussed in this example. Three multiple-point borehole extensometers (MPBX) including two bar electrical transducer and one wire electrical transducer were installed at the roof of the section. Two MPBXs of steel wire electrical transducer were installed at sidewalls. The MPBXs are of 10–13 m in length, for roof and sidewalls, respectively. The monitoring layout is shown in Figure 6.2. The end points are located in the undisturbed zone and have no deformation. In addition, several monitoring holes of 5 m length are set at the location close to the sections for ultrasonic measurements and convergence monitoring.

Figure 6.2 shows the monitored final displacements at different points of the MPBXs. Figure 6.3 shows the displacement-time curves obtained from the MPBX in the left sidewall. From Figure 6.2, it can be seen that the maximum displacement reaches about 1.3 mm on the sidewalls, but less than 0.3 mm at the roof. This may be

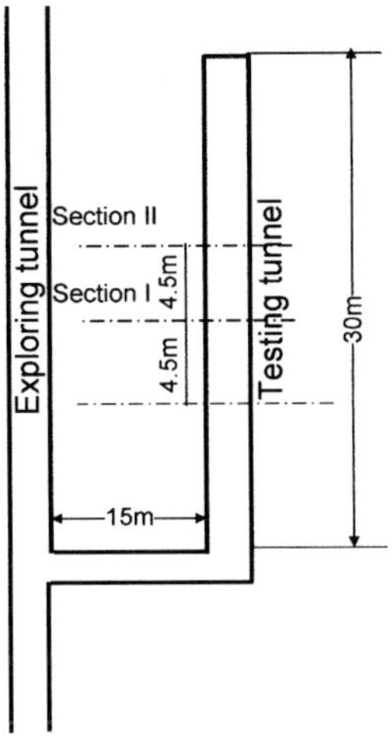


**Figure 6.1.** Layout plan of the monitoring in experimental tunnel.

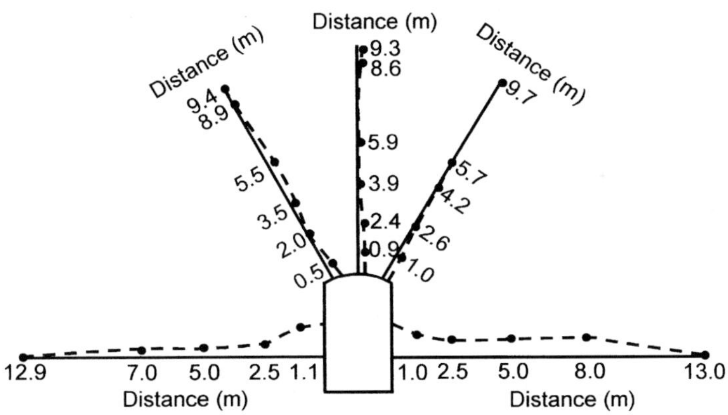


**Figure 6.2.** Displacement measurements by multiple-point borehole extensometers (Section 1 of Figure 6.1).

caused by the high horizontal in situ stress, and high walls of rectangle shape. In addition, the longitudinal ultrasonic wave velocity measured from the boreholes before and after the construction does not change apparently. It indicates that the surrounding rock has not been damaged or disturbed. The rock is at perfectly elastic state. From Figure 6.3, it can be seen that the curve tends to be flat when the excavation approaches Section II ($\Delta v/v = 1$).

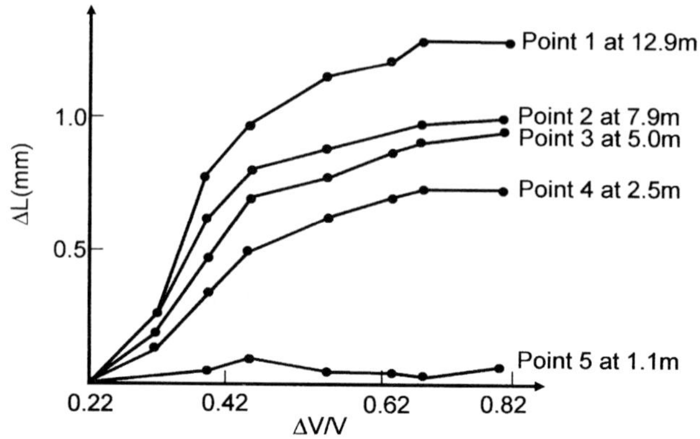


**Figure 6.3.** Changes of displacement with excavation.

The displacement monitored at Section I is the deformation after the 1m excavation from Section I towards Section II. Three-dimensional finite element modelling is carried out to simulate the total displacement and displacements of different excavation stages. The total displacements at left and right walls are computed to be 2.54 mm and 2.45 mm, respectively.

*(ii) Back analysis and results.* The rock mass located around the monitoring sections of the underground complex is a seinite. The rock mass is fractured and the fractures are generally closed. The rock mass is under high in situ stresses. The rock mass is assumed to be uniform, homogeneous and linearly elastic. The in situ stress in rock is assumed to be uniform. Therefore, the linear elastic theory can be used to perform back analysis. If the Young's modulus and the Poisson's ratio are known, it is not difficult to back analyse the in situ stress.

By inputting the monitored displacements and adopting the least square method for solving equation (6.6), the in situ stress can be computed. Taking $E = 40{,}000$ MPa and $\nu = 0.15$ and treating it as a plane strain problem, the maximum principal in situ stresses are computed as $\sigma_1 = 23.7$ MPa and $\sigma_2 = 18.7$ MPa. $\alpha_1$ is at 31.5° to the horizontal plane (Figure 6.4). Figure 6.4 shows comparison of monitored displacement with the back analysis result at left wall. From the figure, it can be seen that the back analysis agrees well with the monitoring. It indicates that back analysis using linear elastic theory is applicable in this particular case. It is also found that the in situ stress back analysed based on the monitored displacements is very close to in situ stress measured. In other words, the 2-D rock deformation back analysis incorporating rock mechanics parameters $E$ and $\nu$ and in situ stresses measured on-site, has verified that the monitored data are reliable.

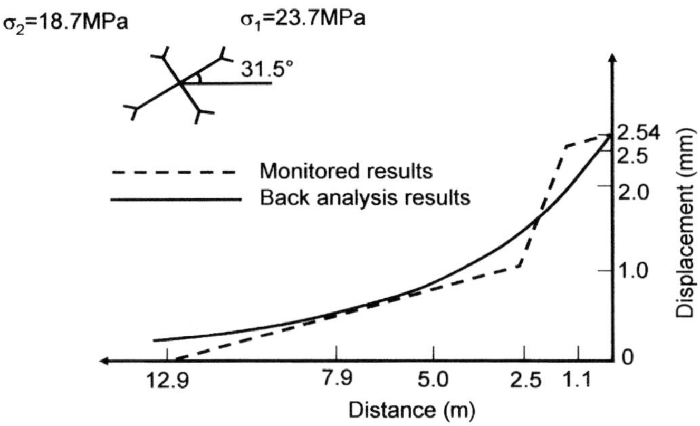


**Figure 6.4.** Monitored and back-analysed displacements at left wall.

### 6.1.2 *Back analysis of in situ stress distribution*

In situ stress is generally a necessary input parameter in the analysis of rock stability. However, for a large underground complex such as the hydroelectric caverns, the in situ stress cannot be treated to be uniform, as the rock covers very large area (often more than 300 m). The in situ stress field changes with depth and distance. A stress function interpolation method [8,52,137] is adopted. Its principle is to use the least square method to obtain the in situ stress field satisfying stress equilibrium and deformation condition, based on-site monitoring data and boundary conditions. The following describes the method for the in situ stress field of the same hydroelectric complex as in Section 6.1.1.

***6.1.2.1. Computational zone and monitored in situ stress.*** The computational zone of the underground complex is from the ground surface to a depth of 800 m and about 200 m radius from the boundary of the opening in horizontal direction. Six sets of measured in situ stresses are taken [8,52].

***6.1.2.2. Determination of stress function.*** The equilibrium equation at a point in rock is a set of inhomogeneous linear differential equations. The solution of the equation set is the sum of its corresponding general solution of the homogenous equation set and specific solution of the inhomogeneous equation set. The specific solution is taken as $\sigma_x = \sigma_y = \tau_{xy} = \tau_{yx} = \sigma_z = 0$, $\tau_{xz} = \rho g x$, and the general solution of the homogeneous equation set is expressed by the stress functions $\phi_1(x, y, z)$, $\phi_2(x, y, z)$ and $\phi_3(x, y, z)$.

Based on equilibrium equation and stress function, following relations can be derived.

$$\left.\begin{aligned} \sigma_x &= \frac{\partial^2\varphi_3}{\partial y^2} + \frac{\partial^2\varphi_2}{\partial z^2} & \tau_{xy} &= -\frac{\partial^2\varphi_3}{\partial x\partial y} \\ \sigma_y &= \frac{\partial^2\varphi_1}{\partial z^2} + \frac{\partial^2\varphi_3}{\partial x^2} & \tau_{yz} &= -\frac{\partial^2\varphi_1}{\partial y\partial z} \\ \sigma_z &= \frac{\partial^2\varphi_2}{\partial x^2} + \frac{\partial^2\varphi_1}{\partial y^2} & \tau_{zx} &= -\frac{\partial^2\varphi_2}{\partial z\partial x} \end{aligned}\right\} \tag{6.7}$$

By considering the fluctuation and deviation of the measurement data of in situ stresses, stress functions are assumed to be polynomial of power 4 and the corresponding stress components to be polynomial of power 2. The general expression of stress function $\varphi_1(x, y, z)$ is:

$$\begin{aligned} \varphi_1(x, y, z) = {} & a_1x^2 + a_2y^2 + a_3z^2 + a_4xy + a_5yz + a_6zx \\ & + a_7x^3 + a_8y^3 + a_9z^3 + a_{10}xyz + a_{11}xy^2 \\ & + a_{12}x^2y + a_{13}yz^2 + a_{14}y^2z + a_{15}zx^2 \\ & + a_{16}z^2x + a_{17}x^4 + a_{18}y^4 + a_{19}z^4 + a_{20}xyz^2 \\ & + a_{21}xy^2z + a_{22}x^2yz + a_{23}xy^3 + a_{24}x^3y \\ & + a_{25}yz^3 + a_{26}y^3z + a_{27}zx^3 + a_{28}z^3x \\ & + a_{29}x^2z^2 + a_{30}y^2z^2 + a_{31}z^2x^2 \end{aligned} \tag{6.8}$$

where $a_i$ is the extrapolation parameter to be determined. The stress components can be expressed as

$$\sigma_x = b_0 + b_1x + b_2y + b_3z + b_4x^2 + b_5y^2 + b_6z^2 + b_7xy + b_8yz + b_9zx \tag{6.9}$$

with

$$\begin{aligned} b_0 &= 2(a''_2 + a'_3) & b_1 &= 2(a''_{13} + a'_{15}) \\ b_2 &= 2(3a''_8 + a'_{16} & b_3 &= 2(a''_{14} + 3a'_9) \\ b_4 &= 2(a''_{26} + a'_{27}) & b_5 &= 2(6a''_{18} + a'_{28}) \\ b_6 &= 2(a''_{28} + a'_{19}) & b_7 &= 2(3a''_{20} + a'_{31}) \\ b_8 &= 6(a''_{25} + a'_{23}) & b_9 &= 2(a''_{30} + 3a'_{21}) \end{aligned}$$

where $a'$ and $a''$ are the corresponding coefficients of $\varphi_2$ and $\varphi_3$.

The expressions of stress functions $\varphi_2$ and $\varphi_3$ and other stress components are similar to equations (6.8) and (6.9) and are not listed here.

By substituting equation (6.7) into six deformation equations (other components are ignored here), we have

$$(1+\nu)\nabla^2\sigma_x + \frac{\partial^2\theta}{\partial x^2} = 0 \tag{6.10}$$

where $\theta = \sigma_x + \sigma_y + \sigma_z$; $\nu$ is the Poisson's ratio.

The six extrapolation functions can be eliminated. By considering the relationship between stress functions and stress components above, each stress function can eliminate some extrapolation coefficients, and the amount of independent extrapolation coefficients to be determined becomes 55.

The corresponding boundary conditions must be determined to obtain the extrapolation functions by using least squares method. They include the stress measurements at the ground surface and at selected locations. The normal and tangent stresses at ground surface with outward normal cosine directions of $l$, $m$ and $n$ are equal to zero, which can be expressed as

$$\left.\begin{aligned} X_0 &= \sigma_x l + \tau_{xy} m + \tau_{xz} n = 0 \\ Y_0 &= \tau_{yx} l + \sigma_y m + \tau_{yz} n = 0 \\ Z_0 &= \tau_{zx} l + \tau_{xy} m + \sigma_z n = 0 \end{aligned}\right\} \tag{6.11}$$

Substituting the stress components with those extrapolation function coefficients as listed in equation (6.9) into equation (6.11), all the boundary conditions can be expressed by extrapolation function coefficients.

Therefore, each stress measurement point below ground has six equations and each measurement at ground surface point has three equations. These equations have a general form as

$$A_1Q_1(x_i) + A_2Q_2(x_i) + \cdots + A_nQ_n(x_i) = B_i \tag{6.12}$$

where $A_n$ is the extrapolation coefficient to be determined ($n=55$); $x_i$ is the co-ordinate $(x, y, z)$ of the point; $Q_i$ is the exponent function of co-ordinate $(j=1, 2, \ldots, n)$; $B_i$ is the stress measurement values at the measurement points or zero values at ground surface points; $i=1, 2, \ldots, n, m=6\times I+3\times K$, $I$ is number of stress measurement points, $K$ is the number of ground surface points.

Based on the above equations, all the coefficients of stress functions can be solved by using least squares method to obtain the co-ordinates of the measurement points. The in situ stress distribution in the whole zone can thus be obtained. The stress distribution computed satisfies force equilibrium conditions, continuous deformation conditions and minimum value in the squares error with actual measurements

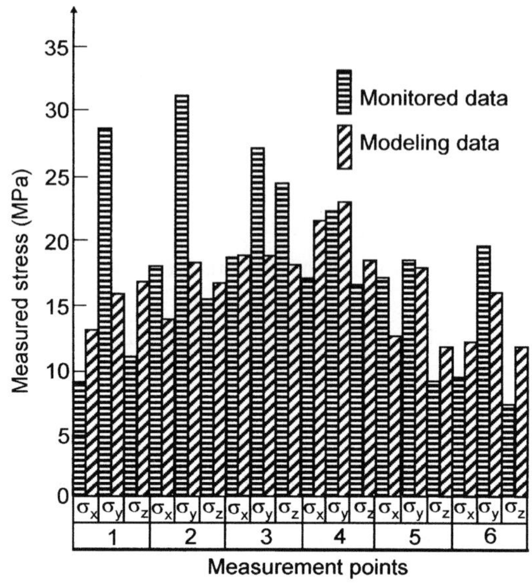


**Figure 6.5.** Comparison of stresses measured and computed.

and the given boundary conditions. Figure 6.5 shows the comparison of stresses between the actual measurements and the computed results using the above method. It can be seen that in general they agree well with each other.

The results obtained by this method were compared with the results obtained by multiple-variable regression method. It is found that both methods provide similar results, and the result obtained by this method generally agree better with the actual measurements. This method is simpler and has the advantage in computer running time and the preparation work. This method also eliminates the human error in determining stress field.

Figure 6.6 shows the stress distribution at a section in the middle of the underground complex.

## 6.2. VISCO-ELASTIC BACK ANALYSIS AND ITS ENGINEERING APPLICATIONS

### *6.2.1 Method of site deformation monitoring and its application results*

This section introduces the back analysis method for visco-elastic media using a case study on a shallow tunnel.

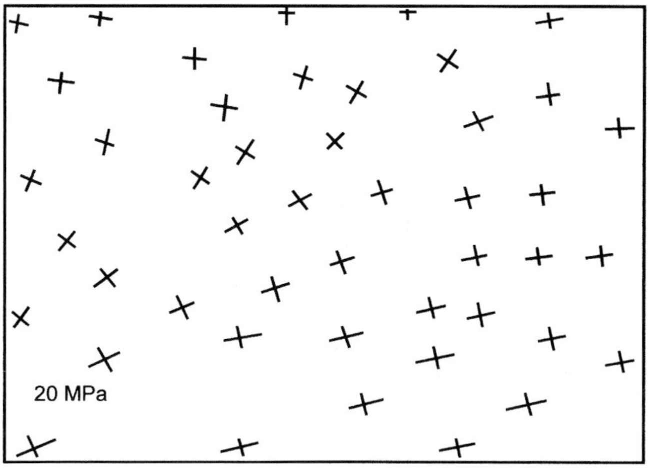


**Figure 6.6.** Stress distribution at a section in the middle of the underground complex.

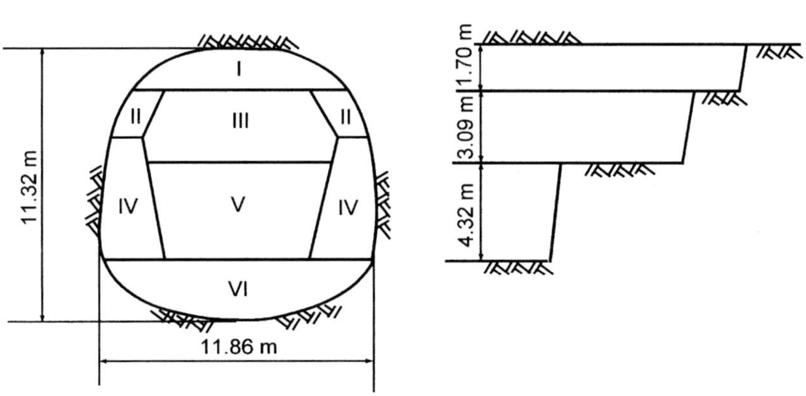


**Figure 6.7.** Tunnel section and multiple-stage excavation procedure.

A large railway tunnel is located in a loess layer. The tunnel is a double-lane heavy vehicle tunnel of 11.5 m high, 12 m wide and with 12 m thick overburden. The tunnel is constructed by the New Austrian Tunnelling Method (NATM) and has an experimental section to conduct monitoring studies. Monitored vertical deformation at the crown due to excavation is presented here to illustrate the application of back analysis. Figure 6.7 shows the tunnel section and multiple-stage excavation procedure.

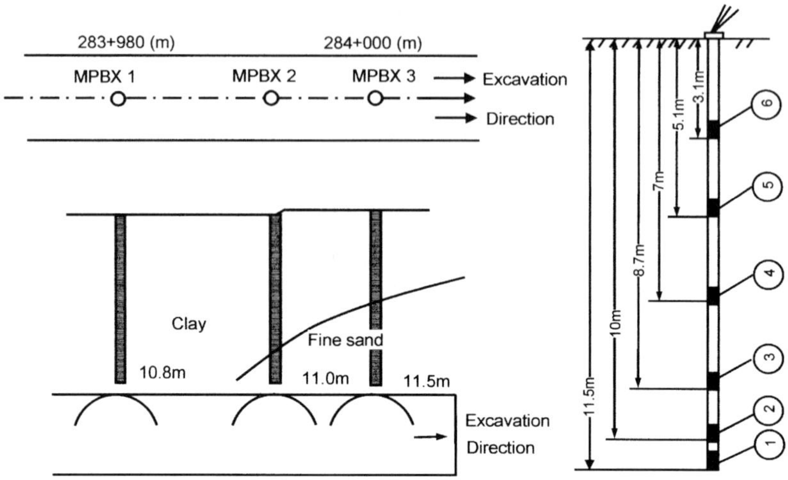


**Figure 6.8.** Ground condition and instrumentation layout.

Figure 6.8 shows the installation of instrumentation. It can be seen that the multiple-point borehole extensometers (MPBX) are installed in the boreholes before the excavation to monitor the absolute deformation in the soil during the tunnel construction. Relationship between ground settlement and MPBX monitored displacement are shown in Figure 6.9.

The monitoring results and analysis of MPBX 3 with a period of 65 days during the excavation of the upper part of the tunnel are shown in Figure 6.10. The largest deformation of 15.29 mm is at the end point (point 1). The deformation can be divided into three stages:

(a) Initial deformation stage (compression deformation stage) in which the deformation occurs during the tunnel excavation from the beginning to the monitoring section, with a period of 24 days. The maximum compression deformation is −1.91 mm and displacement rate is −0.08 mm/day. All the six points at different depths have negative deformations. Initially the monitoring data has small fluctuation in the first 10 days. Then the negative deformation increases and the curves tend to be close together, indicating that deformations occurs mainly due to the ground settlement.
(b) Sharp deformation stage. As the tunnel excavation passes through the monitoring section, the deformations become positive immediately and increase sharply for 10 days before reaching a maximum of 15.19 mm. The average displacement rate is 1.72 mm/day.

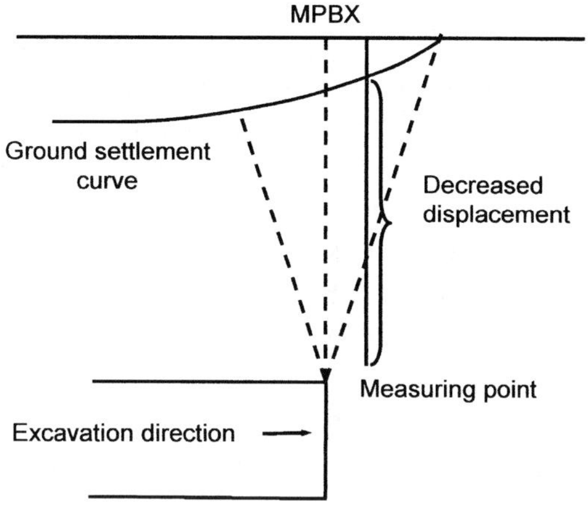


**Figure 6.9.** Notation of negative displacement in pre-installed extensometers.

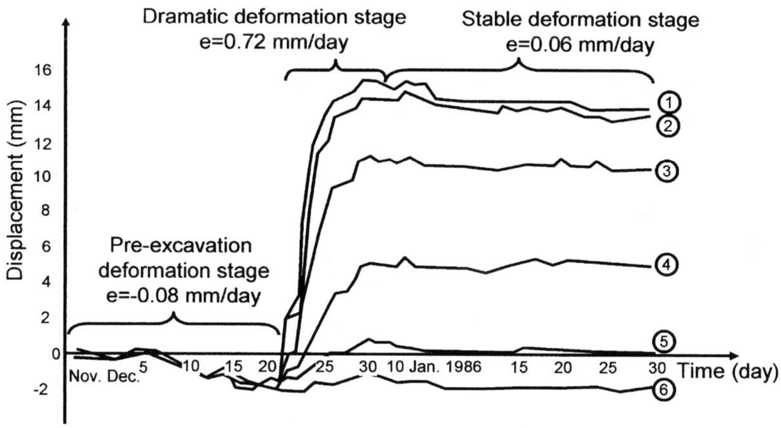


**Figure 6.10.** Monitored displacement in borehole 3 with a period of 65 days during the excavation.

(c) Stable deformation stage. When the tunnel excavation is 14 m away from the monitoring sections, the deformations increase slightly (displacement rate was 0.06 mm/day) for 1 month and then become stable. However, the displacements at two deeper points (points 1 and 2) decrease in the first several days, indicating that there may be saturated sand-lens at these locations.

While the lower part is excavated completely, the displacement increased again about 1–5 mm and the average displacement rate is 0.14 mm/day.

### 6.2.2 *Visco-elastic back analysis*

***6.2.2.1. Computation method.*** As noted in the observation, the deformation in ground above the tunnel develops with the excavation of the tunnel. Therefore, theoretically it is a 3-D problem. To save the computation time, it is often simplified as a virtual 3-D problem or a plane problem, but uses the change of a virtual support force to simulate the unloading due to excavation. Because the monitoring data were obtained from the excavation of the upper part of the tunnel, analysis is performed only for the upper part of the excavation. Finite element elastic analysis is conducted. Figure 6.11 shows the generated mesh. Because of the symmetry, only half of the domain is taken as the computational model with 144 elements (4-node isotropic element) and 170 nodes. The in situ stress is obtained from gravity of the overburden soil. The top of the tunnel is 12 m below the ground surface. The elastic solution of the stress boundary condition is

$$u(x, y, z) = \frac{f(x, y, z)}{E} P \tag{6.13}$$

where $u(x, y, z)$ is the displacement at a point in the 3-D space; $f(x, y, z)$ is the displacement induced by unit load at unit elastic modulus which can be solved by the

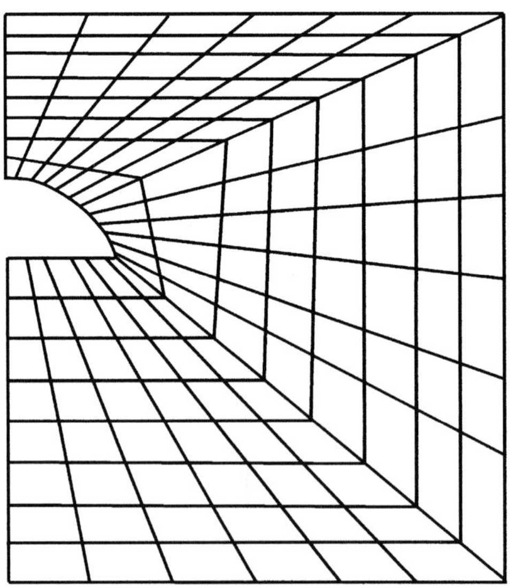

**Figure 6.11.** Generated mesh in modelling.

finite element analysis; $P$ is the excavation load as a function of time.

$$P = P(x,y,z)\left[0.5 - \frac{1}{\pi}\arctan(t - t_0 - 2)\right] \tag{6.14}$$

***6.2.2.2. Visco-elastic analysis results.*** Using three-unit visco-elastic model shown in Figure 6.12 and assuming the Poisson's ratio as a constant, the visco-elastic displacement solution can be expressed as follows by transforming equation (6.14) with visco-elastic responding principle:

$$U(t,x,y,z) = f(x,y,z)M(t) \tag{6.15}$$

where,

$$M(t) = \frac{1}{E_1}\left[\frac{1}{2} - \frac{1}{\pi}\arctan(t - t_0 - 2)\right]\left[1 + \frac{E_0}{E_2}e^{-E_2/\eta(t-t_0)}\right]$$

$f(x,y)$ can be computed using 2-D FEM. Subsequently back analysis regression can be performed for the monitored data by using equation (6.15). Figure 6.13 shows the

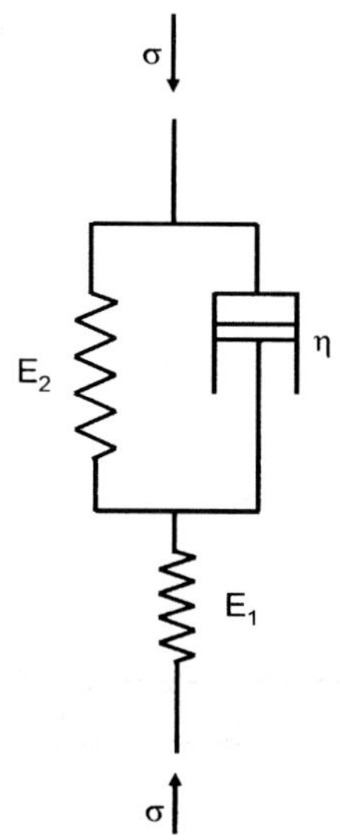


**Figure 6.12.** The three-unit visco-elastic model.

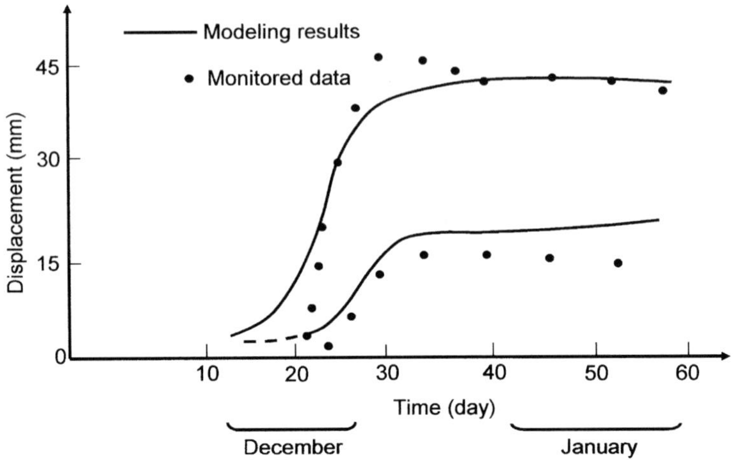


**Figure 6.13.** Regression curves of measurements at points 1 and 4.

regression curves of measurement points 1 and 4. The monitored data are the absolute displacements combining the ground surface settlements and relative displacements. It can be seen that the regression data agree well with the monitored data after the excavation face passed the monitoring sections. Nevertheless, the overall visco-elastic parameters of the soil can also be obtained from the regression of the latter part of the curves. The analysis gives $E_1 = 15.2$ MPa and $E_2 = 45.8$ MPa, and are seemingly reliable compared with the empirical values.

## 6.3. BACK ANALYSIS AND OPTIMISED METHODS IN TRANSVERSE ISOTROPIC ROCK

The layered rocks such as sedimentary formation are often encountered, so it is very important to investigate the back analysis method for such media.

### *6.3.1 Basic formulae of transverse isotropic mechanics*

In continuum mechanics, if the media satisfies the assumptions of uniformity, continuity, small deformation and linear elasticity, the constitutive equations can be expressed as

$$\sigma_{ij} = D_{ijkl}\varepsilon_{kl} \tag{6.16}$$

Considering the symmetry of the strain tensor and stress tensor, the elastic parameters can be reduced from 81 to 36. The matrix form is

$$\{\sigma\} = [D]\{\varepsilon\} \tag{6.17}$$

Due to the symmetry of the elastic matrix, the elastic independent parameters can be reduced to 21. This can be used in general for anisotropic material that has three perpendicular elastic symmetric planes. If the axes are located in the elastic symmetric planes, the general Hooke's law can be expressed in the form of matrix as

$$\begin{Bmatrix} \sigma_x \\ \sigma_y \\ \sigma_z \\ \tau_{xy} \\ \tau_{xz} \\ \tau_{yz} \end{Bmatrix} = \begin{bmatrix} d_{11} & d_{12} & d_{13} & & & \\ d_{21} & d_{22} & d_{23} & & 0 & \\ d_{31} & d_{32} & d_{33} & & & \\ & & & d_{66} & & \\ & 0 & & & d_{55} & \\ & & & & & d_{44} \end{bmatrix} \begin{Bmatrix} \varepsilon_x \\ \varepsilon_y \\ \varepsilon_z \\ \gamma_{xy} \\ \gamma_{xz} \\ \gamma_{yz} \end{Bmatrix} \tag{6.18}$$

containing only nine independent elastic parameters.

The stress–strain relationships in the components of the anisotropic media are

$$\left.\begin{aligned} \sigma_x &= d_{11}\varepsilon_x + d_{12}\varepsilon_y + d_{13}\varepsilon_z \\ \sigma_y &= d_{21}\varepsilon_x + d_{22}\varepsilon_y + d_{23}\varepsilon_z \\ \sigma_z &= d_{31}\varepsilon_x + d_{32}\varepsilon_y + d_{33}\varepsilon_z \\ \tau_{xy} &= d_{66}\gamma_{xy} \\ \tau_{xz} &= d_{55}\gamma_{xz} \\ \tau_{yz} &= d_{44}\gamma_{yz} \end{aligned}\right\} \tag{6.19}$$

Assuming that there is a plane parallel to *xoz* plane at each point in the elastic body as shown in Figure 6.14, the elastic properties in any direction are equivalent. The body is called orthotropic medium. The general Hooke's law has a form of:

$$\begin{Bmatrix} \sigma_x \\ \sigma_y \\ \sigma_z \\ \tau_{xy} \\ \tau_{xz} \\ \tau_{yz} \end{Bmatrix} = \begin{bmatrix} d_{11} & d_{12} & d_{13} & & & \\ d_{21} & d_{22} & d_{23} & & 0 & \\ d_{31} & d_{32} & d_{33} & & & \\ & & & \frac{1}{2}(d_{11} - d_{12}) & & \\ & 0 & & & d_{44} & \\ & & & & & d_{44} \end{bmatrix} \begin{Bmatrix} \varepsilon_x \\ \varepsilon_y \\ \varepsilon_z \\ \gamma_{xy} \\ \gamma_{xz} \\ \gamma_{yz} \end{Bmatrix} \tag{6.20}$$

In this case, only five independent elastic parameters are unknowns. Therefore, equation (6.17) can be modified to be

$$\{\varepsilon\} = [A]\{\sigma\} \tag{6.21}$$

where $[A]$ is the compliance matrix.

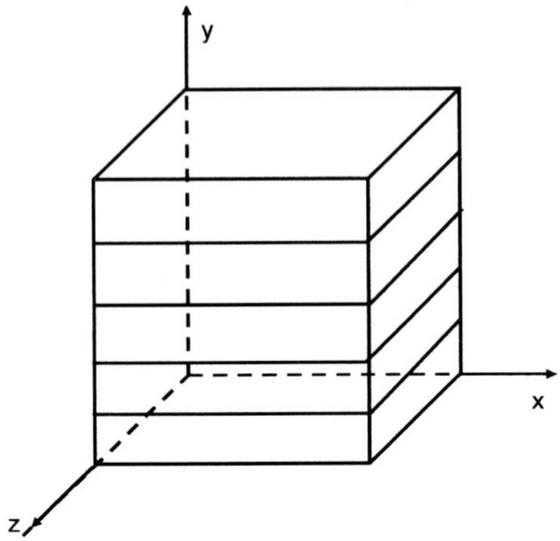


**Figure 6.14.** Transverse isotropic elastic body.

As the $xz$ plane is an elastic isotropic plane, thus

$$E_x = E_z = E_s \qquad E_y = E_n$$
$$\nu_{xz} = \nu_s \qquad \nu_{xy} = \nu_{yz} = \nu_n$$
$$G_{xz} = G_s = 0.5 \times E_s/(1 + \nu_s)$$
$$G_{xy} = G_{yz} = G_{sn}$$

where $[A]$ becomes

$$[A] = \begin{bmatrix} \frac{1}{E_s} & -\frac{\nu_n}{E_n} & -\frac{\nu_s}{E_s} & & & \\ -\frac{\nu_n}{E_n} & \frac{1}{E_n} & -\frac{\nu_n}{E_n} & & 0 & \\ -\frac{\nu_s}{E_s} & -\frac{\nu_n}{E_n} & \frac{1}{E_s} & & & \\ & & & \frac{1}{G_{sn}} & & \\ & 0 & & & \frac{1}{G_{sn}} & \\ & & & & & \frac{2(1+\nu_s)}{E_s} \end{bmatrix} \tag{6.22}$$

If the main axes of the medium are not the same as the coordinate axes, then $[A]$ needs to be transformed in coordinate and new compliance matrix is

$$[A'] = [T][A][T]^T \tag{6.23}$$

where $[T]$ is the coordinate transforming matrix.

For plane strain problems, $[A]$ can be further simplified as

$$[A] = \frac{1}{E_s}\begin{bmatrix} 1-\nu_s^2 & -n\nu_n(1+\nu_s) & 0 \\ -n\nu_n(1+\nu_s) & n(1-n\nu_n^2) & 0 \\ 0 & 0 & \dfrac{E_s}{G_{sn}} \end{bmatrix} \tag{6.24}$$

The stiffness matrix is

$$[D] = [A]^{-1} = \frac{E_n}{m}\begin{bmatrix} n(1-n\nu_n^2) & n\nu_n(1+\nu_s) & 0 \\ n\nu_n(1+\nu_s) & 1-n\nu_n^2 & 0 \\ 0 & 0 & \dfrac{mG_{sn}}{E_n} \end{bmatrix} \tag{6.25}$$

where $n = E_s/E_n$, $m = (1+\nu_s)(1-\nu_s-2n\nu_n^2)$

Let us discuss a special and useful case. Assuming the layer plane orientation is parallel to the tunnel direction ($z$ axis), the excavation face is $xoy$ plane, $x$ axis is horizontal and is parallel or oblique to the layer plane, the analysis can be conducted using equations (6.24) and (6.25). If there is a joint set parallel to the layer plane and with the same mechanical properties, and shear stiffness is $K_s$, normal stiffness is $K_n$ and joint spacing is $b$, the elastic parameters in equations (6.24) and (6.25) can be determined by using the following equation:

$$\left.\begin{aligned} E_s &= E, \nu_s = \nu \\ E_n &= \left(\frac{1}{1/E_s + 1/K_n b}\right) \\ G_{sn} = G_n &= \left[\frac{1}{2((1+\nu_s)/E_s) + 1/K_s b}\right] \\ \nu_n &= E_n\frac{\nu}{E_s} \end{aligned}\right\} \tag{6.26}$$

### 6.3.2 *Optimisation analysis method*

As stated earlier, the inverse approach of back analysis method can be used in a number of cases with simple conditions. In most cases with complicated conditions, only the direct approach of back analysis method can be used. This method is actually a regression method. The most important issue of this direct approach method is the appropriate understanding of the subject to be back analysed. It depends extensively on the experience and knowledge. For example, in underground excavation analysis, the mechanical model (elastic isotropic or anisotropic, visco-elastic or elasto-plastic models), mechanical analysis method or computational program have to be determined. When the mechanical model and analytical method are determined, the next step is to rationalise the regression computation to make it scientific and fast. This is the optimisation of the direct back analysis to be discussed in this section. The objective of optimisation, based on monitored convergence displacement, is to obtain the best sets of material parameters and stress parameters.

Assuming the computational displacement is $u_i(X)$, monitored convergence displacement is $u_i^0$ $(i = 1, 2, \ldots, n)$, the objective function is defined as

$$F(X) = \sum_{i-1}^{n} \left[u_i(X) - u_i^0\right]^2 \tag{6.27}$$

where

$$X = [\sigma_x, \sigma_y, \tau_{xy}, E, \nu, c, \phi]$$

where $\sigma_x$, $\sigma_y$ and $\tau_{xy}$ are three components of in situ stress, $E$ is the elastic modulus, $\nu$ is the Poisson's ratio, $c$ is the cohesion and $\phi$ is the friction angle.

The optimisation method is to make the objective function approach gradually to minimum in the direct approach. Generally, giving one set of initial parameters $X_0$ and their allowable ranges, the optimisation program searches automatically one set of parameters $X$ to make the objective function satisfying the given accuracy requirement.

There are many optimisation methods with or without restraints available [49–51]. Brief outlines of the common optimisation methods are presented below.

(a) Pure shape speeding method: Pure shape acceleration method was developed from the pure shape method. The pure shape method is to compare the function values at $(n + 1)$ peak points of an initial pure shape in $n$-D space, replacing the points of maximum function value, with a new pure shape, and approaching gradually to the points of minimum value through iterations.

(b) Composite shape acceleration method: This method is used to solve optimisation problems of multiple variables (normally not more than 20) with constraints of unequal equations. The optimisation procedure is to take $2n$ peak points from non-linear constraints in $n$-D space to form a pure shape. It is then to compare function values at each peak point one by one, replace the worst point with a new point to improve the objective function and satisfy constraint conditions, and approach gradually to the optimised point.

(c) Mixing penalty function method: This is also termed as sequential unconstrained minimisation technique (SUMT). It is to add constraint function with equal equations and unequal equations in penalty, respectively, to the objective function to form a new objective function with no constraint (penalty function). So that it converts the original minimisation of objective function with constraints into a new minimisation of penalty function with no constraints.

(d) New Powell method: This is an advanced optimisation method to solve the minimisation of complex functions. The iterative procedure is to solve the minimisation of the objective function along a series of conjugate directions starting from the initial point ($X_0$). The forming of the conjugate directions uses only the objective function values at the iteration points. Therefore, it is a direct method. This method is useful when the objective function is a non-convex function. In the case of multiple minimum value points within a certain range, this method can find the optimal point.

For complicated non-linear problems, the objective function is often non-convex function, and the minimum point is not unique. To obtain the optimal solution, the optimal searches in one dimension and multiple dimensions are added to seek minimum points of objective function along conjugate directions. Meanwhile, the step should be decreased in the area with sharp slope of objective function. After obtaining the minimum point, the searches will be performed for its two sides to obtain second and third class minimum points. Comparing them with each other gives the optimal solution. An optimisation generally has three loops to execute different functions: optimisation methods, model types (linear, non-linear or isotropic models) and selection of different back analysis parameters section. Figure 6.15 illustrates a typical program flow.

### *6.3.3 Examples of engineering applications*

The following example is to illustrate the application of the optimisation method in back analysis, for a large hydroelectric project. The power complex is located underground in sedimentary rocks. The surrounding rocks are mainly siltstone interlayered with clay slate. Finite element displacement back analysis coupled with an optimal program was used to examine the displacement distribution, in situ stresses and rock parameters.

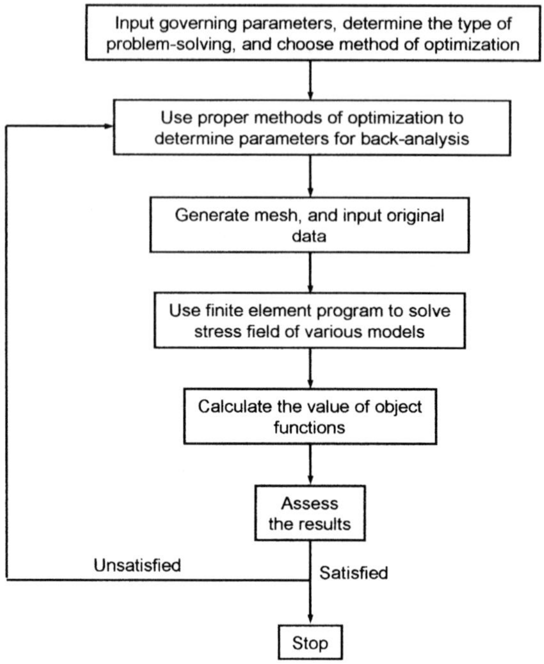


**Figure 6.15.** A typical program flow.

As shown in Figure 6.16, Zones 1 and 3 are the siltstone and Zones 2 and 4 are the clay slate. The siltstone has the Young's modulus of 28 GPa and 35 GPa in directions perpendicular to and parallel to the rock layer, and Poisson's ratio of 0.25. The clay slate has the Young's modulus of 21 GPa and 27 GPa in directions perpendicular to and parallel to the rock layer, and Poisson's ratio of 0.27. The monitored convergence curves are shown in Figure 6.17. The tunnel was constructed by using one-stage excavation. The in situ stresses are uniformly distributed in the surrounding rock masses.

This example is treated as a plane strain problem. The finite element mesh is generated into four-node four-side isotropic parameter elements with 217 elements and 252 nodes. The rock masses are assumed to be isotropic. The modelling assumes that the Young's modulus between the measured modulus in the two directions, and the Poisson's ratio does not change. The back analysis results, obtained using four optimisation methods, are summarised in Table 6.1.

The analysis results of in situ stresses obtained by the Powell method are $\sigma_x = 9.27$ MPa, $\sigma_y = 4.80$ MPa. The measured stresses are $\sigma_x = 9.39$ MPa,

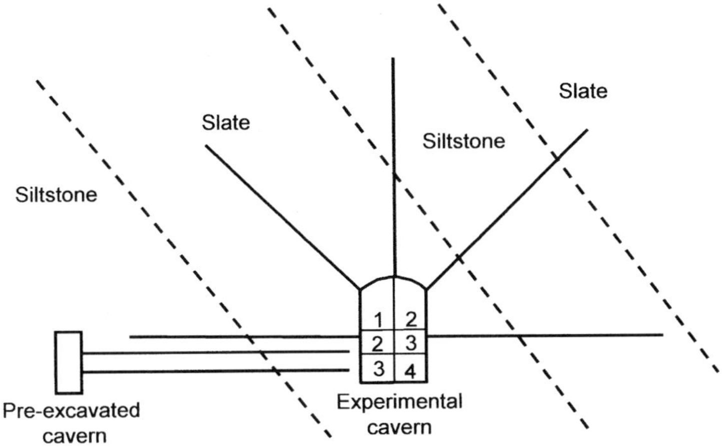


**Figure 6.16.** Cross section of the tunnel and monitoring holes.

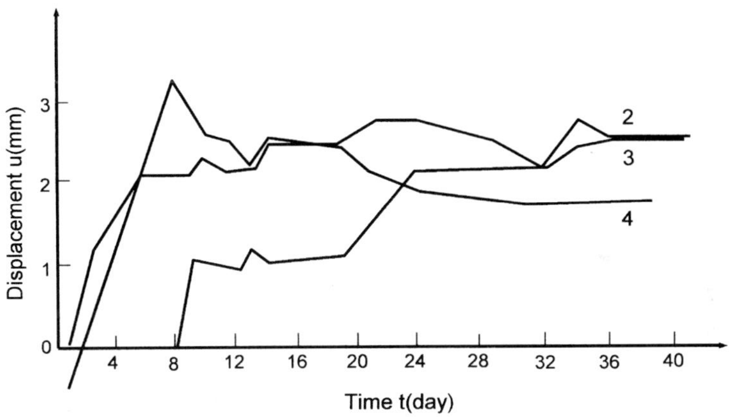


**Figure 6.17.** Monitored displacement histories.

$\sigma_y = 4.75$ MPa. The results from back analysis with optimisation are very close to the actual measured values.

### *6.3.4 Discussions*

The discussions below are based on the experience in using the four optimisation methods to a variety of engineering problems involving visco-elastic and

**Table 6.1.** Results for different analysis methods.

| Method | No of iteration steps | Target function ($10^{-5}$) | Back analysis results (MPa) | | | |
|---|---|---|---|---|---|---|
| | | | $\sigma_x$ | $\sigma_y$ | $E_1$ | $E_2$ |
| Pure shape | 15 | 3.2 | −9.04 | −4.74 | 21829.6 | 30294.0 |
| Composite shape | 50 | 2.5 | −9.00 | −4.80 | 21700.0 | 30560.0 |
| Mixing penalty | 15 | 2.2 | −9.04 | −4.74 | 21721.6 | 30224.5 |
| Powell method | 43 | 3.6 | −9.27 | −4.80 | 21700.0 | 30800.0 |

elasto-plastic back analysis.

(a) Pure shape acceleration method: It has a fast convergence and a high accuracy, especially for elastic, isotropic problems with more than two parameters. But initial value of back analysis parameters must be properly determined in advance.

(b) Composite shape acceleration: Similar to the pure shape acceleration method, it has a fast convergence and a high accuracy, but has constraint functions. Giving constraint functions and an allowable value range of parameters, the program is able to perform back analysis automatically under the constraint conditions.

(c) Mixing penalty function method: It has a constraint optimisation function after the Powell method. It does not require to set properly initial values of back analysis parameters in advance and can find most optimal solutions satisfying constraint conditions automatically. It converges faster than the Powell method because of the 1-D search method feature.

(d) Powell method: It converges slower than that of pure shape and composite shape, but is applicable when the objective functions are complex non-convex functions. It can find the most optimal point when minimum points are not unique in a certain range, especially in the cases of non-linear problems with less than three parameters.

In summary, for the back analysis of general elastic, anisotropic problems, the pure shape acceleration method and composite shape acceleration method are the best choices since they have a fast convergence and a high accuracy. However, for complex elasto-plastic problems, the Powell method or the mixing penalty function method should be used to conduct optimisation search, and then use either pure shape or composite shape methods to perform global optimisation back analysis to obtain the final results. So the unique result can be obtained with fast convergence.

The criteria to select optimisation methods with or without constraint conditions are:

(a) For the cases when the parameters cannot be determined in range, but can be estimated as a possible value, the optimal back analysis approach with no constraint conditions can be used.
(b) However, when the parameters have constraints (equal or unequal constraints), the optimal back analysis approach with constraints will be a better alternative.

## 6.4. BACK ANALYSIS OF JOINTED ROCK MASS AND STABILITY PREDICTION

The sections earlier introduced the back analyses of elastic, visco-elastic, anisotropic and elasto-plastic problems, but does not attempt the back analysis of jointed rock mass. This section discusses the back analysis of jointed rock mass. It is illustrated by an engineering example of a pumped storage hydroelectric complex.

### 6.4.1 Description of the project and monitoring

***6.4.1.1. Description of the project.*** The large-scale pumped storage hydroelectric facilities include main powerhouse, transformer house, ventilation and transport tunnels located underground 200–400 m in depth.

The surrounding rock mass is mainly a medium-grained granite. There are mainly two joint sets at the crown of the powerhouse, with dip direction of 020–030 and 300–320. The joint spacing is usually 1–2 m. The average persistence is about 50%.

Two cases are considered in the analysis. One is to treat the rock mass as an isotropic medium. Another is to consider the joint sets and treat the rock mass as an approximately perpendicularly anisotropic medium.

During construction, displacements are monitored at three sections of ventilation tunnel and transport tunnel.

The positions of measurement Sections I and II in a ventilation tunnel and the layout of MPBX monitoring is shown in Figure 6.18.

***6.4.1.2. Data processing and modification.*** Due to the fluctuation of the monitored data, data were processed. In addition, as most instruments are installed after the face of excavation, the data do not include the displacements occurring before the installation. The lost displacements are determined using 3-D finite element analysis for all the points and corresponding modification coefficients are derived. Therefore, the final displacement based on the 2-D back analysis should be $d = d_m + d_l = \lambda d_m$,

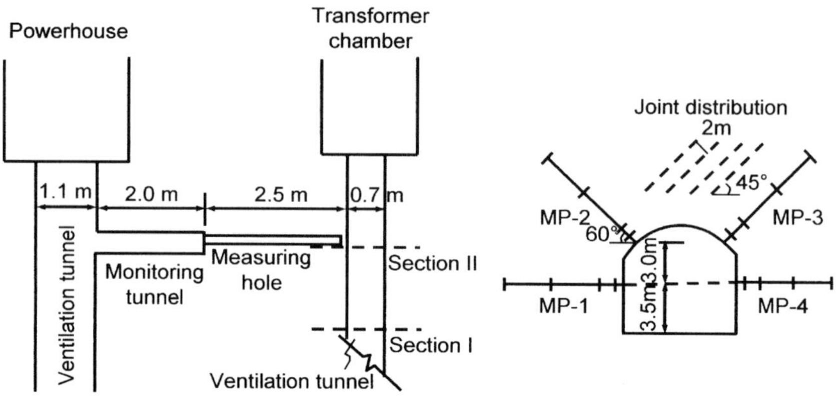


**Figure 6.18.** Monitoring layout in ventilation tunnel.

where $d_m$ is monitored displacements, $d_l$ is lost displacement, $\lambda$ is modification coefficient which is determined from 3-D finite element analysis.

The instruments (multiple-point extensometers or convergence meters) were installed behind the excavation face about 1 m in the two measurement tunnels. Therefore, the displacement measurements reflect partial displacement during the tunnel construction. Three-dimensional finite analysis is to obtain the modification coefficient and determine total displacement at each measurement point.

The overburden thickness of the tunnels is 405 m, so the in situ stresses are calculated as

$$\sigma_x = 6.89\,\text{MPa} \qquad \tau_{xy} = -0.49\,\text{MPa}$$
$$\sigma_y = 9.16\,\text{MPa} \qquad \tau_{yz} = -0.65\,\text{MPa}$$
$$\sigma_z = 13.1\,\text{MPa} \qquad \tau_{zx} = -0.37\,\text{MPa}$$

The Young's modulus is taken as $3 \times 10^4$ MPa, and the Poisson's ratio as 0.20.

Three-dimensional finite element modelling is carried out for each monitoring point to estimate the total displacement. Details of the modelling can be found in [8,9].

Based on the computed total displacement, the monitored displacement will be modified. It was found that the modification coefficient between the monitored and the computed displacements was 1.27 (average). The monitored displacements and modified displacements are presented in Table 6.2.

### 6.4.2 *Back analysis using pure shape acceleration method*

The approach of ordinary back analysis method is to: (a) create constitutive model of the rock mass, (b) assume an initial value of parameter to be back analysed,

**Table 6.2.** The monitored and modified displacements.

| Measurement lines | Measurement points (mm) | | | | Remark |
|---|---|---|---|---|---|
| | 1 | 2 | 3 | 4 | |
| MP-4 | 0.29 | 0.34 | 0.54 | 0.62 | monitored |
| | 0.35 | 0.50 | 0.82 | 0.99 | modified |
| MP-3 | 0.21 | 0.33 | 0.38 | 0.40 | monitored |
| | 0.25 | 0.39 | 0.49 | 0.55 | modified |
| MP-2 | 0.25 | 0.39 | 0.41 | – | monitored |
| | 0.30 | 0.52 | 0.62 | – | modified |
| MP-1 | 0.45 | 0.60 | 0.77 | 0.85 | monitored |
| | 0.58 | 0.77 | 0.99 | 1.09 | modified |

(c) conduct numerical modelling, (d) solve displacements at measurement points, and (e) compare the computed displacement with monitored displacements. The final parameter obtained from the back analysis is the parameter giving minimum difference between the computed displacement and the monitored displacement.

Normally the difference between the computed displacement and the monitored displacements is expressed by the error objective functions as shown in equation (6.27). The pure shape acceleration method is used here to conduct optimisation analysis, because it is an effective method to make the error objective function be minimum.

***6.4.2.1. Computational procedure.*** Displacement analysis procedure consists of optimisation analysis and positive modelling. The optimisation analysis is discussed earlier and its program flow is listed in [403]. The ordinary positive modelling program is modified from a commercial program, capable to model anisotropic material. The modelling program flow is given in [404]. The whole program flow is presented in Figure 6.19, to explain the concept of back analysis.

***6.4.2.2. Computational results.*** The back analysis considers two mechanic models. One is to assume rock as uniform and isotropic medium with no joint. The material parameters are $E$, $\nu$, $c$ and $\phi$. The other is to treat rock as an anisotropic media with one joint set. The parameters are $E$, $K_n$ and $K_s$ to be obtained by back analysis using corresponding constitutive relationship.

*(a) Access tunnel.* In case of isotropic rock, two sets of input displacements are taken in back analysis. Upon optimisation back analysis, the relevant parameters are

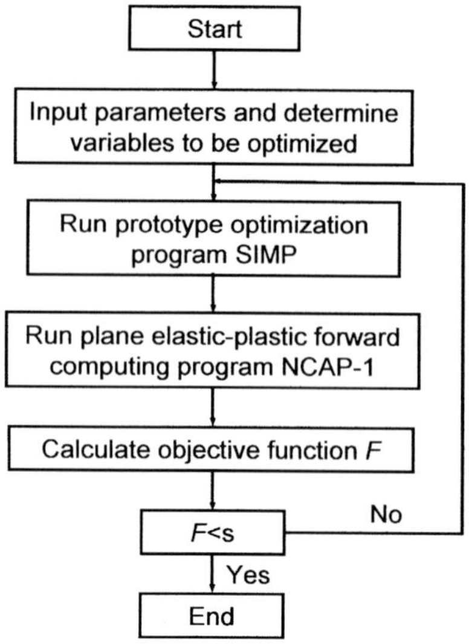


**Figure 6.19.** The complete cycle of program flow.

obtained as listed below:

$$E = 3.70 \times 10^4\,\text{MPa} \qquad \nu = 0.15 \qquad F = 0.6 \times 10^{-3}$$
$$\sigma_1 = -12.87\,\text{MPa} \qquad \sigma_3 = -5.47\,\text{MPa} \qquad \alpha = 50^\circ$$

In case of anisotropic rock, in situ stresses are assumed constant, the back analysis is conducted by inputting in situ stress and corresponding displacement set. The empirical formulae $K_s = 1/5 - 1/10K_n$ (Belytschko *et al.* 1984), is taken as a constraint condition. The back analysis results are list below:

$$E = 6.419 \times 10^4\,\text{MPa} \qquad K_n = 7.25 \times 10^5\,\text{MPa/cm}$$
$$F = 0.59 \times 10^{-3} \qquad K_s = 0.91 \times 10^5\,\text{MPa/cm}$$

*(b) Ventilation tunnel* In case of isotropic rock, the back analysis uses modified displacements at measurement holes MP-4, MP-2. Assuming the Poisson's ratio $\nu = 0.15$, optimised back analysis gives:

$$E = 3.77 \times 10^4\,\text{MPa}$$
$$F = 0.1 \times 10^{-2}$$

$$\sigma_1 = -10.02\,\text{MPa}$$
$$\sigma_3 = -6.78\,\text{MPa}$$
$$\alpha = 25^\circ$$

In case of anisotropic rock, taking the above in situ stress as input data, the back analysis results are:

$$E = 5.91 \times 10^4\,\text{MPa}$$
$$F = 0.2 \times 10^{-2}$$
$$K_n = 48.65 \times 10^4\,\text{MPa/cm}$$
$$K_s = 6.08 \times 10^4\,\text{MPa/cm}$$

The measured displacements and back analysis results for the access tunnel and the ventilation tunnel are compared as shown in Figures 6.20 and 6.21. It can be seen that the results agree well with each other.

Inputting the above back analysis results for each case into the ordinary finite element modelling, the results are compared with the modified displacement measurements as listed in Tables 6.3 and 6.4. It can be seen that the results obtained from back analysis by combining the access tunnel and the ventilation tunnel are better. For each MPBX, the correlation of measured and back-analysed results is better at greater depth than that at shallower depth. This may be because the rock at shallower depth is greatly disturbed by blasting vibration.

### *6.4.3 Stability prediction of powerhouse and transformer chamber*

The powerhouse was constructed by using multiple-stage excavation. In the finite element modelling, however, only the first and last stages are involved. In this section, only the latter is discussed.

Two rock mechanical models are considered in the modelling: uniform linear elastic rock with no joint and anisotropic rock mass with joints. In adopting with joints, the joint influences are estimated by equivalent method discussed in Chapter 3. Two joint sets are considered and equivalent deformation modulus and equivalent strength parameters are obtained.

Three key faults are involved in each modelling. The damage distribution is shown in Figure 6.22. The mesh of the computational model has 596 elements and 589 nodes.

The in situ stresses of ventilation tunnel are obtained by using finite element modelling as: $\sigma_x = 9.42\,\text{MPa}$, $\sigma_y = 7.369\,\text{MPa}$, $\tau_{xy} = 1.245\,\text{MPa}$.

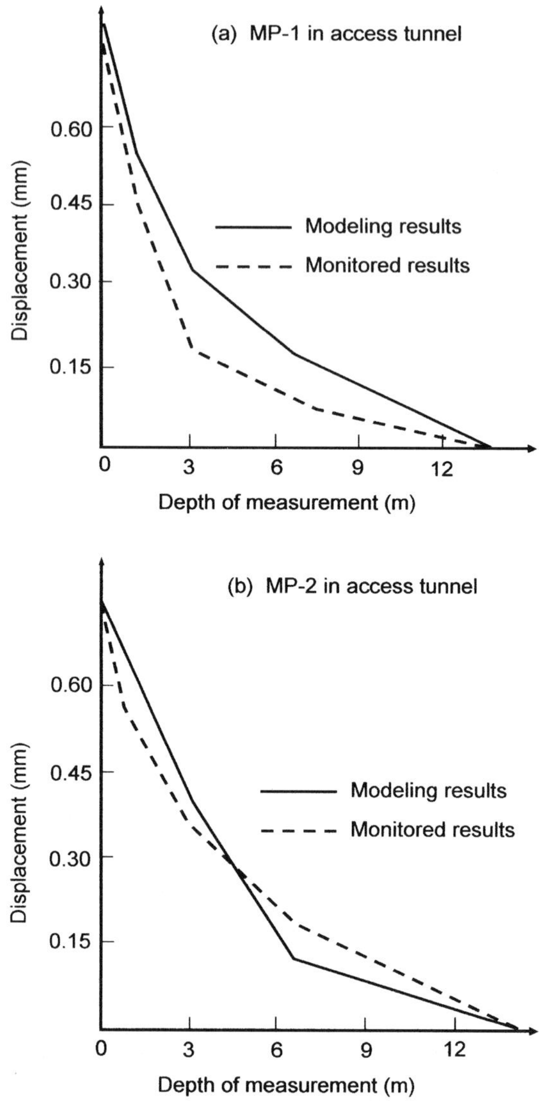


**Figure 6.20.** Comparison of monitored and back-analysed displacement of the access tunnel.

Material parameters of rock are: $E = 3.7 \times 10^4$ MPa; $\nu = 0.20$; $c = 22.27$ MPa; $\phi = 48.1°$; $\sigma_t = 5.0$ MPa.

Joint stiffness are estimated from back analysis, and joint strength parameters are estimated from a similar project. The joint parameters are: $K_n = 6.0 \times 10^5$ MPa/cm; $K_s = 7.5 \times 10^3$ MPa/cm; $c = 0.5$ MPa; $\phi = 35°$.

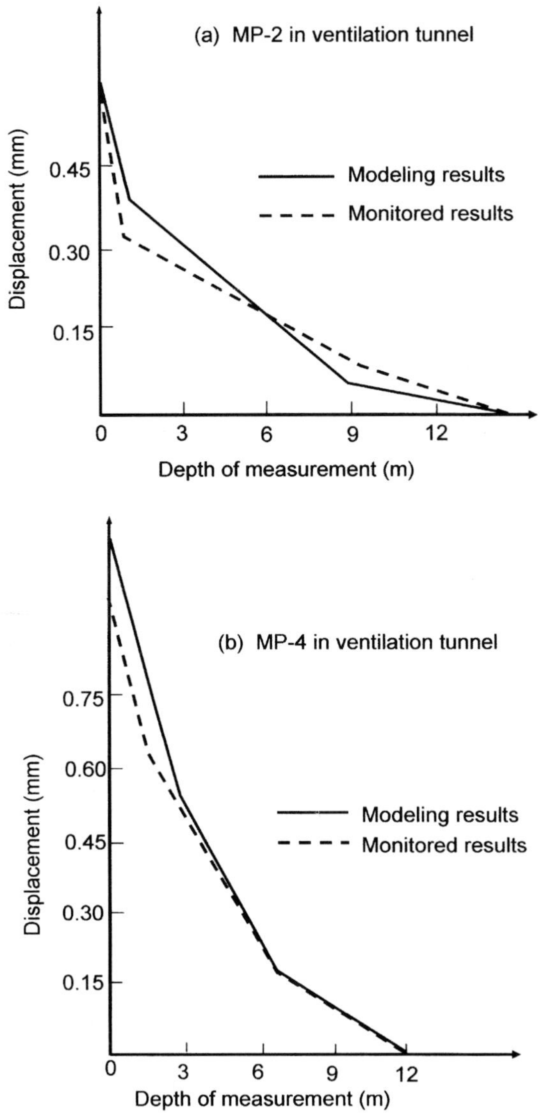


**Figure 6.21.** Comparison of monitored and back-analysed displacements for the ventilation tunnel.

Faults material parameters are: $E = 9.0 \times 10^3$ MPa; $\nu = 0.25$; $c = 0.4$ MPa; $\phi = 30^\circ$; $\sigma_t = 1.0$ MPa.

The layouts of convergence measurements in the powerhouse and transformer chamber are shown in Figure 6.23. The displacement prediction results are listed in

**Table 6.3.** Comparison of measured and computed results in access tunnel.

| Measurement holes | Measurement points (mm) | | | | Remark |
|---|---|---|---|---|---|
| | 1 | 2 | 3 | 4 | |
| MJ-1 | 0.28 | 0.56 | 0.65 | 0.73 | monitored |
| | 0.38 | 0.64 | 0.72 | 0.76 | anisotropic |
| | 0.22 | 0.47 | 0.64 | 0.77 | group (1) |
| | 0.25 | 0.46 | 0.61 | 0.78 | group (2) |
| MJ-2 | 0.13 | 0.35 | 0.63 | 0.76 | monitored |
| | 0.08 | 0.25 | 0.54 | 0.93 | anisotropic |
| | 0.20 | 0.39 | 0.56 | 0.75 | group (1) |
| | 0.34 | 0.79 | 1.29 | 1.82 | group (2) |
| MJ-3 | 0.02 | 0.12 | 0.16 | 0.45 | monitored |
| | 0.03 | 0.08 | 0.18 | 0.29 | anisotropic |
| | 0.21 | 0.60 | 0.97 | 1.34 | group (1) |
| | 0.11 | 0.21 | 0.28 | 0.31 | group (2) |
| MJ-4 | 0.08 | 0.11 | 0.74 | 0.84 | monitored |
| | 0.35 | 0.62 | 0.74 | 0.75 | anisotropic |
| | 0.26 | 0.51 | 0.71 | 0.91 | group (1) |
| | 0.25 | 0.47 | 0.68 | 0.84 | group (2) |

**Table 6.4.** Comparison of measured and computed results in ventilation tunnel.

| Measurement holes | Measurement points (mm) | | | | Remark |
|---|---|---|---|---|---|
| | 1 | 2 | 3 | 4 | |
| MP-4 | 0.35 | 0.50 | 0.82 | 0.99 | monitored |
| | 0.45 | 0.75 | 0.95 | 1.02 | anisotropic |
| | 0.38 | 0.63 | 0.95 | 1.13 | group (1) |
| | 0.36 | 0.60 | 0.92 | 1.10 | group (2) |
| MP-3 | 0.25 | 0.39 | 0.49 | 0.55 | monitored |
| | 0.07 | 0.28 | 0.64 | 0.79 | anisotropic |
| | 0.15 | 0.60 | 1.08 | 1.07 | group (1) |
| | 0.13 | 0.50 | 0.87 | 1.02 | group (2) |
| MP-2 | 0.30 | 0.52 | 0.62 | – | monitored |
| | 0.10 | 0.39 | 0.50 | – | anisotropic |
| | 0.23 | 0.57 | 0.63 | – | group (1) |
| | 0.20 | 0.53 | 0.59 | – | group (2) |
| MP-predrilled hole | 0.58 | 0.77 | 0.99 | 1.09 | monitored |
| | 0.12 | 0.34 | 0.77 | 0.98 | anisotropic |
| | 0.10 | 0.30 | 0.69 | 1.06 | group (1) |
| | 0.09 | 0.29 | 0.67 | 1.04 | group (2) |

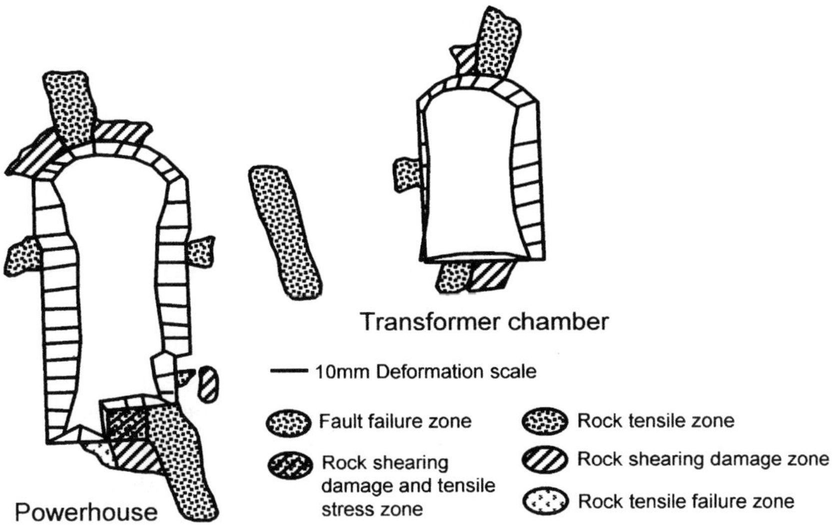


**Figure 6.22.** Distribution of damage zones.

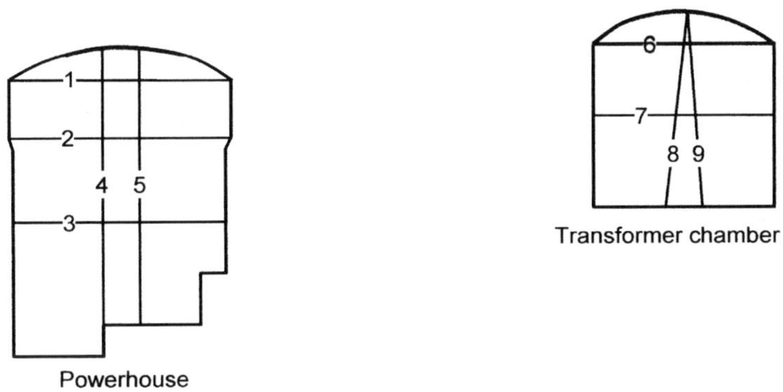


**Figure 6.23.** Layout of convergence measurements.

Table 6.5. From the table, it can be seen that the convergence in the case of considering the joint influence is about 10% greater than that in the case of no joint.

The displacements on tunnel walls and rock failure zone by equivalent model are shown in Figure 6.22. It can be seen that in addition to fault zone failure, shear failure occurs at top and bottom of the tunnel, and tensile failure on the sidewalls.

**Table 6.5.** Predicted displacement of convergence (mm).

| Measurement lines | 1 | 2 | 3 | 4 | 5 | 6 | 7 | 8 | 9 |
|---|---|---|---|---|---|---|---|---|---|
| Equivalent approach | 11.4 | 18.5 | 19.5 | 6.0 | 6.9 | 7.3 | 10.8 | 7.4 | 6.9 |
| Linear elastic | 9.7 | 15.1 | 16.3 | 5.0 | 5.8 | 6.1 | 9.0 | 6.3 | 5.8 |

## 6.5. THREE-DIMENSIONAL BACK ANALYSIS OF ANISOTROPIC ROCK

The example using 3-D back analysis of anisotropic rock mass is shown in this section, in a large-scale hydroelectric power cavern complex. The hydroelectrical power is 70 m high and 30 m wide and constructed underground at the downstream of the right side of a river. The overburden depth is 30–107 m. Two sides of the river are symmetric with average slope of 45–50°. The project is very complicated with the following problems.

(a) Poor engineering geology conditions: The rocks encountered are mainly killas and chorismite with intrusion of other types of rock which are heavily jointed. Breaking of rock in the tunnel occurs easily due to the excavations. The rocks are divided into three types. There are several faults intersecting the powerhouse that may induce instability to the tunnel walls.
(b) High horizontal in situ stress: The site investigation shows that the horizontal in situ stress is more than twice of the vertical in situ stress.
(c) Faults: There are a number of faults intersecting the complex and in the vicinity.

Experimental and numerical analyses show that the faults intersecting the powerhouse cavern may be slipped during excavation.

In summary, this project is constructed under high in situ stresses and complicated ground conditions with non-uniformity, anisotropy and discontinuities. Therefore, displacement monitoring and back analysis are performed to understand the rock mass behaviour and in situ stresses [8,52].

### *6.5.1 Displacement monitoring in trial tunnel and results*

***6.5.1.1. Set-up of displacement monitoring.*** In order to understand the behaviour of underground rock masses during excavation, a pilot tunnel of 1/6 of the cavern size was excavated at the cavern position. The pilot tunnel is 5 m wide, 9.5 m high and 54 m long. The pilot tunnel crosses all the three rock types, and three sections are set up for displacement monitoring. Seven boreholes for MPBX installation are drilled in Section I and II separately, and four are drilled in Section III. Each section has a

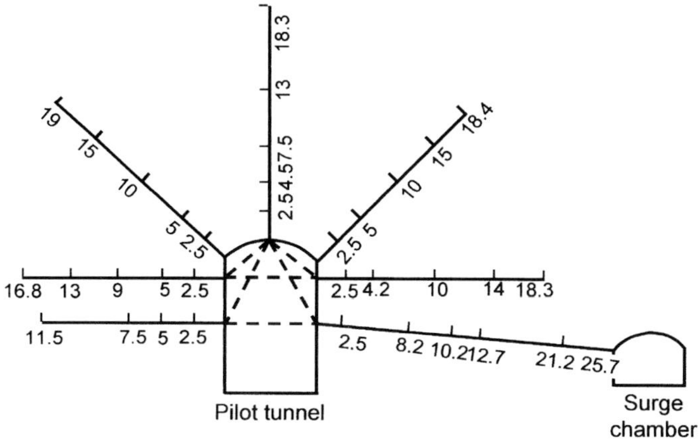


**Figure 6.24.** Setting up of MPBX in Section II.

pre-embedded MPBX to monitor whole deformation process. The pre-embedded MPBX is drilled from the surge chamber to the wall of the pilot tunnel as shown in Figure 6.24. Each monitoring section has 5–7 measurement points on the wall to form more than six convergence monitoring lines to correlate with results obtained by extensometers. In addition, some tests are conducted in the pilot tunnel including measurements of shotcrete and rock bolt stresses, rock bolt pulling test and seismic measurement of rock disturbance zone.

***6.5.1.2. Monitoring results.*** The displacement monitoring results are given here with a period of 250 days until all the displacements become stable. It shows:

(a) Displacement curve tends to be stable in 60–80 days. A typical displacement set is shown in Figure 6.24. A convergence measurement set is shown in Figure 6.25.
(b) The smallest deformation occurs in Section II due to the good rock quality, while the largest displacement is observed in Section III due to the weak rock. Deformation in Section I is remarkable due to the faults. However, in overall, all deformations are quite uniform and not extensive.
(c) The disturbance zone of surrounding rock is about 1–1.5 times of the tunnel size.

### 6.5.2 *Back analysis*

Three-dimensional back analysis is conducted by considering anisotropic rock, complex ground condition and that the in situ stress axis does not coincide with the

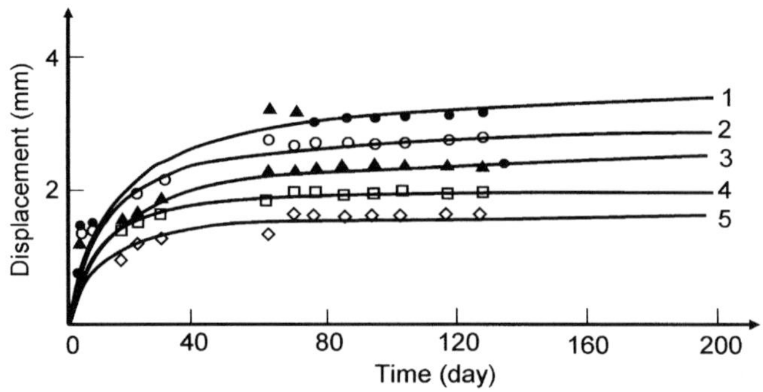


**Figure 6.25.** A typical displacement set.

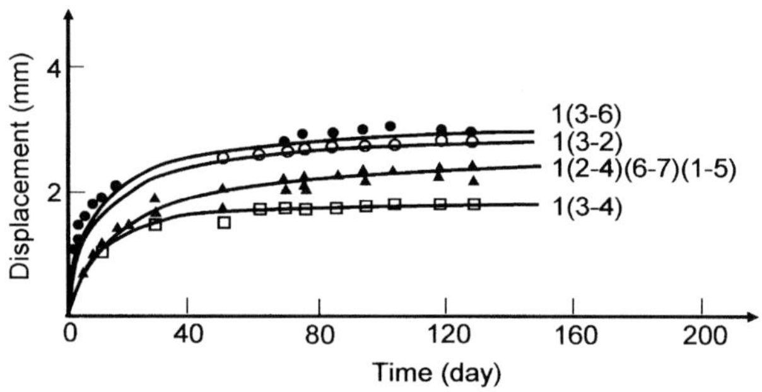


**Figure 6.26.** A convergence measurement set.

tunnel axis. Only the monitored displacements at middle measurement points are used in the back analysis, because it is found that the monitored data at two ends of MPBXs have relatively large error and low reliability. The procedure of back analysis modelling is similar to that outlined in Figure 6.19 of Section 6.4.2. The comparison of modelled results and monitored data for Section II are summarised in Table 6.6.

The back-analysed material parameters of the anisotropic rock mass and in situ stresses are presented in Tables 6.7 and 6.8. The results indicate that the back-analysed results are lower than the results obtained from laboratory tests. The results also reflect the scale effects. In general, large-size rock masses have lower strength and modulus.

The back-analysed in situ stresses agree well with those measured, as shown in Table 6.7. From the above results, it can be seen that satisfactory results are obtained from 3-D non-linear regression based on monitoring displacements. Figure 6.27 shows modelled displacement curves compared with monitored

**Table 6.6.** Comparison of calculated and monitored results in Section II.

| Measurement lines | | | II-1 | II-2 | II-3 | II-4 | II-5 | II-6 | II-7 |
|---|---|---|---|---|---|---|---|---|---|
| Relative displacement | Calculated | $u_1 - u_0$ | 0.18 | 0.14 | 0.10 | 0.15 | 0.15 | 0.24 | 0.23 |
| | Monitored | | 0.19 | 0.18 | 0.56 | 0.15 | 0.16 | 0.17 | – |
| | Calculated | $u_2 - u_0$ | 0.44 | 0.36 | 0.58 | 0.76 | 0.68 | 0.67 | 0.74 |
| | Monitored | | 0.46 | 0.40 | 0.86 | 0.70 | 0.37 | 1.59 | – |
| | Calculated | $u_3 - u_0$ | 0.68 | 0.76 | 0.90 | 1.09 | 0.49 | 1.64 | 1.58 |
| | Monitored | | 0.64 | 0.67 | 1.22 | 1.03 | 0.67 | 1.78 | – |
| | Calculated | $u_4 - u_0$ | 1.13 | 1.21 | 1.53 | 1.67 | 2.03 | 2.36 | 2.34 |
| | Monitored | | 1.13 | 1.23 | 1.48 | 1.57 | 2.06 | 2.25 | – |
| | Calculated | $u_5 - u_0$ | 2.46 | 2.66 | 2.98 | 2.73 | 2.90 | 3.57 | 3.36 |
| | Monitored | | 2.78 | 3.14 | 3.27 | 2.50 | 2.99 | 3.21 | – |

**Table 6.7.** Mechanical parameters of rock mass obtained by back analysis.

| Classification of rock mass | $E_{\|\|}(\times 10^3\,\text{MPa})$ | $E_{\perp}(\times 10^3\,\text{MPa})$ | $u_{\|\|}$ | $u_{\perp}$ |
|---|---|---|---|---|
| A | 17.0 | 11.5 | 0.27 | 0.27 |
| $B_1$ | 13.5 | 9.0 | 0.28 | 0.28 |
| $B_2$ | 11.0 | 7.4 | 0.28 | 0.28 |
| C | 8.5 | 4.3 | 0.29 | 0.29 |
| $f_{27}$ | 1.0 | – | 0.3 | – |
| $f_{25}$ | 2.0 | – | 0.3 | – |

**Table 6.8.** Comparison of calculated and measured in situ stresses.

| In situ stress | $\sigma_x$ | $\sigma_y$ | $\sigma_z$ | $\tau_{xy}$ | $\tau_{yz}$ | $c_a$ |
|---|---|---|---|---|---|---|
| Measured | 5.9 | 7.2 | 7.7 | 0.3 | 0.2 | 0.6 |
| Calculated | 5.4 | 6.7 | 8.0 | –0.9 | 0.4 | –0.5 |

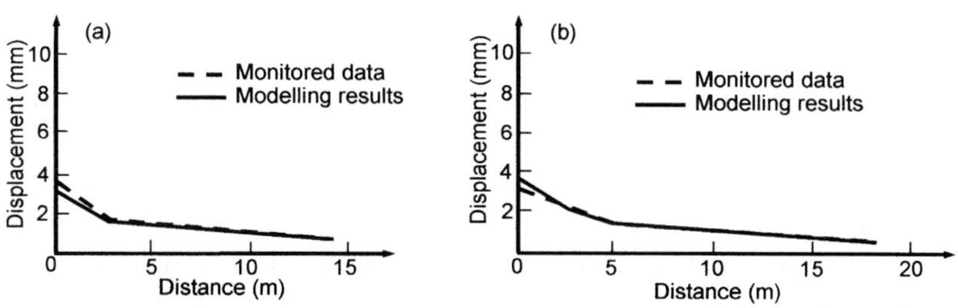


**Figure 6.27.** Modelled displacement curves compared with monitored displacement curves in two typical MPBXs.

displacement curves obtained from multiple-point extensometers for two typical MPBXs. Again, good matches are observed.

## 6.6. THREE-DIMENSIONAL BACK ANALYSIS OF JOINTED ROCK MASS AND STABILITY ANALYSIS

A 2-D back analysis of jointed rock mass for a pumped storage hydroelectric power station was conducted in Section 6.4. The analysis was based on the monitoring information of two branch tunnels, before the excavation of the powerhouse cavern. This section is to illustrate, for the same project, the 3-D back analysis based on monitoring of the powerhouse cavern. Brief concept of the method will be introduced and some main results will be discussed in this section. The theory has been discussed in the previous sections.

### *6.6.1 Mechanic model*

Similar to the example in Section 6.4, the equivalent model of jointed rock is also used in this 3-D back analysis case. For the joint persistence, the 3-D back analysis uses area equivalence, while 2-D uses line equivalence. The material parameters of rock and joints are obtained from approximate weighted average based on the joint arrangement. The joint persistence is taken as 80% from site investigation results.

The material parameters of rock element used in the 3-D modelling are $E$, $\nu$, $c_r$, $\phi_r$ and $\sigma_t$, where $\sigma_t$ is tensile strength of rock. The parameters used for joints are $K_n$, $K_s$, $c_j$, $\phi_j$ and $\alpha_j$, where $\alpha_j$ is the joint dip angle.

Equivalently transferring jointed rock mass into quasi-continuous rock mass, new stiffness matrix $[D]$ can be obtained. It is a $6 \times 6$ full matrix for 3-D ($3 \times 3$ matrix for 2-D). The new matrix will replace elastic matrix in the finite element modelling for the rock mass.

### *6.6.2 Summary of site monitoring data*

Five monitoring sections are set up in the powerhouse cavern, while two monitoring sections are set in the transformer chamber cavern. Altogether, a total number of 36 monitoring MPBXs and 144 monitoring points are installed in those sections. Upon the completion of powerhouse and transformer chamber constructions, it is found that only data at 90 monitoring points in 28 MPBXs give effective data. After data sorting, the data from 15 points in 3 MPBXs have clean trends and are considered reliable, and these data are used in the 2-D and 3-D back analysis.

Figure 6.28 shows the multiple-stage excavation of the powerhouse and the transformer chamber and the layout of instruments. The instrument details and monitoring data in powerhouse are shown in Tables 6.9 and 6.10. Data in Table 6.9 are used for 2-D analysis and data in Table 6.10 are used for 3-D analysis. From the monitoring data it can be seen that the displacements are small with a maximum value of 6 mm close to the right side of powerhouse cavern.

In Table 6.10, the $n$ horizontal lines refer to $(n-1)$ stage excavation. The vertical lines across the horizontal lines represent excavation face position and

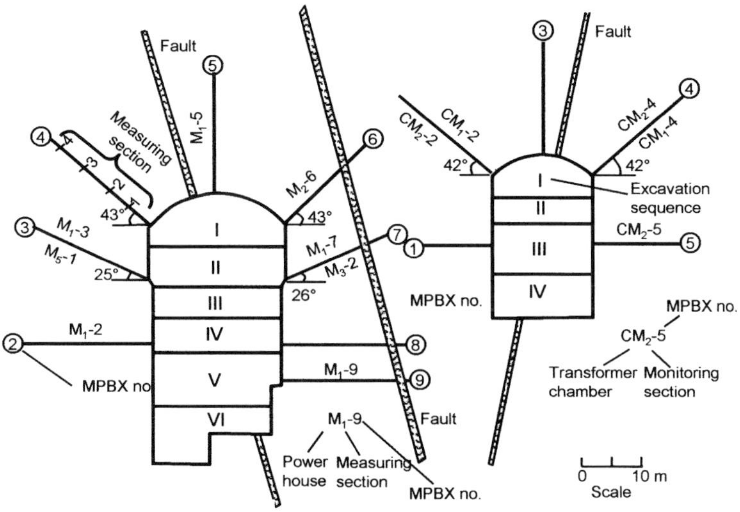


**Figure 6.28.** Excavation stage of the caverns and layout of instrumentation.

**Table 6.9.** Measured displacement by MPBX (mm).

| | 1 | 2 | 3 | 4 | Original status |
|---|---|---|---|---|---|
| $M_1$-3 | – | 2.73 | 1.12 | 3.11 | After excavation of Stage II |
| $M_1$-5 | – | – | −2.44 | −4.12 | 5 m away from excavation face in Stage I |
| $M_1$-7 | 0.14 | 1.01 | 1.04 | 3.51 | After excavation of Stage I |
| $M_1$-9 | 0.17 | – | 0.90 | 1.72 | After excavation of Stage V |
| $M_2$-2 | 1.31 | 1.47 | – | 2.28 | After excavation of Stage IV |
| $M_2$-6 | 3.57 | – | 5.25 | 6.05 | 1.5 m away from excavation face in Stage I |
| $M_5$-1 | – | 0.11 | 0.30 | 0.60 | After excavation of Stage II |
| $M_5$-2 | – | 0.72 | 2.44 | 5.05 | |

**Table 6.10.** Instrumentation set-up and measured displacement.

| | Arrangement | Point | Depth (m) | Measured value (mm) | Location |
|---|---|---|---|---|---|
| 1 |  | $M_1^3 - 5$ | 8 | −2.44 | 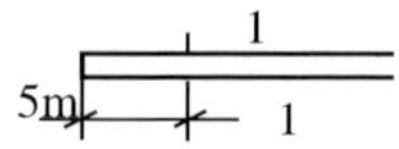  |
| | | $M_1^4 - 5$ | 23 | −4.12 | |
| 2 |  | $M_2^1 - 6$ | 1 | 3.57 | 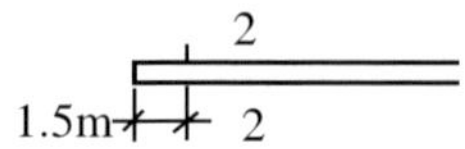  |
| | | $M_2^3 - 6$ | 8 | 5.25 | |
| | | $M_2^4 - 6$ | 23 | 6.05 | |
| 3 |  | $CM_1^1 - 2$ | 1 | 0.09 | 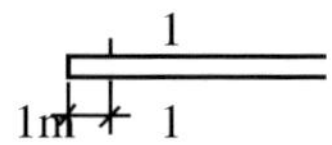  |
| | | $CM_1^2 - 2$ | 3 | 0.63 | |
| | | $CM_1^4 - 2$ | 19 | 0.85 | |
| 4 |  | $CM_1^4 - 3$ | 19 | 0.71 | |
| 5 |  | $CM_1^1 - 4$ | 1 | 0.42 | |
| | | $CM_1^3 - 4$ | 8 | 1.14 | |
| | | $CM_1^4 - 4$ | 19 | 2.41 | |
| 6 | 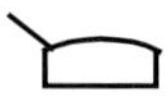 | $CM_2^2 - 2$ | 3 | 35 | 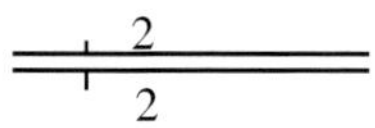  |
| 7 | 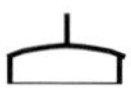 | $CM_2^3 - 3$ | 8 | 0.34 | |
| 8 | 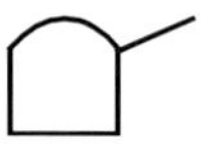 | $CM_2^1 - 4$ | 1 | 0.91 | |
| | | $CM_2^3 - 4$ | 8 | 1.72 | |
| 9 |  | $M_1^1 - 7$ | 1 | 0.14 | 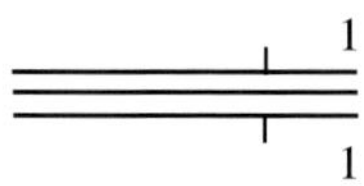  |
| | | $M_1^3 - 7$ | 3 | 1.01 | |
| | | $M_1^3 - 7$ | 8 | 1.04 | |
| | | $M_1^4 - 7$ | 23 | 3.51 | |
| 10 |  | $M_1^2 - 3$ | 3 | 2.73 | |
| | | $M_1^3 - 3$ | 8 | 1.12 | |
| | | $M_1^4 - 3$ | 23 | 3.11 | |
| 11 |  | $M_5^2 - 1$ | 3 | 0.11 | 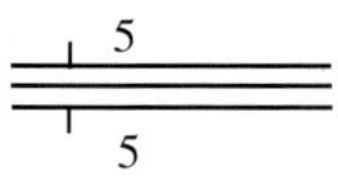  |
| | | $M_5^3 - 1$ | 8 | 0.30 | |
| | | $M_5^4 - 1$ | 23 | 0.60 | |

(*continued*)

**Table 6.10.** Continued.

| | Arrangement | Point | Depth (m) | Measured value (mm) | Location |
|---|---|---|---|---|---|
| 12 |  | $M_5^2 - 2$ | 3 | 0.72 | |
| | | $M_5^3 - 2$ | 8 | 2.44 | |
| | | $M_5^4 - 2$ | 23 | 5.05 | |
| 13 |  | $CM_2^1 - 5$ | 1 | 0.09 | 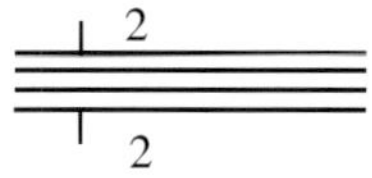 |
| | | $CM_2^3 - 5$ | 6 | 0.34 | |
| | | $CM_2^4 - 5$ | 12 | 1.05 | |
| 14 | 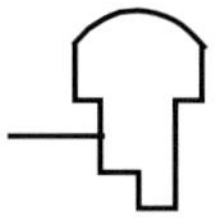 | $M_2^1 - 2$ | 1 | 1.31 | 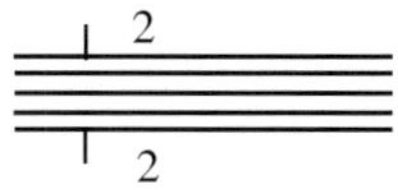 |
| | | $M_2^3 - 2$ | 3 | 1.47 | |
| | | $M_2^4 - 2$ | 12 | 2.28 | |
| 15 | 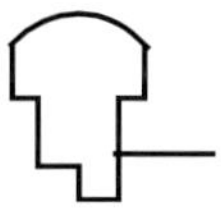 | $M_1^1 - 5$ | 1 | 0.17 |  |
| | | $M_1^3 - 5$ | 6 | 0.90 | |
| | | $M_1^4 - 5$ | 12 | 1.72 | |

that non-crossing the horizontal lines mean that the excavation stage has been completed.

### *6.6.3 Finite element back analysis of underground powerhouse complex*

Two-dimensional modelling uses 630 nodes and 635 elements, while 3-D modelling uses 4304 nodes and 3994 elements. Rock material parameters adopted in the 2-D back analysis are given in Section 6.4. To reduce the mesh preparation work, all the computations use the same mesh arrangement as shown in Figure 6.29. However, mesh relating to the excavation will be changed to simulate the multiple-stage excavation process.

Upon the back analysis, the final rock material parameters are obtained as following:

$$E = 3.7 \times 10^4 \, \text{MPa}$$

$$n = 0.24$$

$$c = 1.29 \, \text{MPa}$$

$$\phi = 41^\circ$$

$$\sigma_t = 3.5 \, \text{MPa}$$

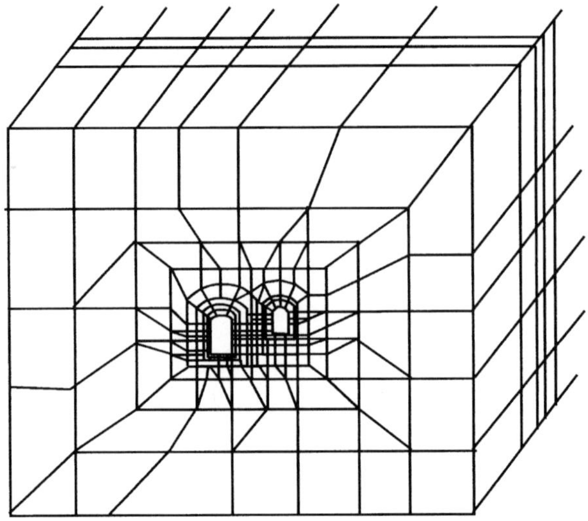

**Figure 6.29.** 3-D FEM model mesh.

Joint parameters are:

$$K_n = 6.0 \times 10^5\,\text{MPa}$$

$$K_t = 7.5 \times 10^4\,\text{MPa}$$

$$c = 0.5\,\text{MPa}$$

$$\phi = 35^\circ$$

In situ stresses are:

$$\sigma_x = 8.95\,\text{MPa}$$
$$\sigma_y = 12.84\,\text{MPa}$$
$$\sigma_z = 6.89\,\text{MPa}$$
$$\tau_{xy} = 0.58\,\text{MPa}$$
$$\tau_{yz} = 0.38\,\text{MPa}$$
$$\tau_{zx} = 0.41\,\text{MPa}$$

Comparing above results with results obtained in Section 6.4, there is no much difference, except the rock material strength parameters.

The results from the 2-D back analysis and from the monitoring in the powerhouse and the transformer chamber are compared in Figures 6.30 and 6.31.

The 3-D back analysis results and the monitoring data in the powerhouse are compared in Figure 6.32.

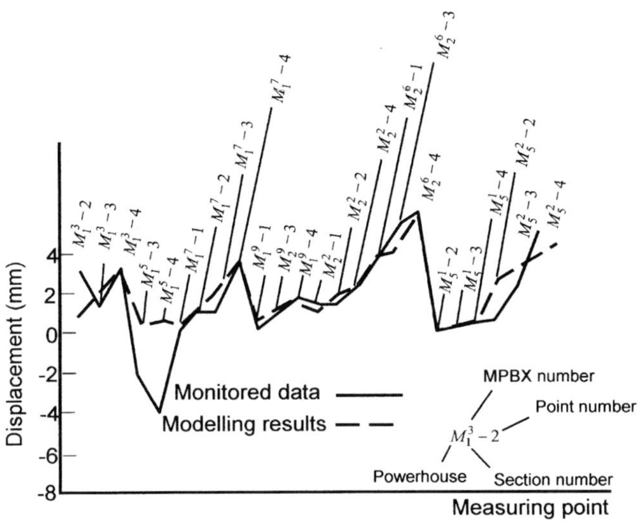


**Figure 6.30.** 2-D back analysis and monitoring in the powerhouse.

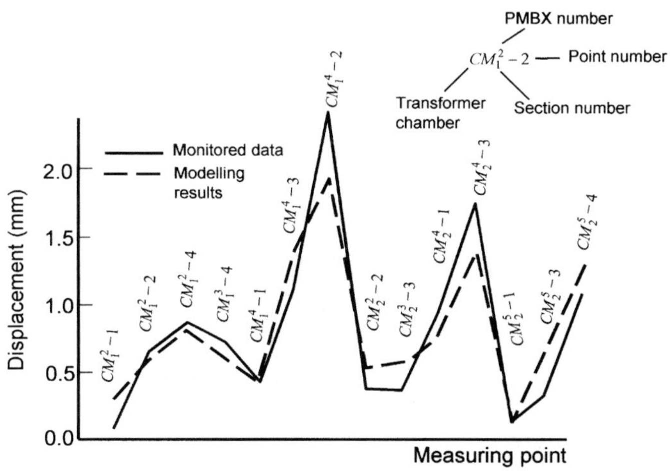


**Figure 6.31.** 2-D back analysis and monitoring in the transformer chamber.

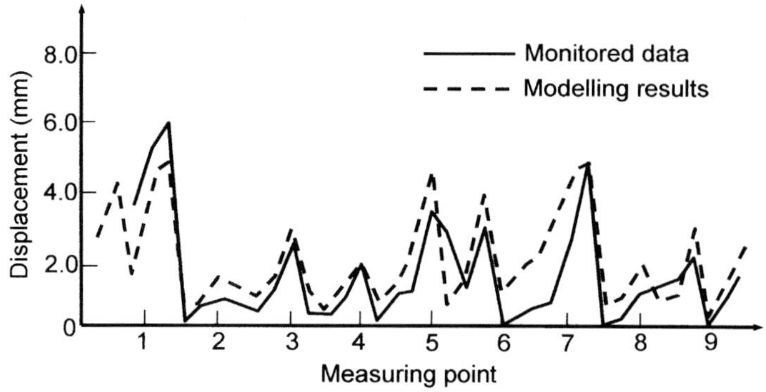


**Figure 6.32.** 3-D back analysis and monitoring data in the powerhouse.

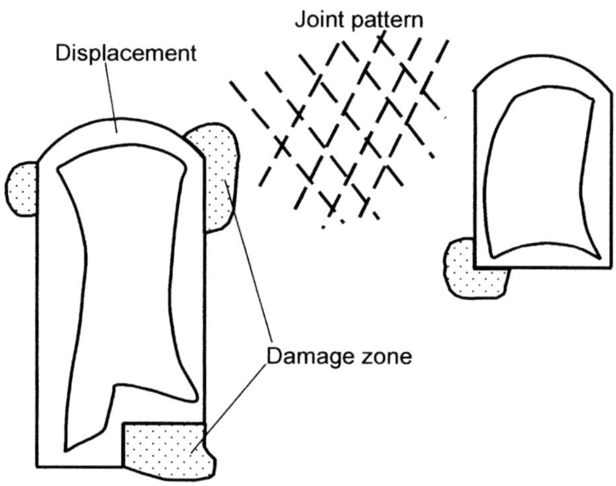


**Figure 6.33.** Displacement distribution and failure zone in the rock mass around the cavern.

### *6.6.4 Stability of powerhouse and transformer chamber*

Non-linear analysis of powerhouse and transformer chamber using above back-analysed parameters provides the displacement distribution in the surrounding rock mass. The displacement distribution and failure zone in the rock mass around the cavern at a particular section are shown in Figure 6.33. The displacements at roof are smaller than those at the walls and the bottom of the powerhouse. The upward displacement at the lower corner of powerhouse is the largest. Shear failure occurs locally. The above results are obtained in the case of no supports. The rock mass can be stabilised by reinforcement and support.

From the above analysis results, it can be concluded that the mechanics models, especially the equivalent continuous model for the jointed rock mass proposed, are applicable and validated. In addition, the rock mechanic parameters from the back analysis using those analytic models have given good results. Therefore, the analytical methods are further verified.

## 6.7. APPLICATIONS OF STATISTICS MODEL IN DEFORMATION PREDICTION

In underground constructions, monitoring of rock mass deformation becomes more and more important and is widely applied. Especially when the observational construction method is adopted in tunnelling. The deformation monitoring provides the basic information for back analysis. The information obtained from deformation monitoring can be used in at least three aspects. Firstly, it is used to predict the safety during construction. The deformation monitoring can reflect the total deformation and deformation rate. Secondly, ordinary conditions including in situ stress and material parameters can be back analysed. Lastly, the monitored deformation can be used to predict the rock behaviour in subsequent construction [53–56,398,405,406]. This section is to discuss the last application.

There are several methods to predict rock parameters and behaviour in subsequent construction, for example, the mathematical analytical method. In usual analytical method, the rock model with material parameters and in situ stresses must be determined in advance and provided to the model as input data. In back analysis method, the in situ stresses and material parameters are obtained from back analysis with monitored data and computational work. However, the statistic method is aimed at saving computational work. The method relies on the observed rock behaviour and monitored data, develops the inter-relation between the properties and behaviour, and predicts the behaviour of the rock [407–413]. This method can accordingly avoid influences of many mechanical uncertainties and provides results in short time and with high accuracy.

There are many methods to be used for data regression in statistics approach, such as Laglongi, power sliding, spline, regression model, time series analysis and grey system theory [408,410,412,414–417]. In this section, a method combining non-linear regression method and modified grey system model is introduced through an example to predict displacement in a project.

### 6.7.1 Non-linear regression model

As a statistical model, the regression model requires to make judgement on the data distribution and to assume a model in advance based on the user's experience and skills. Then the parameters of the given model are recognised by least square estimation of error square sum function. Finally, statistic checking is conducted and prediction is made. Presently, linear regression model is well developed. For general non-linear regression model, the least square estimation of parameters is often carried out by using numerical modelling [24,30,33]. It is transformed to general linear model by mathematical transformation under certain conditions, and then obtain least square estimation of parameters.

From the features of rock deformation curves, the regression model is assumed as hyperbolic function

$$u = \frac{t}{A + Bt} \tag{6.28}$$

where $\{t_i\}$ denotes time sequence, $\{u_i\}$, $(i = 1, 2, \ldots, n)$ is the measured displacements.

Let $y = t/u$, then,

$$Y = A + Bt \tag{6.29}$$

So it is transformed to be a linear regression model.

Setting

$$\left.\begin{aligned}
\bar{t} &= \frac{1}{n}\sum_{i=1}^{n} t_i \\
\bar{y} &= \frac{1}{n}\sum_{i=1}^{n} y_i \\
L_{tt} &= \sum_{i=1}^{n} (t_i - \bar{t})^2 \\
L_{yy} &= \sum_{i=1}^{n} (y_i - \bar{y})^2 \\
L_{yt} &= \sum_{i=1}^{n} (t_i - \bar{t})(y_i - \bar{y})
\end{aligned}\right\} \tag{6.30}$$

It is known that

$$B = L_{yt}/L_{tt}$$
$$A = \bar{y} - B\bar{t}$$

The linear relationship of variable $y$ and $t$ can be expressed by using relation coefficient,

$$r = \frac{L_{yt}}{\sqrt{L_{tt}L_{yy}}}$$

Obviously $|r| \leq 1$. When $|r| = 1$, it becomes fully linearly related. When $|r| = 0$, it becomes fully linearly unrelated. The nearer $|r|$ to 1, the more linearly it is related.

Maximum displacement is given as:

$$u_{\infty} = \lim_{t \to \infty} \frac{t}{A + Bt} = \frac{1}{B} \tag{6.31}$$

It is the predicted final displacement.

The initial displacement rate is given as:

$$u'_{t \to 0} = 1/A$$

From the above two formulae, the physical meaning of $A$ and $B$ can be recognised.

### 6.7.2 *Grey system theory model*

Because all the grey models GM have the same basic conditions and principles, the basic grey model GM(1,1) is used here for analysis [414,415]. The grey theory assumes discretised source data series as $u^{(0)} = \{u_1^{(0)}, u_2^{(0)}, \ldots, u_n^{(0)}\}$, by conducting one accumulation forming treatment (AGO), generates a forming series $u^{(1)} = \{u_1^{(1)}, u_2^{(1)}, \ldots, u_n^{(1)}\}$, and then creates a one-stage deviation equation, GM(1,1) is created as

$$\frac{\mathrm{d}u^{(1)}}{\mathrm{d}t} + \alpha u^{(1)} = b \tag{6.32}$$

The solution of GM(1,1) is:

$$\hat{u}_{t+1}^{(1)} = \left(u_1^{(0)} - \frac{b}{a}\right)e^{-at} + \frac{b}{a} \tag{6.33}$$

Equation (6.33) is the grey prediction formulae. Deviating $u_{t+1}^{(1)}$, we can get the prediction formulae of source data as:

$$\hat{u}_{t+1}^{(0)} = \left(-au_1^{(0)} + b\right)e^{-at} \tag{6.34}$$

GM(1,1) requires equal time step. However, monitoring data do not satisfy this requirement. Local internal insertion or smooth treatment can be used to form equal time step.

The GM(1,1) model of equal time step is subsequently improved, to make it applicable for solving problems of non-equal time step. The improved model is applied to predict single pile capacity and showed that it was effective. The key feature of non-equal time step model is to replace grey deviation by the difference form:

$$\frac{\mathrm{d}x^{(1)}}{\mathrm{d}t} = \frac{x^{(1)}(k+1) - x^{(1)}(k)}{t^0(k+1) - t^0(k)} = \frac{x^{(0)}(k+1)}{t^{(-1)}(k+1)} \tag{6.35}$$

The problem now is how to take the background value of $dx^{(1)}/dt$. It was noted that this value does not change significantly when $x$ changes from $x^{(1)}(k)$ to $x^{(1)}(k+1)$ if $t^{(0)}(k+1)-t^{(0)}(k)$ is sufficiently small. Therefore, we have

$$z^{(1)}(k+1) = 0.5x^{(1)}(k+1) + 0.5x^{(1)}(k)$$

as the background value of $\mathrm{d}x^{(1)}/\mathrm{d}t$ between $t^{(0)}(k)$ and $t_{(0)}(k+1)$.

The background value of $\mathrm{d}x^{(1)}/\mathrm{d}t$ can further be improved based on the features of rock deformation curves that change sharply in the initial stage. Usually $u$–$t$ curves is up convex, the background value of $\mathrm{d}x^{(1)}/\mathrm{d}t$ can be taken as:

$$z^{(1)}(k+1) = x^{(1)}(k) + x^{(1)}(k+1) - \sqrt{x^{(1)}(k)x^{(1)}(k+1)} \tag{6.36}$$

From equations (6.31), (6.34) and (6.36), we have,

$$x^{(0)}(k+1) + at^{(-1)}(k+1)z^{(1)}(k+1) = bt^{(-1)}(k+1) \tag{6.37}$$

Based on least squares method, we have,

$$\begin{Bmatrix} a \\ b \end{Bmatrix} = \left(B^T B\right)^{-1} B^T Y_n \tag{6.38}$$

where

$$[B] = \begin{bmatrix} t^{(-1)}(2) & & & & 0 \\ & t^{(-1)}(3) & & & \\ & & \cdot & & \\ & & & \cdot & \\ & & & \cdot & \\ 0 & & & & t^{(-1)}(n) \end{bmatrix}$$

$$\times \begin{bmatrix} -\left[x^{(1)}(1) + x^{(1)}(2) - \sqrt{x^{(1)}(1)x^{(1)}(2)}\right] & 1 \\ -\left[x^{(1)}(2) + x^{(1)}(3) - \sqrt{x^{(1)}(2)x^{(1)}(3)}\right] & 1 \\ \cdot & \\ \cdot & \\ -\left[x^{(1)}(n-1) + x^{(1)}(n) - \sqrt{x^{(1)}(n-1)x^{(1)}(n)}\right] & 1 \end{bmatrix}$$

$$Y_n = \begin{bmatrix} x^{(0)}(2) \\ x^{(0)}(3) \\ \cdot \\ \cdot \\ \cdot \\ x^{(0)}(n) \end{bmatrix}$$

The measured displacements in underground excavation consist of time series $\{t_i\}$ and corresponding displacement series $\{u_i\}$. The displacement prediction formulae is obtained from the monitored data as

$$\ddot{u}_{k+1} = \left(u_1 - \frac{b}{a}\right) e^{a(t_1 - t_{k+1})} + \frac{b}{a} \tag{6.39}$$

where

$$[B] = \begin{bmatrix} t^{(-1)}(2) & & & 0 \\ & t^{(-1)}(3) & & \\ & & \ddots & \\ 0 & & & t^{(-1)}(n) \end{bmatrix} \times \begin{bmatrix} -(u_1 + u_2 - \sqrt{u_1 u_2}) & 1 \\ -(u_2 + u_3 - \sqrt{u_2 u_3}) & 1 \\ \vdots & \\ -(u_{n-1} + u_n - \sqrt{u_{n-1} u_n}) & 1 \end{bmatrix}$$

$$Y_n = \begin{bmatrix} u_2 - u_1 \\ u_3 - u_2 \\ \vdots \\ u_n - u_{n-1} \end{bmatrix}$$

The stability of the model is often examined by backward error ratio $c$ and small error ratio $p$ [111].
Error is given as:

$$\varepsilon(i) = u(i) - \hat{u}(i) \quad i = 1, 2, \ldots, n$$

Mean error is given as:

$$\bar{\varepsilon} = \frac{1}{n} \sum_{i=1}^{n} \varepsilon(i)$$

Variance is given as:

$$s_1^2 = \frac{1}{n} \sum_{i=1}^{n} [\varepsilon(i) - \bar{\varepsilon}]$$

Mean of source data is calculated by:

$$\bar{u} = \frac{1}{n} \sum_{i=1}^{n} u(i)$$

Variance of source data is calculated by:

$$s_2^2 = \frac{1}{n} \sum_{i=1}^{n} [u(i) - \bar{u}]$$

**Table 6.11.** Classification of accuracy grade.

| Accuracy | Excellent | Good | Fair | Bad |
|---|---|---|---|---|
| $c$ | $<0.35$ | $<0.50$ | $<0.65$ | $\leq 0.65$ |
| $p$ | $>0.95$ | $>0.80$ | $>0.70$ | $\geq 0.70$ |

Posterior error ratio is estimated by:

$$c = \frac{s_1}{s_2}$$

Probability of small error is estimated by:

$$p = p\{|\varepsilon(i) - \bar{\varepsilon}| < 0.6745 s_2\}$$

A small value of $c$ represents a good model. A small value of $c$ generally leads to a large $s_2$ and a small $s_1$. Small $s_1$ represents small error deviation, and small $c$ means small difference between the computational data and the actual data. A large value of $p$ represents a good grey model, and means more points with small difference between the error and the mean of error ($< 0.6745 s_2$).

The accuracy can be divided into several grades based on $c$ and $p$ value as shown in Table 6.11. If the $c$ and $p$ values are within the allowable range, the model can be used to make prediction. Otherwise, the model needs to be modified until the accuracy is in a satisfactory range.

### *6.7.3 Engineering application*

The grey model is applied to a trial tunnel of a hydroelectric power station. The layout of extensometers is shown in Figure 6.34. Some regression displacement curves and monitoring curves are shown in Figure 6.35. The monitoring displacements and prediction results in later days are listed in Table 6.12.

From Figure 6.35 and Table 6.12, it can be seen that hyperbolic model and grey model all agree well with monitoring data. The final monitoring displacements are between the two predictions. Therefore, the use of both models is suggested.

### *6.7.4 Discussion*

The grey model can be created only when the source series is a smooth discrete function. It requires the source series determined or has a determinable trend.

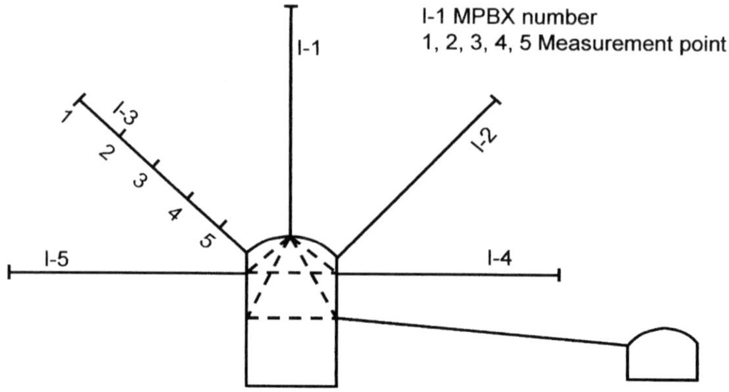


**Figure 6.34.** Layout of extensometers.

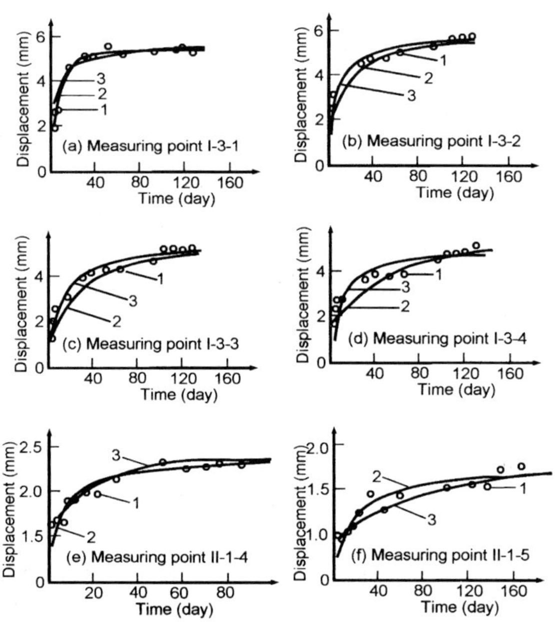


**Figure 6.35.** Regression displacement curves and monitoring curves.

However, the feature of the grey model is to treat source series and to generate forming series, and to describe $x(1)$ by using the grey amplitude and indirectly describe $x^{(0)}$. Therefore, in practice, the grey model does not require $x^{(0)}$ to be a smooth discrete function, but allows $x^{(0)}$ to have certain flexibility. This extends

**Table 6.12.** Comparison of measured and predicted displacements.

| Time (day) | I-3-1 | | | I-3-2 | | | I-3-3 | | | I-3-4 | | | I-3-5 | | |
|---|---|---|---|---|---|---|---|---|---|---|---|---|---|---|---|
| | M | HM | GM | M | HM | GM | M | HM | GM | M | HM | GM | M | HM | GM |
| 140 | 5.28 | 5.29 | 5.21 | 5.46 | 5.38 | 5.26 | 5.16 | 5.07 | 4.92 | 5.10 | 4.81 | 4.94 | 4.17 | 4.04 | 4.15 |
| 149 | 5.31 | 5.30 | 5.21 | 5.47 | 5.39 | 5.27 | 5.18 | 5.10 | 4.93 | 5.11 | 4.84 | 4.98 | 4.16 | 4.05 | 4.19 |
| 167 | 5.33 | 5.32 | 5.21 | 5.49 | 5.42 | 5.28 | 5.21 | 5.13 | 4.95 | 5.13 | 4.87 | 5.05 | 4.18 | 4.08 | 4.28 |
| 173 | 5.30 | 5.32 | 5.21 | 5.46 | 5.43 | 5.28 | 5.14 | 5.14 | 4.96 | 5.03 | 4.88 | 5.09 | 4.15 | 4.09 | 4.31 |
| Infinity | | 5.49 | 5.21 | | 5.66 | 5.29 | | 5.47 | 4.98 | | 5.15 | 5.21 | | 4.31 | 4.60 |
| | II-1-1 | | | II-1-2 | | | II-1-3 | | | II-1-4 | | | II-1-5 | | |
| 94 | 2.29 | 2.26 | 2.14 | 2.57 | 2.58 | 2.49 | 1.44 | 1.46 | 1.50 | 2.15 | 2.41 | 2.19 | 1.66 | 1.56 | 1.57 |
| 103 | 2.32 | 2.29 | 2.22 | 2.59 | 2.60 | 2.50 | 1.46 | 1.47 | 1.51 | 2.14 | 2.15 | 2.19 | 1.65 | 1.57 | 1.59 |
| 118 | 2.31 | 2.33 | 2.32 | 2.59 | 2.62 | 2.51 | 1.43 | 1.48 | 1.52 | 2.13 | 2.16 | 2.20 | 1.63 | 1.59 | 1.62 |
| 128 | 2.32 | 2.36 | 2.37 | 2.62 | 2.64 | 2.52 | 1.41 | 1.50 | 1.53 | 2.10 | 2.17 | 2.20 | 1.61 | 1.60 | 1.63 |
| Infinity | | 2.68 | 2.77 | | 2.83 | 2.52 | | 1.62 | 1.54 | | 2.26 | 2.21 | | 1.71 | 1.68 |

Note: (1) M – Measured; HM – Hyperbolic Model; GM – Grey Model; (2) I-3-1 refers to measurement point 1 in Section I MPBX 3, similarly for other points.

the application range of the grey model. The behaviour of rock displacement in underground projects makes the grey model more suitable and applicable.

However, all models are created based on the existing monitoring information. If geological conditions of the excavation are changed, the models need to be re-created or otherwise the expected prediction accuracy will become low.

Generally the grey model is applicable in predicting final displacements in underground engineering. However, hyperbolic model is applicable for describing monitoring displacements that follows a simple function. It is not suitable for other situations.

Chapter 7

# Construction Mechanics and Optimisation of Excavation Schemes

Construction of rock engineering projects, especially large-scale underground excavations in rock masses, usually requires a long period to complete, i.e., a few months to a few years. To the extreme, a mining project may take tens or hundreds of years. The construction of these rock engineering projects will disturb the initial stable state of the rock masses. After that, various rock mass parameters interact in a dynamic interactive process until the rock mass reaches a new stable state [1–3]. The construction is therefore a dynamic interactive process in time and in space. The success in constructing and managing a rock project not only depends on the eventual state of the project, but also on the interim process and the construction methods adopted [63–65,69,70].

Construction of large-scale rock projects is implemented by continual excavation of new working faces. Each newly excavated face interacts dynamically with the existing excavated space in time and in space [63,64,66,69]. Rock mass is a geological medium, which is generally discontinuous, inhomogeneous and anisotropic. The rock mass has uncertain and variable parameters, which are further changed by the engineering activities. It is very important to consider the effects of engineering activities, especially under complex geological conditions such as high rock stress, weakness zones, faults, joints, and ground water [69,180,348,350,406]. For any large-scale rock engineering project, the effects of engineering activities must be taken into account for the design and construction. From the viewpoint of mechanics, the dynamic interactive process of rock engineering constructions is non-inverse and non-linear. Its eventual state (or solution) is not unique but changeable with the interim process [1–3]. In other words, the eventual state is strongly dependent on the stress paths or stress histories. This leads to the necessity of the optimisation of construction process.

It is generally impossible to complete the construction of any large-scale projects just at one stage of full face excavation, especially for large underground caverns. In practice, they are constructed step by step following a certain sequence of excavation [67,68]. The excavation scheme is decided according to the layout of access tunnels, types of tunnelling machines and characteristics of rock masses. In the series of sequential construction, each step of construction corresponds to a certain type of temporary cavern geometry, i.e., different sequences of construction correspond to the different temporary loading conditions. During and after the construction, rock mass parameters are affected by the continually variable cavern geometry and loading conditions. Typical rock mass parameters are rock stress, rock damage and circumferential displacement around the cavern. The disturbance to the surrounding rock mass and the redistribution of rock stress are primarily due to excavation activities. Therefore, excavation sequences and associated methods have widely crucial implication on the deformation and stability of underground excavation [418–437].

Many engineering measures have been proposed to stabilise the large-scale underground excavations. Of them, the most effective measure is to adopt a proper excavation sequence and to install effective rock reinforcement [418–426]. This forms the principle of the construction mechanics. The initiation and development of the construction mechanics are presented and discussed in various literatures [8,9,52,63,68,404,431]. It was proposed that the excavation steps should be reduced to a minimum number, because the rock mass is of very poor strength against cyclic loads and movements [68,404]. Some researchers studied the effects of excavation sequences on the rock stress distribution at different points of surrounding rock mass [68]. Some suggested that the excavated geometry should match the initial distribution and orientation of field stresses [194,212,432]. Zhu *et al.* [52] conducted comparative studies on multiple schemes of different excavation sequences, and obtained an optimum scheme for improving the surrounding rock mass. Yu and Yu [433] presented a new concept of rock memory and emphasised the importance of analyses on the functions of surrounding rock masses. The basic principles and applications of construction mechanics are briefly discussed in the following sections.

## 7.1. BASIC PRINCIPLES OF INTERACTIVE CONSTRUCTION MECHANICS

### *7.1.1 Basic principles*

It has been commonly recognised that the schemes and sequences of rock excavations and supports have significant effects on the stabilisation and cost of underground rock engineering projects. The approach of the New Austrian Tunnelling Method (NATM) actually reflects the significant effects imposed by excavation activities. NATM emphasises the effects of excavation methods and attends to the economical aspects of construction [438]. It does not, however, examine the mechanisms of construction from a wider viewpoint of philosophy. It is necessary to examine a new concept in rock mechanics and engineering – interactive construction mechanics of rock engineering.

The concept of interactive construction mechanics applies the rock mechanics theory to rock engineering practice, by examining the interactive mechanics in the excavation and support process [8,9,434]. The principles of the interactive construction mechanics are outlined in detail in the following paragraphs.

(a) Rock engineering construction under complex rock mass conditions is a complete open system. This open system is affected by the uncertainty in natural geological factors. As a result, analyses of rock mass stability and estimation of construction cost become a systematic work. In order to take the full consideration of the effects of various factors, it is required to study not only the natural factors (e.g., geological conditions, initial stress and mechanical properties of rock mass), but also human factors (excavation sequence, excavation method and geometry). The main idea of this principle is to view a rock engineering project as an open and interactive system rather than a close and static system. The interactive viewpoint emphasises both the natural factors and the human activities. For example, optimisation methods may be adopted in advance at the design stage to minimise the potential problems. At construction stage, pre-measures may be taken to reduce the possibility of anticipated problems. Optimisation analyses should be conducted in advance.
(b) Rock mass stability and economic aspects during and after construction are not only related to the eventual states, but also to the excavation sequences and methods. This is because the boundaries of excavated rocks change in time and in space. From the mechanics point of view, construction is a non-linear process that is related to both the eventual states and stress paths as well as stress histories. The sequences of large cavern construction under complex geological conditions can significantly influence the safety and economy of the caverns during excavation and operation. The cavern excavation is actually a process of loading and unloading the rock masses at different time and positions. As the surrounding rock mass is a non-linear mechanical medium, the different excavation sequences imply different stress paths and histories imposed on the rock mass. Hence, it produces very different damage to the rock mass. The difference in damage produced by the various sequences of excavation may be very significant.
(c) The approach should be adopted before the construction to determine the optimum excavation schemes for final decision-making. In the optimisation analyses, rock supports should be considered as an important factor. Optimisation analyses on the excavation sequences should be taken after the cavern design but before the actual construction, in view of the mechanics of excavation process.
(d) The design and control of construction process have significant effects on the rock engineering projects. It is important to understand the interaction and the response of surrounding rock masses, to adopt proper excavation methods and

rock supporting measures, and to minimise the potential instability. It emphasises the control on the construction process. Careful considerations must be taken to stabilise the rock masses by choosing very suitable construction technologies and methods, such as charge weight of blasting, instant rock bolting and shotcreting, and groundwater drainage. That is, the rock mass damages resulting from the excavating and supporting activities should be minimised.

(e) The optimum construction scheme should be continually modified during the construction based on observation and monitoring of the surrounding rock mass. The information on rock mass conditions shall be updated during the construction, so that the existing scheme can be continually assessed and improved. The actual rock mass conditions may be different from expected conditions during excavation due to the high uncertainty in natural geological environment. The traditional site investigation techniques can only access to a very small quantity of rock mass. Hence, in situ monitoring can be used to verify and modify the predictions of the rock mass. Furthermore, the new information obtained in monitoring can be used for back analyses, or statistical analyses. Accordingly, modifications can be made to the originally proposed mechanical model of rock masses, geological and mechanical parameters, and eventually to the rock structure design and rock supporting measures. The modifications should be continually conducted during the excavation.

(f) Site investigation, design, construction and research shall be fully combined within an integrated system. It is unwise to strictly follow a specific scheme and to ignore the changes in rock mass conditions. Instead, modification of the existing scheme should be continually carried out according to the current conditions. This principle emphasises the policy guarantee for the implementation of the above principles. It is because the construction of large-scale complex rock projects requires interactive management and close cooperation between geologists, design engineers, contractors, researchers, clients and site engineers. One of the differences between rock engineering projects and other civil engineering projects is that rock engineering projects require continual updating of the site data and modifications to the initially proposed design schemes. The excavation methods and schemes should have the flexibility for timely updating and modifications, which need to be reflected in the design codes and the construction specifications.

### *7.1.2 Engineering applications*

Applications of the construction mechanics to an engineering project are illustrated in this section to demonstrate the necessity and importance of optimisation analyses on excavation sequences.

***7.1.2.1 Description of the project.*** The hydroelectric power project is a large underground cavern complex, consisting of a main powerhouse cavern, a transforming cavern and a tailrace cavern. The powerhouse cavern is 31.2 m wide and 72.6 m high, the transformer chamber is 17.4 m wide and 36.0 m high, and the surge chamber is 21.7 m wide and 76.9 m high. Bolts and shotcrete are needed for the rock reinforcements. The rock type is syenite. The in situ horizontal field stress at the main cavern is about 25 MPa. The cavern complex is 200 m deep below ground [52]. The initial in situ stress distribution is analysed by the method described in Chapter 6. The plane strain model is used to analyse the changes in stress fields disturbed by different excavation sequences. The finite element mesh is composed of 815 quadrilateral elements and a few triangular elements. The central part of the mesh around the caverns is shown in Figure 7.1. An improved two-dimensional non-linear finite element program, is used in the computation. The mechanical properties of the rock masses are summarised in Table 7.1. Figure 7.2 shows the constitutive model of the rock mass. The strength criterion shown in Figure 7.3 is mainly based on the

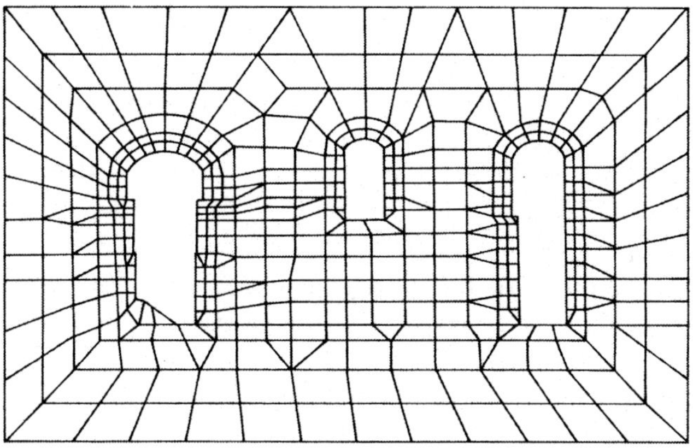

**Figure 7.1.** FEM meshes of the surrounding rock masses.

**Table 7.1.** Mechanical properties of the rock mass.

| Mechanical properties | Elastic properties | Residual properties |
|---|---|---|
| Young's modulus $E$ (MPa) | $3.5 \times 10^4$ | |
| Poisson's ratio $\mu$ | 0.2 | |
| Internal frictional angle $\varphi$ (°) | 60 | 54 |
| Cohesion $C$ (MPa) | 5.0 | 1.25 |
| Tensile strength (MPa) | 1.7 | 0 |

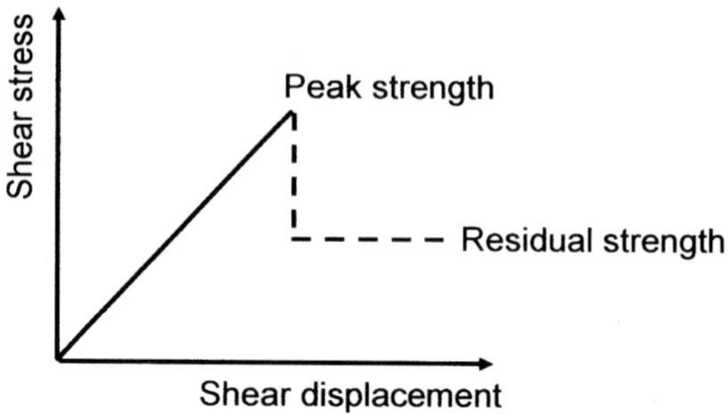


**Figure 7.2.** Constitutive model of rock masses.

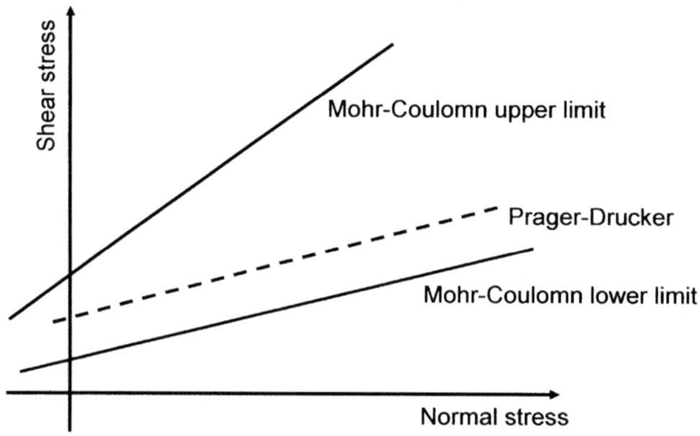


**Figure 7.3.** Strength criterion.

Prager–Drucker criterion, which is further modified by increasing the cohesion to a reasonable value. The distributions of bolts and pre-stressed cables are shown in Figure 7.4 [52]. The contributions of rock bolts to rock mass stiffness are treated by equivalent area and stiffness in the computational model, while the contributions to rock mass strength are treated by increasing the cohesion (internal frictional angle remains unchanged). These treatment methods adopted have been discussed in Section 3.3 of Chapter 3.

***7.1.2.2 Computational implementation and results for different excavation sequences.*** A total number of 11 computational schemes are compared to examine the influences of different excavation sequences on the rock mass stability. This includes the schemes of one-, five- or six-stage excavation. The detailed explanations

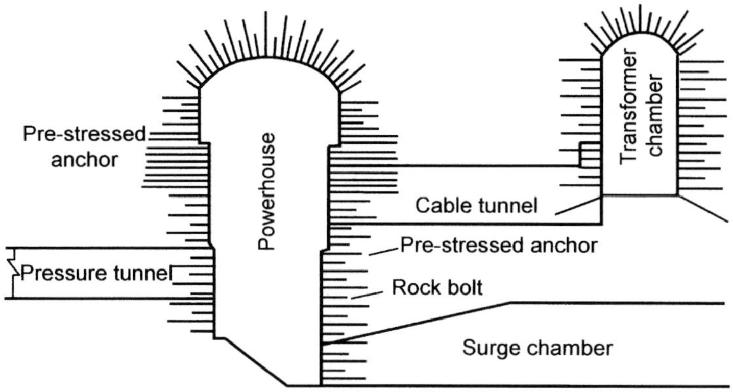


**Figure 7.4.** Installation scheme of rock bolts and cables.

to the different schemes are shown in Table 7.2 and Figure 7.5. The computational results on some of the schemes are compared and discussed in the following sections:

*(i) Non-linear one-stage excavation* As an extreme case, the scheme of non-linear one-stage excavation is firstly considered. The stress in the surrounding rock mass increases, and becomes greater than the strength. The over-stressed area continually expands into the surrounding fields. The damage zone is larger than that obtained from elastic analysis. A large plastic zone connecting the three caverns eventually forms, as shown in Figure 7.6.

*(ii) Non-linear five-stage excavation (5–I and 5–II)* The excavation sequence is given in Table 7.2. The computational results indicate that the damage area is significantly reduced in comparison with the one-stage excavation scheme. The difference in displacements at cavern corners also decreases. However, the damage area around each cavern is still connected, forming a continuous area. After the rock bolts are taken into account in the computation, the results indicate that the damage areas around the left two caverns are disconnected, but those around the right two caverns are still inter-connected, as shown in Figure 7.7. This implies that the results are not perfect.

*(iii) Non-linear six-stage excavation (6–I, 6–II and 6–III)* This is the suggested excavation scheme by the design engineers. It is divided into three sequences of excavation: 6–I, 6–II and 6–III (Table 7.2). The eventual stress state for Scheme 6–I is shown in Figure 7.8. Although the total damage area is smaller than that of one-stage scheme, the damage areas between the three caverns are still connected. In this six-stage scheme, the damage area is in fact slightly larger than that of the five-stage scheme. This implies that more excavation stages do not always lead to less damage

**Table 7.2.** Schedule of different excavation schemes and steps for three caverns.

| | | No 1 | No 2 | No 3 | No 4 | No 5 | No 6 | Note | |
|---|---|---|---|---|---|---|---|---|---|
| 1 | (1–I) | A,B,C,D,E, F,G,H,I,J, K,L,M,N,P | | | | | | Elastic one-step excavation | |
| | (1–II) | | | | | | | | |
| 2 | (1–III) | A,B,C,D,E, F,G,H,I,J, K,L,M,N,P | | | | | | Elasto-plastic one-step excavation | Elasto-plastic (rows 2–11) |
| 3 | (5–III) | A,I | B,F,J | G | H,K,L,M | C,D,E,N,P | | | |
| 4 | (6–I) | A | I | B,F,J | C,D,E | G,K,L,M,N,P | H | The originally proposed scheme | |
| 5 | (6–II) | A,I | B,F,J | G | H,K,L,M | N,P | C,D,E | | |
| 6 | (5–III) | A,I | B,J | C,K | D,F,L | E,G,M,N | H,P | | |
| 7 | (5–II) | A,I | B,F,J | G | H,K,L,M | N,P | C,D,E | No tunnels | |
| 8 | (5–II) | A,I | B,F,J | G | H,K,L,M | C,D,E,N,P | | Having rock bolts (only short bolts render support to rock mass) | |
| 9 | (5–II) | A,I | B,F,J | G | H,K,L,M | C,D,E,N,P | | Having condensed rock bolts (both short and long bolts render support, but they are not pre-stressed) | |
| 10 | (6–III) | A,I | B,J | C,K | D,F,L | E,G,M,N | H,P | Having condensed rock bolts and pre-stressed rock cables | |
| 11 | (6–III) | A,I | B,J | C,K | D,F,L | E,G,M,N | H,P | Having condensed rock bolts and pre-stressed rock cables; subject to additional load induced by crane beams | |

Note: assume that tunnels are completed at the first step of excavation.

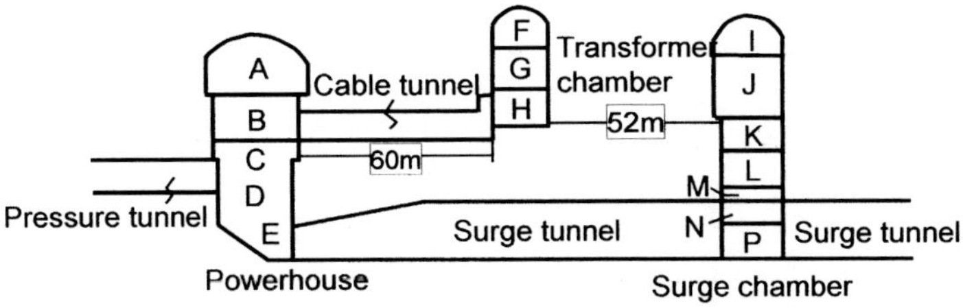


**Figure 7.5.** Different schemes for multiple-stage excavation.

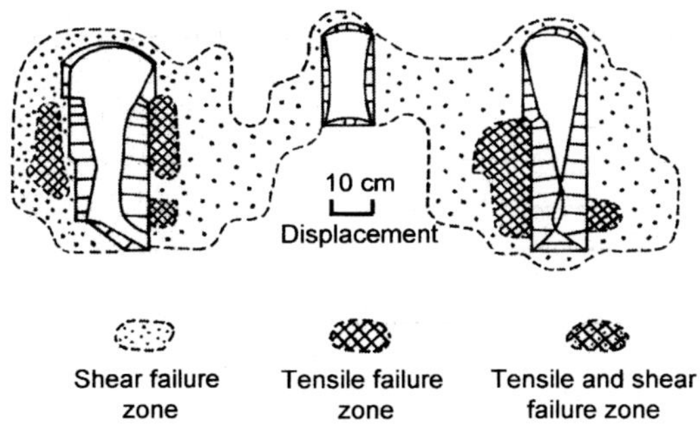


**Figure 7.6.** Damage zones and displacements of one-stage excavation without support (non-linear analysis).

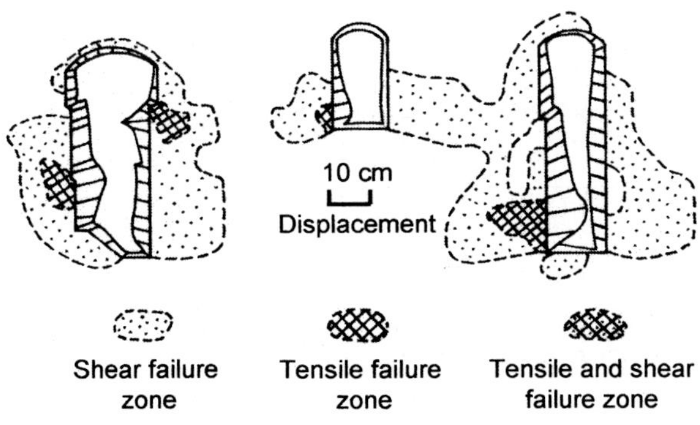


**Figure 7.7.** Damage zones and displacements after five-stage excavation with anchoring of rock bolts (non-linear analysis).

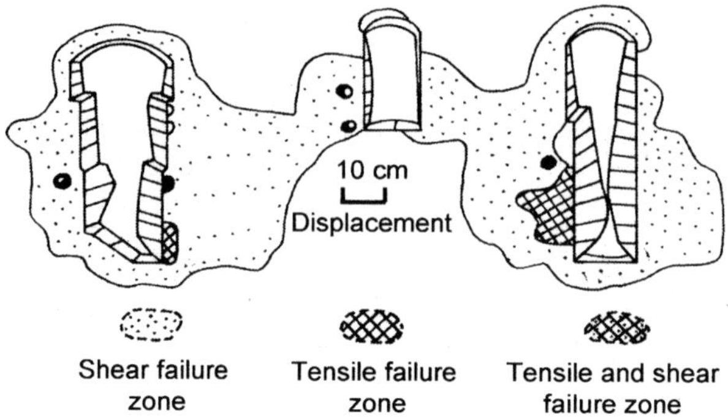


**Figure 7.8.** Damage zones and displacement of last stage of six-stage excavation without anchoring (non-linear analysis).

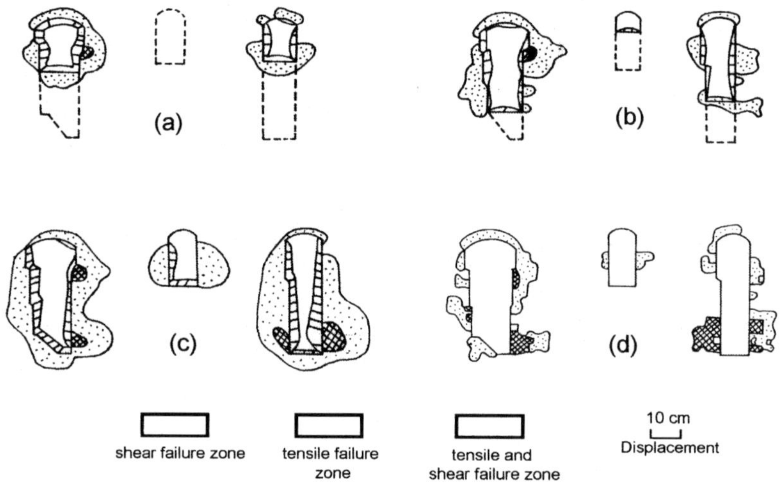


**Figure 7.9.** Damage zones of Scheme 6-III, (a) at excavation stage 2, (b) at excavation stage 4, (c) at excavation stage 6, (d) at excavation stage 6 with reduced bolt spacing.

of the surrounding rock mass. Scheme 6–III is suggested based on the analysis and experience. The two caverns at the left and right sides are first excavated, and the cavern at the centre is excavated later. Figure 7.9 illustrates the damage area and displacement distribution at Stages 2, 4 and 6 in the excavation process. It can be seen that the effects are very significant in isolating the damage areas between the three caverns, reducing the damage zones around each cavern (within 20 m thickness) and reducing the displacements at the cavern perimeter (half the displacement in the previous case). Figure 7.9d indicates that the rock mass stability is further improved after reducing the spacing of rock bolts and cables. The damage areas are therefore significantly reduced such that many of them become isolated individual small areas.

***7.1.2.3 Discussions.***

(a) The results from numerical analyses show that different excavation sequences produce very different effects on the rock mass stability. The one-stage excavation may generate large damage areas and displacements at the cavern perimeter. This implies that the stress history plays an important role, and the optimisation of excavation sequences is necessary.
(b) In the situation of multiple cavern excavations in rock mass with high horizontal stress, in order to minimise the damage zone, excavations should be conducted at different times. In concurrent excavations, the excavations should be kept at far distance to reduce the interaction between the excavations.
(c) The excavated volume at each stage of the excavation should be limited to a reasonable amount in order to reduce the area of over-stressing. In addition, rock support measures such as rock bolt and shotcrete should be applied instantly. All of these have positive effects on the rock mass stability.
(d) At the connections between caverns and tunnels, three-dimensional numerical analyses should be conducted, because the two-dimensional analyses may not give quantitative results with the simplified model.
(e) Since the above method of excavation optimisation cannot consider the energy loss in rock mass, the damage areas are generally over-estimated. So they should be normally used as the reference values for comparisons between different excavation schemes. Back analysis from monitoring can overcome the short-coming of the above method.

## 7.2. APPLICATIONS OF INTERACTIVE PROGRAMMING IN OPTIMISATION OF CAVERN CONSTRUCTION

The concepts and principles of the interactive construction mechanics of rock engineering, together with the applications to an engineering project have been presented in the previous section. However, the method used has its limitations. For example, the schemes for optimisation analyses are selected based on engineering experiences. It is possible that the optimal scheme obtained with this method is a local optimal scheme within the limited number of proposals rather than a global optimal scheme. In this section, the interactive programming principle and associated methods are discussed and applied to construction optimisation.

### *7.2.1 Principles of interactive programming*

The method of interactive programming, originally proposed by Bellman [439] is a specific technique solving the optimisation problems. Generally, optimisation is the process of finding the best solution to the problems from a group of possible solutions according to the given requirements. Correspondingly, the interactive programming is a multistage sequential decision-making process of finding the best solution.

An optimal policy has the property that whatever the initial state and initial decision are, the remaining decisions must constitute an optimal policy with regard to the state resulting from the first decision. This is known as Bellman's Principle of Optimality [439].

The main advantages of the interactive programming method are:

(a) An optimisation problem of multiple ($n$) dimensions can be changed into $n$ problems of one dimension, which may be solved in a sequential manner. This solution-searching procedure cannot be implemented with the traditional optimisation methods.
(b) The global maximum or minimum value can be directly determined. By contrast, the traditional optimisation methods possibly give the local maximum or minimum value, which leads to the difficulty of judging the globality or locality of the solutions obtained. In fact, the interactive programming method is not an 'algorithm' but an approach to break and reconstruct the problems so that a suitable optimisation method can be applied.

The following technical terms and statements are commonly used in the interactive programming method in describing the optimisation problems.

A physical system can always be taken as a hierarchical system in a certain sequence. The interactive programming method divides the optimisation problem into a series of *stages*, which may represent, for example, time or space. At each stage, the *state* of system can be depicted by a relatively small group of variables. These variables are termed as *state variables* or *state vectors* (or simply state). Single or multiple *decisions* need to be made in any state of system at each stage. These decisions may depend on either the stage or the state, or both, while the history during which the system comes to the current state or stage is not important. In other words, the decision-making is only based on the current stage or state. After a decision is made, a *benefit* is obtained correspondingly. At the same time the system changes to the next stage. The benefit is governed by a known single-value function of a certain given state. Similarly, the state of the system after *alteration* is governed by a single-value function of the current state altered. The eventual objective of hierarchical development of the system stages is to find the maximum or minimum values of the function of state variables. The key elements related to the interactive programming are *stage*, *state*, *decision*, *alteration* and *benefit*.

(a) Stage: The concept of stage is required in representing the sequence of decision-making. The function of state is to give the number of the sequences involved in the system developing process. Many problems seemingly without stage nature can still be analysed in a stage manner. For example, during the process of excavation, several excavation phases may be combined into one excavation stage. The optimisation problems of multiple variables can bc treated as a process of multiple decision-making according to a certain sequence. During this process, each step in the sequence can be represented by a stage, which depends on the current state.

(b) State: State space can be represented as a non-nil combination $\Lambda_1$. An element $\lambda \in \Lambda$. This element is called as state, which is the description of a variable (or a group of variables). The state space $\Lambda$ is composed of all the state variables. For example, during the process of excavation, the state variables describe the current step in excavation sequences within the excavation system. The corresponding state space $\Lambda$ is composed of all the possible excavation steps.

(c) Decision: The system state must include all the information required for determining all possible decisions. Therefore, for each state $\lambda \in \Lambda$, there is a non-nil combination $X_\lambda$. It is termed as decision combination of $\lambda$. One of the elements, $X_3(\lambda) \in X_\lambda$ is a decision or a decision variable. It represents an allowable choice when the system is at the state $\lambda$. Decision combination $X_\lambda$ is composed of all the possible choices when the system is in the state $\lambda$. For example, as shown in Figures 7.10 and 7.11, the decision under the state P1T1T2T3 is T4 or P2.

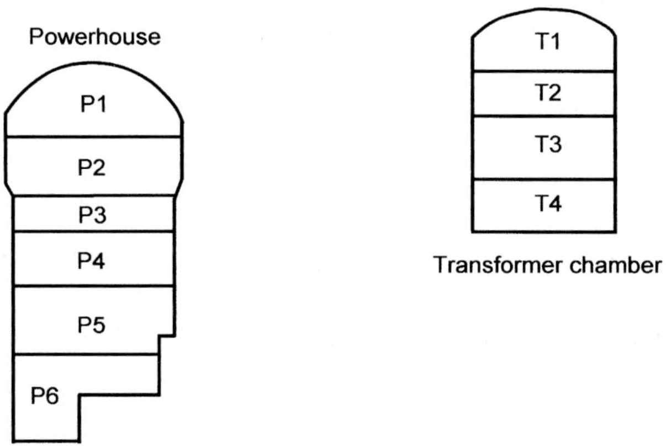


**Figure 7.10.** Excavation stages of the powerhouse and transformer chamber.

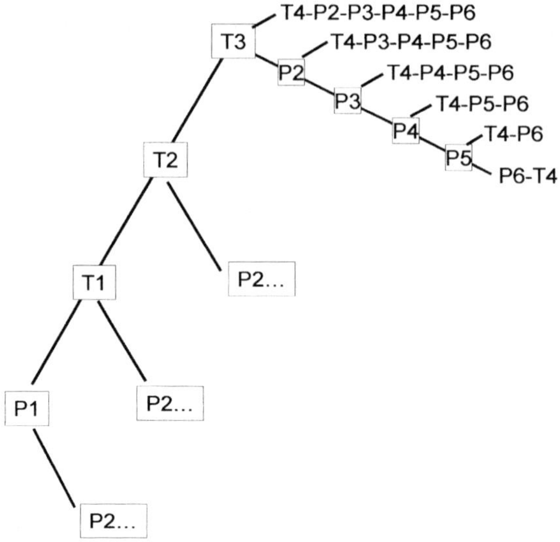


**Figure 7.11.** Possible combinations of excavation sequences.

(d) Benefit: Because a problem of dynamic programming is an optimisation problem, there may be an objective function to assess the given decisions. The total benefit is a combination (sum or product) of benefits at all the stages. It is the accumulation of state alteration (from one stage to another). The benefit function is changeable from one stage to another. There is a requirement for proper definition of the benefit at each stage. For example, for the optimisation of the cavern construction, the benefit function can be defined as the area of rock damage around the caverns.

### *7.2.2 Applications to the optimisation of cavern construction*

A hydropower station is chosen to demonstrate the applications of the interactive programming method. The station mainly consists of a powerhouse cavern (20 m span and 45 m height) and a transformer cavern (18 m span and 28 m height). The design excavation scheme includes six stages (numbered as P1, P2, ..., P6) for the power house cavern and in four stages (numbered as T1, T2, T3, T4) for the transformer cavern, respectively, as shown in Figure 7.10.

To simplify the optimisation problem, some assumptions are made as follows:

(a) Caverns are excavated in the sequence from top to bottom within each stage;
(b) Each excavation stage is considered as one excavation phase; and

(c) Stage P1 corresponds to the first phase of excavation for the powerhouse cavern, and so on.

A total of 168 possible construction schemes can be obtained. However, it is impossible to conduct finite element computations for all the 168 excavation schemes. In the past study, a limited number of schemes were selected for the finite element computations, and the computing results are compared to obtain a local optimal scheme. In the present study, a global optimal scheme is determined from $n$ number of possible schemes, and,

$$n = X_s N_c$$

where $X_s$ is the total number of excavation sequences, and $N_c$ is the number of caverns. For the case in this study, $n = 10 \times 2 = 20$. That is, only 20 finite element (FEM) computations are required to obtain the global optimal scheme.

Figure 7.11 schematically illustrates some of the possible combinations of excavation sequences. In the FEM computation, rock mass is treated as an isotropic elastic-plastic medium. The mechanical properties are shown in Table 7.3.

The in situ field stress is calculated based on the back analysis conducted in Section 6.5 of Chapter 6,

$$\sigma_x = 9.43\,\text{MPa}, \quad \sigma_y = 7.43\,\text{MPa}, \quad \tau_{xy} = 1.24\,\text{MPa}$$

To investigate the effects of in situ stress on selecting the construction sequences, four conditions of stress field are assumed by changing the angle $\alpha$ between $\sigma_1$ and horizontal direction. These are: (a) $\alpha = 25.2°$, (b) $\alpha = 0°$, (c) $\alpha = 90°$, and (d) $\alpha = -25.2°$, as shown in Figure 7.12.

The benefit function is defined by the area of rock damaged around the cavern. Figure 7.13 illustrates the optimisation process for Case (a), where the value in the bracket is the benefit. Figure 7.14 shows the distribution of damage zones around the cavern corresponding to the four cases in Figure 7.12. Table 7.4 shows the computational results for the four conditions of field stresses of Figure 7.12.

**Table 7.3.** Mechanical properties of the surrounding rock mass.

| Parameter | $E$ (MPa) | $\nu$ | $c$ (MPa) | $\phi$ (°) | $R_t$ (Mpa) |
|---|---|---|---|---|---|
| Value | 35000 | 0.22 | 1.30 | 50.20 | 5.0 |

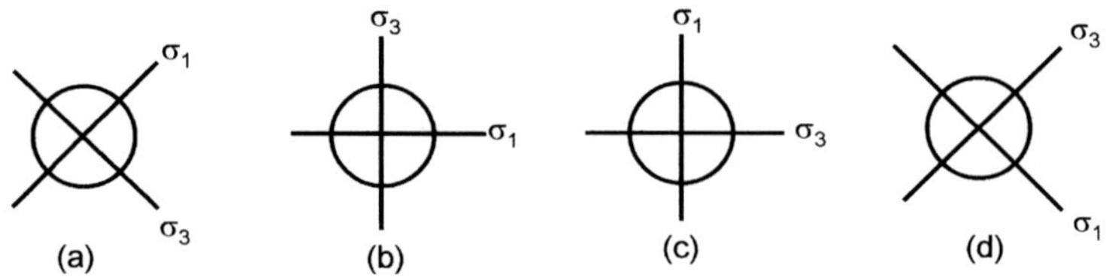


**Figure 7.12.** Different distributions of principal stresses.

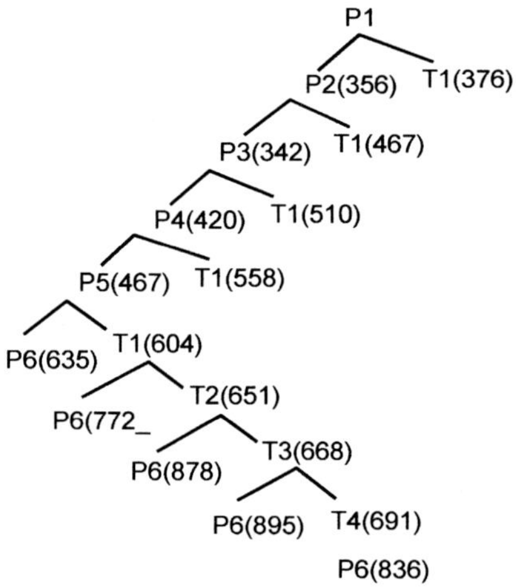


**Figure 7.13.** Optimisation process for excavation case (a) in Figure 7.12.

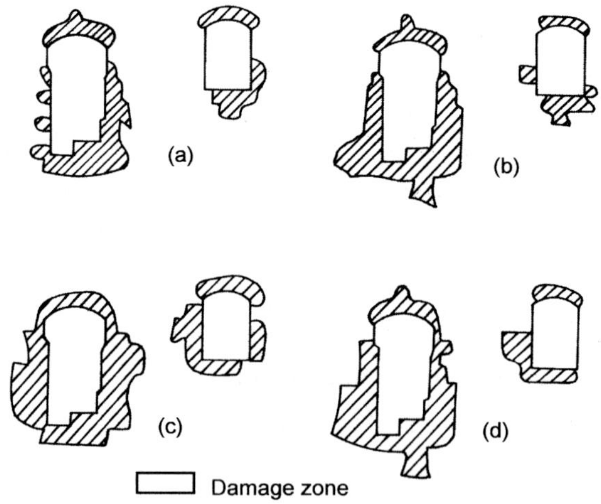


**Figure 7.14.** Damage zones in surrounding rock masses corresponding to cases in Figure 7.12.

**Table 7.4.** Computational results for different stress field in Figure 7.12.

| Stress field | (a) | (b) | (c) | (d) |
|---|---|---|---|---|
| Optimal scheme | P1...P5T1 ...T4P6 | The same as (a) | P1P2T1T2P3 T3T4P5P6 | The same as (a) |
| Damage area ($m^2$) | 836 | 1119 | 1370 | 1282 |
| Maximum horizontal relative displacement | 1.24 | 1.46 | 0.87 | 1.44 |

**Table 7.5.** The eventual damage area and displacements for different construction schemes.

| Schemes | 1 | 2 | 3 | 4 | 5 |
|---|---|---|---|---|---|
| Damage area ($m^2$) | 1730 | 859 | 928 | 938 | 836 |
| Maximum horizontal relative displacement | 1.90 | 1.21 | 1.36 | 1.23 | 1.24 |

To compare the results for different construction schemes under the same stress field (a), the following schemes are first selected:
Scheme 1: PT (one-step full-face excavation)
Scheme 2: P1-P2-P3-P4-P5-P6-T1-T2-T3-T4
Scheme 3: P1-T1-P2-T2-P3-T3-P4-T4-P5-P6
Scheme 4: P1-P2-T1-P3-P4-T2-P5-P6-T3-T4 (actual construction)
Scheme 5: P1-P2-P3-P4-P5-T1-T2-T3-T4-P6 (optimal construction)

The damage area and the displacement at the cavern edges for the above five schemes are compared in Table 7.5.

***7.2.2.1 Discussions.*** The interactive programming method is more efficient than the traditional enumerating method in the optimisation of large volume underground excavation. Although it cannot be totally ensured that the obtained optimal scheme is the real global one due to constraints from the assumptions and the specially defined benefit function, the result can at least be taken as the relatively global optimal scheme under the assumed conditions. Improvement is still required in the future.

The following conclusions can be drawn from this study:

(a) After the excavation steps are determined, the optimal excavation scheme is significantly influenced by the in situ stress magnitude and direction.
(b) The optimal scheme searching method used in the present study largely shortens the computation time in comparison with the traditional enumerating method.

The required computing time is proportional to the product of the total excavation steps and cavern number.

(c) The benefit function needs to be properly pre-determined. For example, it can be the convergent displacement, damage area, or both. On the other hand, because the benefit function adopted in the present study is the damage area at each step of excavation rather than the accumulated damage area after the last excavation step, it may not be ensured to find the global optimal scheme. Global optimal scheme can be obtained by combining the benefit functions.

(d) In the present study, only stress field is taken into consideration. In the cases of presence of major geological structures, further investigation is required.

(e) The results in this study are obtained through two-dimensional numerical computation. They provide qualitative basis for three-dimensional FEM problems of cavern construction. In fact, it is very difficult to fully solve the optimisation problems by three-dimensional numerical modelling. The possible approach to the three-dimensional optimisation problems is to compare the computational results for a few representative construction schemes.

(f) It can be seen from Table 7.5 that the cavern layout and excavation steps have been arranged in such a way that an optimal construction scheme can be achieved. However, the actual construction may not follow the optimal scheme. The damage area in the actual construction is 12% larger than that in the optimal construction scheme, while the maximum horizontal displacement at sidewalls of the cavern is almost the same.

It should be noted that the method in this study provides the optimal scheme for construction sequences in view of rock mass stability of the caverns. Before the method is used, the excavation steps have been pre-determined. Cavern construction optimisation shall be conducted by considering both the rock stability and the economics, based on the site conditions.

## 7.3. ARTIFICIAL INTELLIGENCE TECHNIQUES IN CONSTRUCTION OPTIMISATION

In the previous section, the interactive programming theory is applied and engineering case studies are conducted to illustrate the method of cavern construction optimisation. However, there are still two problems to be solved: (a) determining all the possible construction sequences; and (b) creating the input data files required by the numerical computations according to the selected construction sequences. Traditionally, this work is manually implemented with very low efficiency. The artificial intelligence technique has the capabilities of reasoning, understanding, planning, decision-making and learning [1–3,440–443]. Its applications to cavern construction optimisation can help to solve the above problems. In this section, the artificial intelligence technique is utilised in such aspects as automatic determination of construction sequences, automatic generation of input data files and automatic decision-making for construction optimisation.

### *7.3.1 Artificial intelligence language prolog*

Prolog is a very important language tool in artificial intelligence programming, and is mainly used to deal with logic problems. Prolog language enables computers to have reasoning capability in problem solving. Since it was produced in 1970s, it has quickly become a popular language of artificial intelligence. Prolog is similar to natural languages, and is therefore easy to learn and utilise by the users. The computer program written with this language is simple and understandable. The reasoning capability of prolog has lead to wide applications in the fields of database, logic algorithm, expert system, natural language interpretation, and automatic programming.

One of the typical prolog languages is Turbo Prolog. Turbo Prolog has kept the main advantages of the traditional explanation type prolog, with enhanced running-speed improved functions.

In comparison with other computer languages such as Pascal, Turbo Prolog is a more advanced language. The program written with Turbo Prolog is 10 times shorter than that with Pascal for the same problem. This is because Turbo Prolog has an internal pattern-matching mechanism, and a simple but effective method to deal with recurrent process. These two advantages are utilised in this study for construction sequence determination.

### *7.3.2 Problem solving algorithm in cavern construction optimisation*

***7.3.2.1 Automatic determination of cavern excavation sequences.*** Determining cavern excavation sequence is to find out all the possible excavation sequences. Suppose that a cavern complex comprises Cavern A and Cavern B. Caverns are divided into different excavated faces $a_1, a_2, \ldots, a_6$ and $b_1, b_2, \ldots, b_4$, as shown in Figure 7.15. The basic rules for excavation are (Zhu *et al.* 1992):

(a) Excavation face $a_1$ is the first step;
(b) One face is excavated as one step;
(c) Each cavern is excavated from top to bottom.

A data table is the basic structure in Turbo Prolog language. It is equivalent to the data group in Pascal language. This table consists of a set of sequential elements. The order of data in the table is an important feature of table. In this context, the

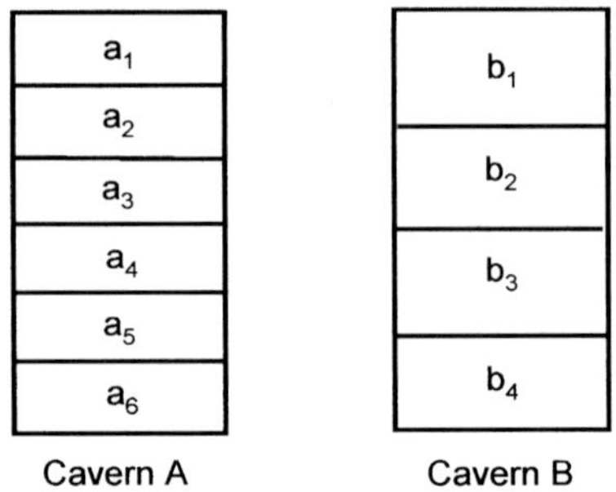


**Figure 7.15.** Stepwise excavation of cavern complex.

excavated faces in each excavation step are represented by the elements of the table, and the excavation sequence is represented by the element order in the table. For example, a table $L = [a_1, b_1, a_2, b_2, a_3, b_3, a_4, b_4, a_5, a_6]$ represents that: (a) the excavated face at first excavation step is $a_1$, (b) the face at the last step is $a_6$, (c) the element sequence in the table is the excavation sequence, and (d) there are a total of 10 excavation steps.

For the initially generated table $[a_1, a_2, a_3, a_4, a_5, a_6, b_1, b_2, b_3, b_4]$, the element order in the table is quickly determined according to the above three rules. The results are written into the output file 'order.out', which includes the combination of all the possible excavation sequences.

To implement the above process of sortation in the table, eleven predicates are defined. It is commonly known that a sentence in computer program is of two-fold characteristics: description and operation. When a computer program is constructed, people always pay attenuation to its operation. However, when the program needs to be verified and explained, people always concern about its description. When a logic algorithm is constructed, one of the methods for combining the two characteristics is to construct the program by considering its operation at first instance, and then translate it into descriptive sentences.

A computer program is generally constructed from top to bottom according to the sequential steps of problem solving. In other words, a general problem is firstly broken into several sub-problems, and then the sub-problems are individually solved. This is a natural process to construct computer programs.

Describing the logic in the table order is to find the sequential position exchange of the elements in the table. The logic module is 'sort (Xs, Ys)', where Ys is the table obtained when all the elements in Xs satisfy the 'ordered' condition.

sort (Xs, Ys): – permutation (Xs, Ys), ordered (Ys)

The statement 'sort (Xs, Ys)' has been broken into 'permutation (Xs, Ys)' and 'ordered (Ys)', which are further explained as below.

Checking whether a table is arranged in the right order according to given requirements can be implemented with the following sentences:

ordered ([-]).
ordered (Ys): —
is_after ($a_2$, $a_1$, Ys), / *$a_2$ and Ys are after $a_1$ */
is_after ($a_3$, $a_2$, Ys),
is_after ($a_6$, $a_5$, Ys),
is_after ($b_1$, $a_1$, Ys),
is_after ($b_2$, $b_1$, Ys),
is_after ($b_4$, $b_3$, Ys),

'Fact' indicates that only the tables consisting of one element have been arranged in order, while 'rule' indicates that the order must satisfy the three assumptions.

Permutation represents a manner of element position exchange in the table. That is, an element is randomly selected, then it is taken as the first element in the table, and then the other elements are iteratively exchanged. Its basic fact indicates that a nil table is the only exchange. Permutation ([$a_1$, ..., $a_n$], L) feeds back the result that L has a total number of $n!$ solutions.

Permutation ($X_s$, [Z|$Z_s$]): —select (Z, $X_s$, $Y_s$), permutation ($Y_s$, $Z_s$)
Permutation ([ ], [ ])

Determining the relative positions of two elements involves three predicates: member, position and is_after. Member is to determine whether an element is involved in the table:

member (X, [X|$X_s$]),
member (X, [Y|$Y_s$]): —member (X, $Y_s$).

The above program means: if X is the first part of the table, or X is the element of the last part of the table, then X is an element in the table.

Position is to calculate the position order of an element in the table:

position (A, [A|_], 1),
position (A, [_|T], N): —
member (A, T),
position (A, T, No),
N = No + 1

It is specified that A is placed at the first position if A is the first element in the table. The meaning of 'is_after ($A_2$, $A_1$, L)' is: if the position number of $A_2$ is bigger than that of $A_1$, $A_2$ is after $A_1$ in the table.

The result of selecting X from [X|$X_s$] is $X_s$, or, if the result of selecting X from $Y_s$ is $Z_s$, then the result of selecting X from [Y|$Y_s$] is [Y|$Z_s$], i.e.,

```
select (X, [X|Xs], Xs)
select (X, [Y|Ys], [Y|Zs]): — select (X, Ys, Zs)
```

'Writelist (L)' and 'write10 (list, N)' mean that a table is exported and each line of the table contains N symbols.
'Run' means that the results of order arrangement are saved in the file 'order.out'.
'Make-cut (L): $-L = L_0$' (cut/fail method) means that the program is repeatedly run until L is equal to $L_0$. $L_0$ can be displayed on the screen in advance.

In general, the above method can be described as: the initial table of excavation sequence is processed through Predicate 'sort', and then a table of order arrangement is produced according to the 'ordered' rule.

The program Autoorder.PRO for generating the excavation sequence is described as below:

---

```
Autoorder.PRO
code = 10000
trail = 4000
errolevel = 0

domains
A = symbol
N = integer
file = myfile
list = symbol*

predicates

run
ordered (List)
member (A, List)
make_cut (List)
write10 (List)
sort (List, List)
writelist (List)
is_after (A, A, List)
position (A, List, N)
select (A, List, List)
permutation (List, List)
```

```
clauses

position (A, [A|_], 1).
position (A, [_|T], N): —
member (A, T),
position (A, T, NO),
N = NO + 1.

is_after (A2, A1, L): —
position (A1, L, K1),
position (A2, L, K2),
K2 > K1.

member (Name, [Name|_]).
member (Name, [_|Tail]): —
member (Name, Tail).

select (P, [P|Xs], Xs).
select (P, [Y|Ys], [Y|Zs]): —
select (P, Ys, Zs).

permutation (Xs, [Z|Zs]): —
select (Z, Xs, Ys),
permutation (Ys, Zs).
permutation ([ ], [ ]).
writelist (NL): — write10(NL, 0), nl.
write10 (TL, 10): — !, write10 (TL, 0).
Write10 ([H|T], N): — !, write (H, " "),
N1=N+1, write10 (T, N1).
Write10 ([ ], _).

ordered ([ _ ]).
ordered (Ys): —
is_after (a2, a1, Ys),
is_after (a3, a2, Ys),
is_after (a4, a3, Ys),
is_after (a5, a4, Ys),
is_after (a6, a5, Ys),
is_after (b1, a1, Ys),
is_after (b2, a1, Ys),
is_after (b3, b2, Ys),
is_after (b4, b3, Ys),
```

```
sort (Xs, Ys): —
permutation (Xs, Ys),
ordered (Ys).

run: —
openwrite (myfile, "order.out"),
sort ([a1, a2, a3, a4, a5, a6, b1, b2, b3, b4], L),
writedevice (myfile),
writelist (L)
make_cut (L), !,
closefile (myfile).

Make_cut (L): —
L = [a1, b1, b2, b3, b4, a2, a3, a4, a5, a6].

goal
run.
```

---

It can be seen that the program is simple and easy to understand. Figure 7.16 shows the computational results on all the possible construction sequences (a total of 126 schemes). It only requires about one minute to run the program in a 486 computer. By comparison, the manual method is time-consuming and some schemes may be missed. Figure 7.17 schematically illustrates partial manual results by manual sorting.

***7.3.2.2 Automatic generation of data files for finite element computation.*** The finite element program used in this study is written by the authors. The structure of data files required for determining the excavation sequences is shown in Figure 7.18.

Each data file above is divided into two files: const.dat and excav.dat. The first one is the same for each excavation scheme, and the second one is changeable with different excavation schemes. The information on multi-stage excavation is saved into data files named according to the excavated faces. For example, Data file a1 contains the information on excavating face a1. Data files a1, a2, a3, a4, a5, a6, b1, b2, b3, b4 are then processed to form a complete file excav.dat, which represents excavation scheme [a1, a2, ..., a6, b1, b2, ..., b4]. Two predicates are used in the processing: 'file_str (Filename, Text)' and 'form_fort.file (L)'. The 'file_str (Filename, Text)' reads symbols from a data file and sends them into a variable or a file. The 'form_fort file (L)' defines a data file as the symbol at the top of the table (the first excavation step); and then sends the content of data file to a variable termed as 'text'. The 'text' is finally written into the data file 'excav.dat'. This process continues until

| | | |
|---|---|---|
| a1 a2 a3 a4 a5 a6 b1 b2 b3 b4 | a1 a2 b1 b2 a3 b3 a4 b4 a5 a6 | a1 a2 a3 a4 a5 b1 a6 b2 b3 b4 |
| a1 a2 b1 b2 b3 a3 a4 a5 b4 a6 | a1 a2 a3 a4 a5 b1 b2 a6 b3 b4 | a1 a2 b1 b2 b3 a3 a4 a5 a6 b4 |
| a1 a2 a3 a4 a5 b1 b2 b3 a6 b4 | a1 a2 b1 b2 b3 a3 a4 a5 b4 a6 | a1 a2 a3 a4 a5 b1 b2 b3 b4 a6 |
| a1 a2 b1 b2 b3 a3 a4 b4 a5 a6 | a1 a2 a3 a4 b1 a5 a6 b2 b3 b4 | a1 a2 b1 b2 b3 a3 b4 a4 a5 a6 |
| a1 a2 a3 a4 b1 a5 b2 a6 b3 b4 | a1 a2 b1 b2 b3 b4 a3 a4 a5 a6 | a1 a2 a3 a4 b1 a5 b2 b3 a6 b4 |
| a1 b1 a2 a3 a4 a5 a6 b2 b3 b4 | a1 a2 a3 a4 b1 a5 b2 b3 b4 a6 | a1 b1 a2 a3 a4 a5 b2 a6 b3 b4 |
| a1 a2 a3 a4 b1 b2 a5 a6 b3 b4 | a1 b1 a2 a3 a4 a5 b2 b3 a6 b4 | a1 a2 a3 a4 b1 b2 a5 b3 a6 b4 |
| a1 b1 a2 a3 a4 a5 b2 b3 b4 a6 | a1 a2 a3 a4 b1 b2 a5 b3 b4 a6 | a1 b1 a2 a3 a4 b2 a5 a6 b3 b4 |
| a1 a2 a3 a4 b1 b2 b3 a5 a6 b4 | a1 b1 a2 a3 a4 b2 a5 b3 a6 b4 | a1 a2 a3 a4 b1 b2 b3 a5 b4 a6 |
| a1 b1 a2 a3 a4 b2 a5 b3 b4 a6 | a1 a2 a3 a4 b1 b2 b3 b4 a5 a6 | a1 b1 a2 a3 a4 b2 b3 a5 a6 b4 |
| a1 a2 a3 b1 a4 a5 a6 b2 b3 b4 | a1 b1 a2 a3 a4 b2 b3 a5 b4 a6 | a1 a2 a3 b1 a4 a5 b2 a6 b3 b4 |
| a1 b1 a2 a3 a4 b2 b3 b4 a5 a6 | a1 a2 a3 b1 a4 a5 b2 b3 a6 b4 | a1 b1 a2 a3 b2 a4 a5 a6 b3 b4 |
| a1 a2 a3 b1 a4 a5 b2 b3 b4 a6 | a1 b1 a2 a3 b2 a4 a5 b3 a6 b4 | a1 a2 a3 b1 a4 b2 a5 a6 b3 b4 |
| a1 b1 a2 a3 b2 a4 a5 b3 b4 a6 | a1 a2 a3 b1 a4 b2 a5 b3 a6 b4 | a1 b1 a2 a3 b2 a4 b3 a5 a6 b4 |
| a1 a2 a3 b1 a4 b2 a5 b3 b4 a6 | a1 b1 a2 a3 b2 a4 b3 a5 b4 a6 | a1 a2 a3 b1 a4 b2 b3 a5 a6 b4 |
| a1 b1 a2 a3 b2 b3 b4 a4 a5 a6 | a1 a2 a3 b1 a4 b2 b3 a5 b4 a6 | a1 b1 a2 a3 b2 b3 a4 a5 a6 b4 |
| a1 a2 a3 b1 a4 b2 b3 b4 a5 a6 | a1 b1 a2 a3 b2 b3 a4 a5 b4 a6 | a1 a2 a3 b1 b2 a4 a5 a6 b3 b4 |
| a1 b1 a2 a3 b2 b3 a4 b4 a5 a6 | a1 a2 a3 b1 b2 a4 a5 b3 a6 b4 | a1 b1 a2 a3 b2 b3 b4 a4 a5 a6 |
| a1 a2 a3 b1 b2 a4 a5 b3 b4 a6 | a1 b1 a2 b2 a3 a4 a5 a6 b3 b4 | a1 a2 a3 b1 b2 a4 b3 a5 a6 b4 |
| a1 b1 a2 b2 a3 a4 a5 b3 a6 b4 | a1 a2 a3 b1 b2 a4 b3 a5 b4 a6 | a1 b1 a2 b2 a3 a4 a5 b3 b4 a6 |
| a1 a2 a3 b1 b2 a4 b3 b4 a5 a6 | a1 b1 a2 b2 a3 a4 b3 a5 a6 b4 | a1 a2 a3 b1 b2 b3 a4 a5 a6 b4 |
| a1 b1 a2 b2 a3 a4 b3 a5 b4 a6 | a1 a2 a3 b1 b2 b3 a4 a5 b4 a6 | a1 b1 a2 b2 a3 a4 b3 b4 a5 a6 |
| a1 a2 a3 b1 b2 b3 a4 b4 a5 a6 | a1 b1 a2 b2 a3 b3 a4 a5 a6 b4 | a1 a2 a3 b1 b2 b3 b4 a4 a5 a6 |
| a1 b1 a2 b2 a3 b3 a4 a5 b4 a6 | a1 a2 b1 a3 a4 a5 a6 b2 b3 b4 | a1 b1 a2 b2 a3 b3 a4 b4 a5 a6 |
| a1 a2 b1 a3 a4 a5 b2 a6 b3 b4 | a1 b1 a2 b2 a3 b3 b4 a4 a5 a6 | a1 a2 b1 a3 a4 a5 b2 b3 a6 b4 |
| a1 b1 a2 b2 b3 a3 a4 a5 a6 b4 | a1 a2 b1 a3 a4 a5 b2 b3 b4 a6 | a1 b1 a2 b2 b3 a3 a4 a5 b4 a6 |
| a1 a2 b1 a3 a4 b2 a5 a6 b3 b4 | a1 b1 a2 b2 b3 a3 a4 b4 a5 a6 | a1 a2 b1 a3 a4 b2 a5 b3 a6 b4 |
| a1 b1 a2 b2 b3 a3 b4 a4 a5 a6 | a1 a2 b1 a3 a4 b2 a5 b3 b4 a6 | a1 b1 a2 b2 b3 b4 a3 a4 a5 a6 |
| a1 a2 b1 a3 a4 b2 b3 a5 a6 b4 | a1 b1 b2 a2 a3 a4 a5 a6 b3 b4 | a1 a2 b1 a3 a4 b2 b3 a5 b4 a6 |
| a1 b1 b2 a2 a3 a4 a5 b3 a6 b4 | a1 a2 b1 a3 a4 b2 b3 b4 a5 a6 | a1 b1 b2 a2 a3 a4 a5 b3 b4 a6 |
| a1 a2 b1 a3 b2 a4 a5 a6 b3 b4 | a1 b1 b2 a2 a3 a4 b3 a5 a6 b4 | a1 a2 b1 a3 b2 a4 a5 b3 a6 b4 |
| a1 b1 b2 a2 a3 a4 b3 a5 b4 a6 | a1 a2 b1 a3 b2 a4 a5 b3 b4 a6 | a1 b1 b2 a2 a3 a4 b3 b4 a5 a6 |
| a1 a2 b1 a3 b2 a4 b3 a5 a6 b4 | a1 b1 b2 a2 a3 b3 a4 a5 a6 b4 | a1 a2 b1 a3 b2 a4 b3 a5 b4 a6 |
| a1 b1 b2 a2 a3 b3 a4 a5 b4 a6 | a1 a2 b1 a3 b2 a4 b3 b4 a5 a6 | a1 b1 b2 a2 a3 b3 a4 b4 a5 a6 |
| a1 a2 b1 a3 b2 b3 a4 a5 a6 b4 | a1 b1 b2 a2 a3 b3 b4 a4 a5 a6 | a1 a2 b1 a3 b2 b3 a4 a5 b4 a6 |
| a1 b1 b2 a2 b3 a3 a4 a5 a6 b4 | a1 a2 b1 a3 b2 b3 a4 b4 a5 a6 | a1 b1 b2 a2 b3 a3 a4 a5 b4 a6 |
| a1 a2 b1 a3 b2 b3 b4 a4 a5 a6 | a1 b1 b2 a2 b3 a3 a4 b4 a5 a6 | a1 a2 b1 b2 a3 a4 a5 a6 b3 b4 |
| a1 b1 b2 a2 b3 a3 b4 a4 a5 a6 | a1 a2 b1 b2 a3 a4 a5 b3 a6 b4 | a1 b1 b2 a2 b3 b4 a3 a4 a5 a6 |
| a1 a2 b1 b2 a3 a4 a5 b3 b4 a6 | a1 b1 b2 b3 a2 a3 a4 a5 a6 b4 | a1 a2 b1 b2 a3 a4 b3 a5 a6 b4 |
| a1 b1 b2 b3 a2 a3 a4 a5 b4 a6 | a1 a2 b1 b2 a3 a4 b3 a5 b4 a6 | a1 b1 b2 b3 a2 a3 a4 b4 a5 a6 |
| a1 a2 b1 b2 a3 a4 b3 b4 a5 a6 | a1 b1 b2 b3 a2 a3 b4 a4 a5 a6 | a1 a2 b1 b2 a3 b3 a4 a5 a6 b4 |
| a1 b1 b2 b3 a2 b4 a3 a4 a5 a6 | a1 a2 b1 b2 a3 b3 a4 a5 b4 a6 | a1 b1 b2 b3 b4 a2 a3 a4 a5 a6 |

**Figure 7.16.** Sorting results of cavern construction (126 schemes).

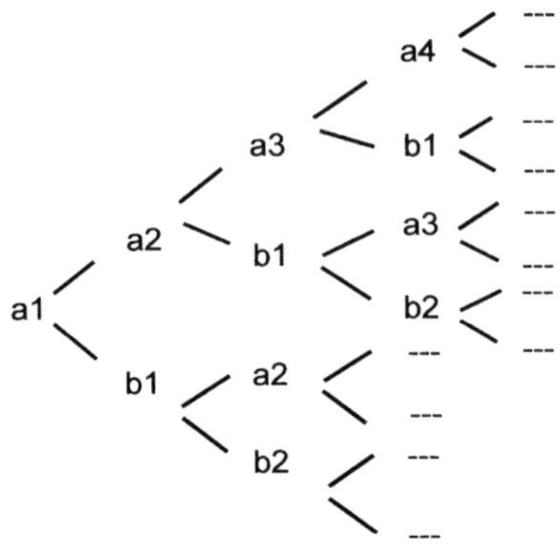


**Figure 7.17.** Hand-sorted construction sequences.

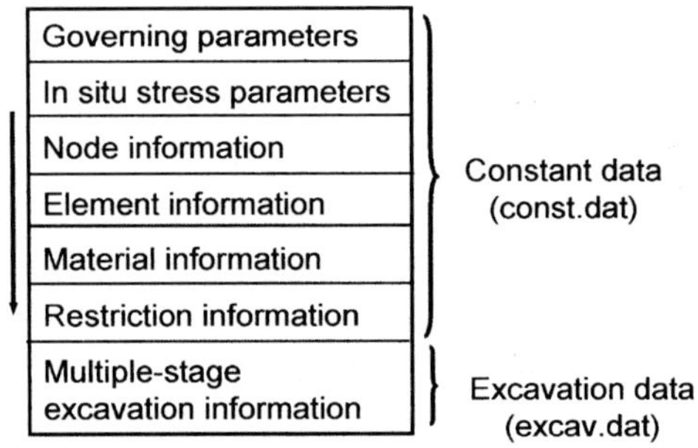


**Figure 7.18.** Structure of data files.

the bottom of the table (the last excavation step) is reached. The program for automatic generation of data files is shown below:

---

```
Autoform.Pro
domains

file=myfile

L=symbol*

A=symbol
predicates

form_fort_file (L)
```

```
form_fort_file (H|T): —

filename=H,

file_str (Filcname, Text),

write (Text).
run (L): —
openwrite (myfile, 'exc.dat'),
writedevice (myfile),
form_fort_file (L),
closefile (myfile).
```

---

*7.3.2.3* ***Implementation of excavation scheme optimisation.*** The automatic optimisation of cavern excavation schemes is implemented based on the results of an order arrangement on excavation sequence and data files generated in the last section. The whole process of excavation optimisation is schematically shown in Figure 7.19. The process can be simply described as: (a) make known the first step $a_1$ through reading file 'order.out'; (b) make known the second step $a_2$ or $b_1$; (c) form two tables $L_1=[a_1, a_2]$ and $L_2=[a_1, b_1]$; (d) call Auto_form.PRO and the finite element program; and (e) search for the optimal scheme according to the benefit function. By repeating this process, an optimal scheme is obtained, that is, $L_{opt}=[a_1, a_2, a_3, a_4, a_5, T_1, T_2, T_3, T_4, a_6]$.

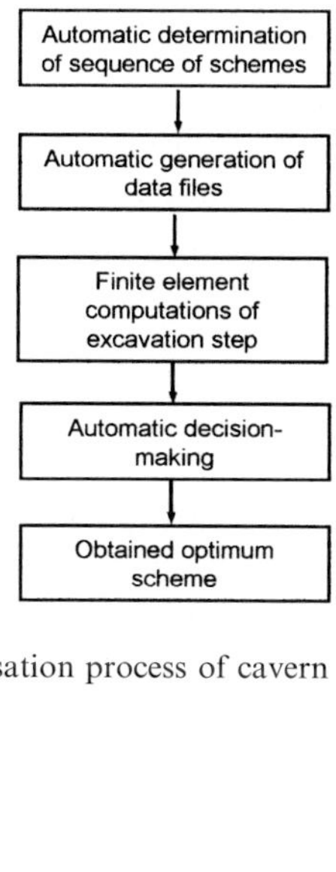


**Figure 7.19.** Optimisation process of cavern complex construction.

### 7.3.3 *Discussions*

Earlier in this chapter, a new method for the optimisation of cavern construction schemes is outlined. The interactive programming theory is used for cavern construction optimisation, based on the concepts of interactive construction mechanics associated with rock engineering. The traditional optimisation approach, which is based on the simple comparisons among multiple schemes from finite element computations, is improved in this method based on multi-stage decision-making. To further improve the efficiency of optimisation process, the artificial intelligence Prolog language is utilised. Accordingly, the optimisation is automatically implemented. The computer program is simple and easy to understand.

Decision-making in cavern construction optimisation is to select the optimal scheme from all the possible schemes according to the given requirements. After all the possible schemes are evaluated, an optimal scheme is then determined as the final option [8,9]. If there is only one index for evaluating the schemes in a system, the system is called single index decision-making system. Optimisation in this system is relatively easy, and can be implemented by comparing the indices of all the possible schemes. However, cavern construction is a complex system with multiple indices for evaluation. In other words, it is a multiple indices decision-making system. In such a system, the optimal scheme cannot be obtained directly from the evaluation results on a single index. The comprehensive evaluation on the multiple indices is required. In the study illustrated in this chapter, damage area around the cavern is taken as the benefit function (evaluation index). This means that the optimisation of cavern construction is still based on a single index. Further improvement on the interactive optimisation method can adopt multiple indices for construction optimisation.

## 7.4. ENGINEERING APPLICATIONS OF ARTIFICIAL INTELLIGENCE OPTIMISATION METHODS

An engineering example is given to demonstrate the effectiveness of applying the construction mechanics and artificial intelligence methods in cavern construction optimisation. The project is an underground hydroelectrical power station excavated in rocks.

### 7.4.1 *Description of the project*

The power station is excavated in a Triassic sandstone. The geological structure is uniclines. The dip direction of rock strata is NEE-SE, and dip angle is about 10°. No large fault is found except for bedding fractures. Groundwater level is deep below ground surface, and the permeability of the rock masses is low. The cavern complex

**Table 7.6.** Mechanical parameters of rock layers.

| Layer | $T_1^2$ | $T_1^{3-1}$ | $T_1^{3-2}$ | $T_1^4$ | $T_1^{5-1}$ | $T_1^{5-2}$ | $T_1^{5-3}$ |
|---|---|---|---|---|---|---|---|
| Young's modulus (GPa) | 11.5 | 12.8 | 12.0 | 13.5 | 9.5 | 12.8 | 10.5 |
| Poisson's ratio | 0.22 | 0.21 | 0.22 | 0.2 | 0.24 | 0.21 | 0.22 |
| Cohesion (MPa) | 1.0 | 1.17 | 1.0 | 1.5 | 0.83 | 1.25 | 1.0 |
| Friction angle (°) | 35 | 35 | 35 | 35 | 35 | 35 | 35 |

is approximately 70–100 m below the ground level. According to the in situ measurement, the horizontal stress is about 3 MPa; and the vertical stress is equal to the overburden stress. Underground cavern complex is housed in fractured mudstones ($T_1^{3-1}$, $T_1^{3-2}$ and $T_1^4$). Among them, $T_1^{3-1}$ comprises thick layers (about 30 m) of calc-silicon fine-grained quartz sandstone and calc sandstone with small amount of mudstone or siltstone. $T_1^{3-2}$ comprises thick layers (about 30 m) of calc or argillaceous-calc siltstone and fine-grained sandstone. $T_1^4$ comprises very thick layers (about 60 m) of silicon or calc sandstone containing little amount of mudstone or siltstone. The mechanical properties of each layer of rock formations are presented in Table 7.6.

### 7.4.2 *Layout of cavern group and arrangement of step excavation*

The layout of the cavern complex is shown in Figure 7.20, together with the distribution of rock layers. Excavation was designed to be completed in 10 stages. This excavation scheme is used in the study of cavern construction optimisation. In addition, several assumptions are made:

(a) $C_0$ is the first excavated;
(b) Excavation sequence is from top to bottom;
(c) Each stage is excavated in one run.

### 7.4.3 *Optimisation of excavation sequence*

Two stages of study have been performed. At the first stage, the surrounding rock mass is assumed isotropic and the same mechanical properties are used. At the second stage, the rock mass is divided into seven layers of different mechanical properties (locally isotropic medium) according to the geological conditions (Figure 7.20 and Table 7.6). The results obtained from the two different studies are different and outlined in the following sections.

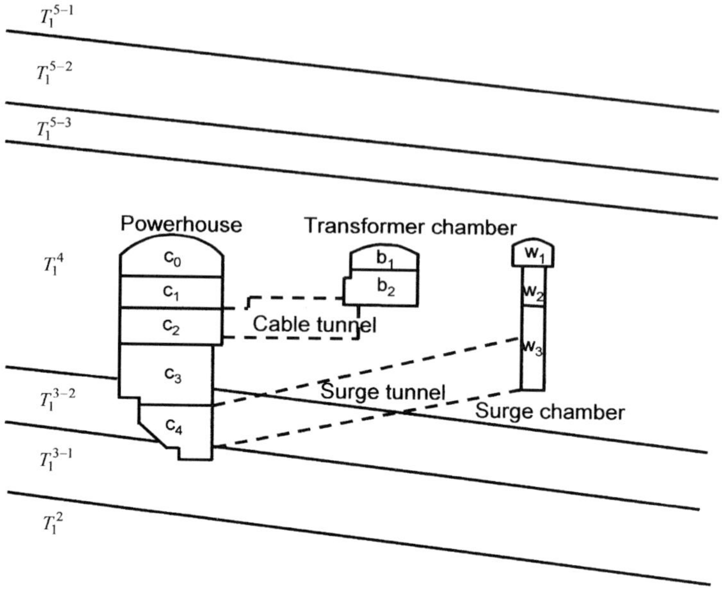


**Figure 7.20.** Cavern layout and excavation stage.

***7.4.3.1 Rock mass assumed as isotropic medium.*** The studies are conducted using the method outlined in Section 7.3.2. The analysis is based on two dimensional finite element modelling. The average depth of cavern complex to ground surface is 85 m. The material parameters for modelling are:

$\gamma = 2610\ \text{kg/m}^3$,
$E = 11000$ MPa,
$\nu = 0.20$,
$c = 0.3$ MPa,
$\phi = 33°$.

Two cases of in situ stress fields are modelled, one based on the elastic theory and the other based on the in situ stress measurement.

The in situ stresses are estimated as:

Case I: elastic theory
$\sigma_v = \sigma_1 = 2.61 \times 85/10 = 2.22$ MPa (vertical)
$\sigma_h = \sigma_3 = \frac{\nu}{1-\nu}\sigma_1 = 0.56$ MPa (horizontal)

Case II: In situ stress measurement
$\sigma_v = \sigma_1 = 2.22$ MPa,
$\sigma_h = \sigma_3 = 3.0$ MPa.

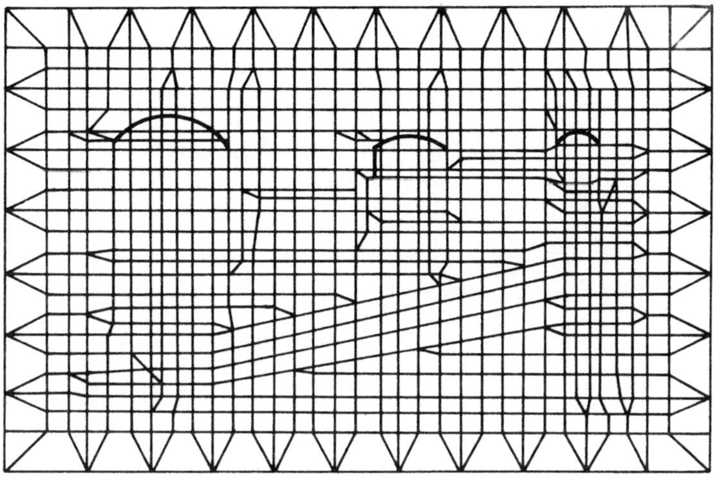

**Figure 7.21.** Mesh of the computation model.

To simulate the excavation process, the whole cavern area is divided into a mesh of triangle and quadrilateral shapes. The total number of elements is 1326, and the total number of nodes is 1284. The central part of the model is shown in Figure 7.21.

The artificial intelligence language Turbo Prolog 2.0 Version is used (see Section 7.32).

The results present 1260 possible excavation sequence schemes. It is impossible and unnecessary to carry out finite element computation for so many schemes. However, by using the construction mechanics theory, optimum schemes can be quickly sorted out. The searching processes for the optimum scheme with two different cases of in situ stress fields are shown in Figure 7.22.

The optimum excavation sequences using optimisation method, are different from the original scheme, $C_0 \rightarrow C_4 + W_3 \rightarrow b_1 + W_1 + C_1 \rightarrow C_2 + b_2 + W_2 \rightarrow C_3$, where $C_4 + W_3$ means that $C_4$ and $W_3$ are simultaneously excavated (Figure 7.20). Assessment of damage areas around the caverns are conducted, and comparisons between the different schemes are shown in Figures 7.23 and 7.24, and Table 7.7.

Usually, in a large underground cavern complex, there are many possible excavation schemes. However, by using the construction mechanics together with interactive programming theory and artificial intelligence method, the optimum scheme can be obtained. Compared with the originally proposed scheme, the damage area from the original scheme is 3–10 times larger than that of the optimum scheme. Although this is obtained with certain assumptions, it however indicates the importance of optimisation of cavern construction scheme.

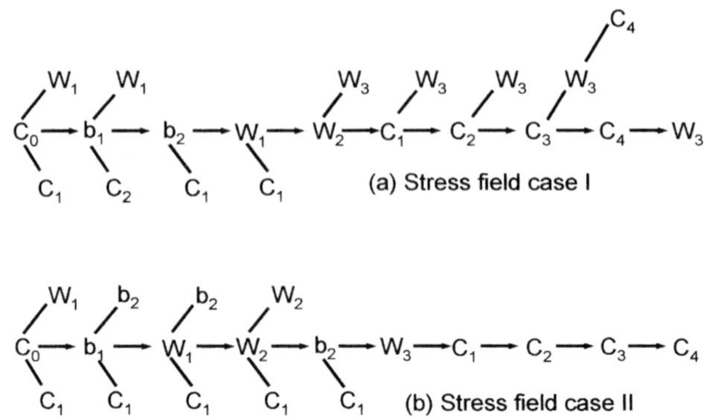


**Figure 7.22.** Optimisation search processes for two cases of in situ stress fields, (a) Case I, and (b) Case II.

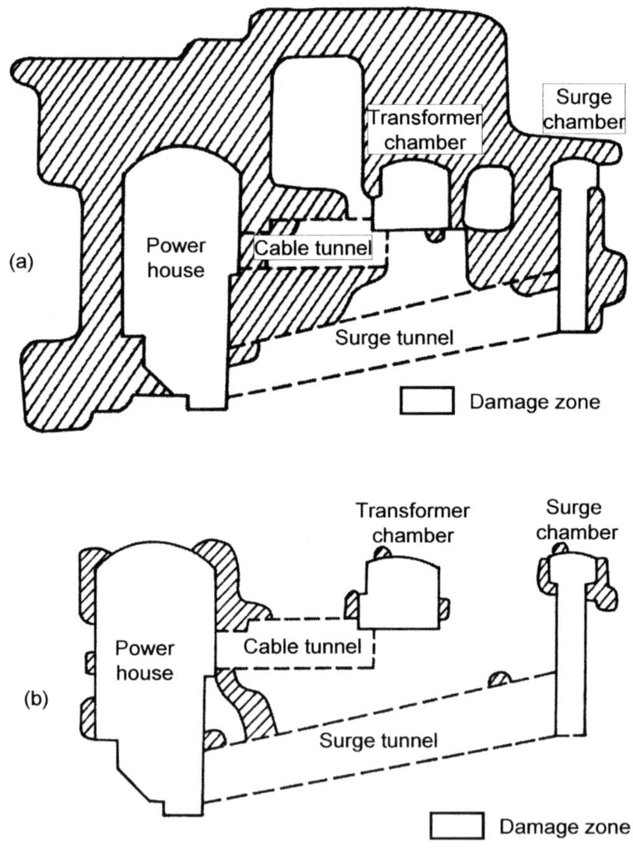


**Figure 7.23.** Damage distribution of the stress field case I, (a) original excavation scheme, and (b) optimised excavation scheme.

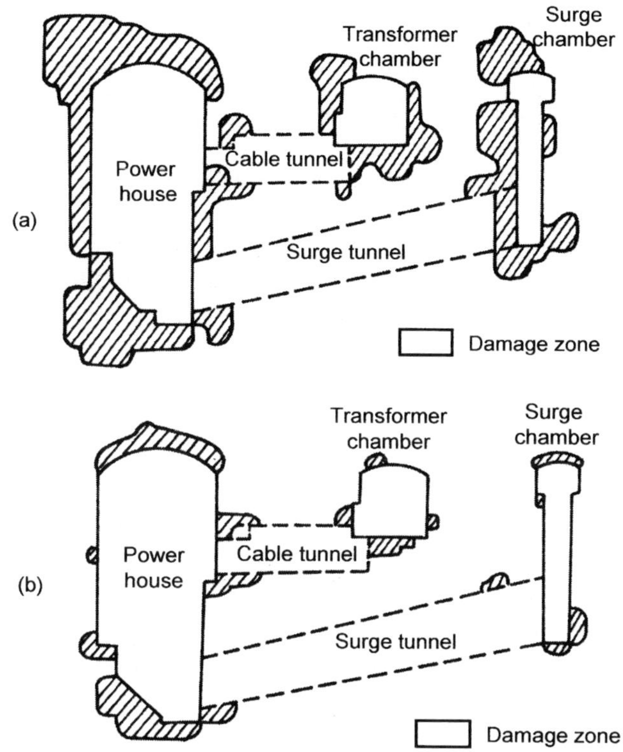


**Figure 7.24.** Damage distribution of the stress field case II, (a) original excavation scheme, and (b) optimised excavation scheme.

**Table 7.7.** Comparisons between the optimised and original excavation schemes.

| Scheme | | Total damage area ($m^2$) | Vertical convergence of powerhouse (mm) | Horizontal convergence of powerhouse (mm) |
|---|---|---|---|---|
| Stress case I | Optimised | 513 | 9.64 | 1.68 |
| | Original | 5139 | 11.75 | 0.69 |
| Stress case II | Optimised | 690 | 4.41 | 20.88 |
| | Original | 2540 | 6.67 | 12.51 |

***7.4.3.2 Rock mass assumed as layered isotropic medium.*** According to the geological conditions, the rock mass is divided into seven layers. Each layer is treated as an isotropic medium. The density of rock mass in all the layers is $2610\,kg/m^3$. The mechanical properties adopted for each rock layer are the same as those shown in Table 7.6.

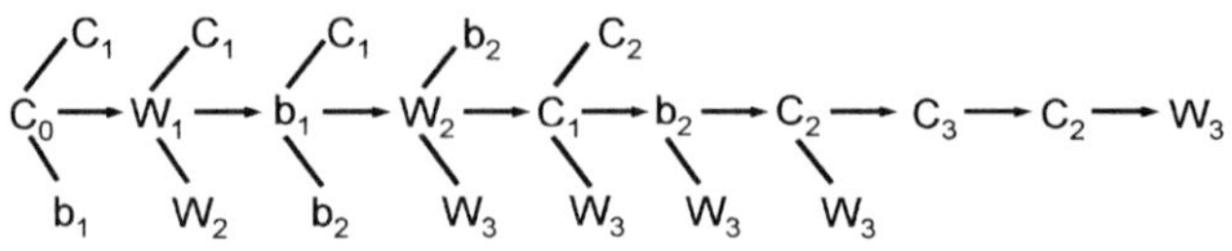


**Figure 7.25.** Optimisation searching process of construction schemes.

The average depth of cavern complex to ground surface is 85 m. The average rock mass density is 2610 kg/m$^3$. The initial stress field is determined from the in situ stress measurement. The horizontal stress obtained from measurement is 3.0 MPa. Therefore,

$\sigma_v = \sigma_1 = 2.61 \times 85/10 = 2.22$ MPa (vertical)
$\sigma_h = \sigma_3 = 3.0$ MPa.

The finite element computation is based on plane strain elastic-plastic model. The total number of finite elements is 1506, and the node number is 1211.

The program and the method are the same as those in the previous modelling. 1260 possible schemes are obtained, and subsequently reduced to 9 schemes. If the benefit function is defined by the damage area of surrounding rock mass, the smaller damage area is then taken as the optimum scheme. The optimum scheme searching process is shown in Figure 7.25.

The excavation sequence in the original scheme is

$$C_0 \to C_4 + W_3 \to b_2 + W_2 + C_1 \to C_2 + b_2 + W_2 \to C_3,$$

where $C_4 + W_4$ represents that $C_4$ and $W_3$ are simultaneously excavated. The comparisons between the optimised scheme and original scheme are shown in Figure 7.26 and Table 7.8.

It can be seen that the damage area of the optimised excavation scheme is reduced by 90% in comparison with the original scheme. A substantial improvement can be achieved by the optimisation.

It should be noted that the optimised scheme obtained in this study is different from that for rock mass assumed to be isotropic in the previous study. This implies that the optimisation results depend on the rock mass conditions.

*7.4.3.3* ***Summaries.***

(a) Applications of the interactive construction mechanics theory together with the artificial intelligence method can optimise the construction schemes of cavern complex scientifically and automatically, as a result high efficiency is achieved.

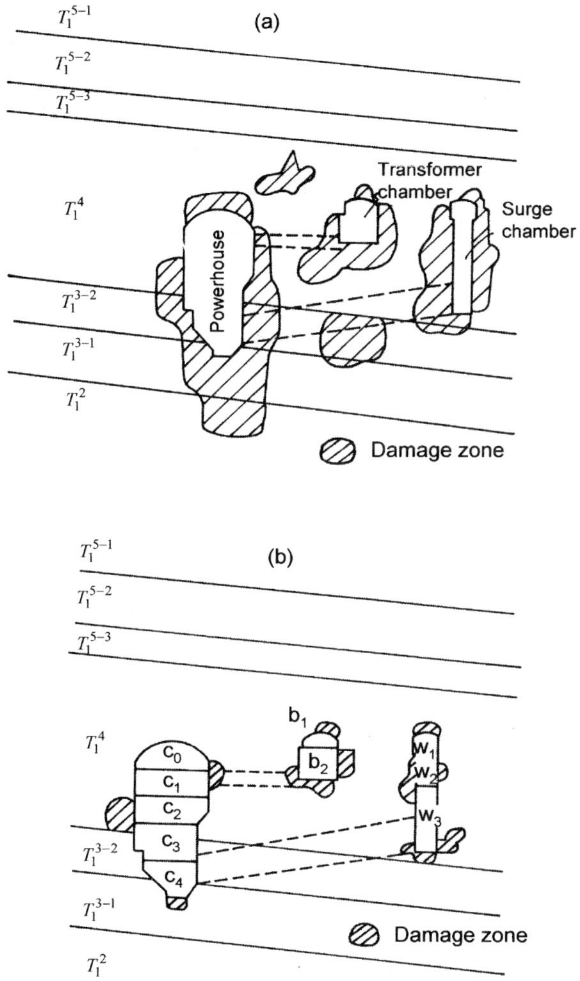


**Figure 7.26.** Damage distribution of (a) original excavation scheme, and (b) optimal excavation scheme.

**Table 7.8.** Comparisons between the optimised and the original excavation schemes.

| Scheme | Total damage area ($m^2$) | Vertical convergence of powerhouse (m) | Horizontal convergence of powerhouse (mm) |
|---|---|---|---|
| Optimised | 329 | 0.16 | 9.6 |
| Original | 4533 | 1.13 | 1.9 |

(b) Advantages of the optimisation for construction are significant for large-scale rock engineering projects. The damage area of the optimised scheme can sometimes be reduced to 1/10 of that of the unoptimised scheme.
(c) The application of the optimisation method of combining interactive construction mechanics theory and artificial intelligence technique to the heavily fractured rock mass and to the three-dimensional problems needed to be studied.

Chapter 8
# Reinforcement Mechanism of Rock Bolts

Rock bolts have been used in rock engineering from end of the 19th century [445]. In 1890, reinforced steel was used in reinforcing the rock mass in mines in north Wales. After the 1940s, this support technique was adopted worldwide. The use of shotcrete together with rock bolts further promotes the application of bolt as a main supporting method. Based on the rock bolt technique, a new construction method combining the use of rock bolts and shotcrete gradually formed in Europe in the 1950s and 1960s. This method sets a milestone in the underground construction technology. In recent years, many researchers and engineers have studied the effects of bolts in rock reinforcement. This chapter summarises some of the research findings.

## 8.1. EFFECT OF BOLTS ON SUPPORTING THE ROCK MASS

### *8.1.1 Effects of rock bolts*

Two main types of rock bolts are widely employed in the underground excavation in rocks. One is the end-anchored bolts varying from wedge to resin anchoring. The other one is the full-length anchored type varying from fully grouted to split-set bolts.

The effects of rock bolts on the surrounding rock are mainly two-fold: providing the reaction force and the reinforcement to the surrounding rock mass. The former prevents deformation occurring in the surrounding rock mass and provides the tensile force to resist the movement of rock blocks. The latter reinforces the rock mass to be an integrated medium and to easily form arches around the excavation [446–477].

***8.1.1.1 Reinforcement.*** The efforts of other support methods such as timber frame, segment lining, and cast concrete lining are mainly external supports. As the rock bolts are inserted into the rock masses, the effects of rock bolts on the rock mass are similar to that of steel bars in reinforced concrete. They increase the integrity and overall strength of the rock masses. The bolts prevent the rock masses from sliding and failing and increase the load-bearing capacity effectively. The rock mass is reinforced by the bolts and becomes an effective self-support medium [446–473].

***8.1.1.2 Post effect of pre-stressing.*** The deformation modulus of the rock bolt (usually steel) is generally greater than that of the surrounding rock masses. The difference between the moduli restricts the surrounding rock mass to deform (usually tensile deformation towards the centre of the opening) after excavation. The restriction occurs at the ends of anchored rock bolts and along the full length of grouted rock bolts. This restriction is mobilised once deformation starts to occur and it can be regarded as a complementary pre-stress on the excavation surface after the excavation. Rock bolts significantly increase the integrity of the rock mass due to this pre-stressing effect [446–448,474].

***8.1.1.3 Prompt prevention.*** Rock bolts and shotcrete are often applied immediately after excavation. In some cases, rock bolts can also be implemented prior to the excavation, to restrain the relaxation of the surrounding rock and to improve the rock mass strength early. The early application of rock bolts provides prompt prevention of failure.

***8.1.1.4 Good match to deformation.*** Rock bolts possess better match ability to the deformation of rock mass than other support methods. It is often referred as 'flexible support'. This support method is also very suitable for tunnels in soft rock.

***8.1.1.5 Flexibility in construction.*** The size and distribution of rock bolts can be adjusted according to the geological conditions of the surrounding rock. Bolting and shotcreting can be completed in several steps [475,476]. Additional bolts can be installed at a later stage. The application of bolts has flexibility during the construction.

### *8.1.2 Reinforcement mechanism of rock bolts*

The reinforcement mechanisms of rock bolts are very complex. In this section, they are discussed by examining the shear strength and other properties of the bolted rock masses.

A series of studies on the reinforcement mechanism of rock bolts have been carried out. Test results indicated that application of rock bolts increases the peak compressive strength of the rock mass by 50~100%, compared with that of the rock mass without bolts.

For a single opening, the shear strength of the surrounding rock mass can be expressed by the Mohr–Coulomb strength condition, as shown in Figure 8.1.

The reaction force contributed by rock bolts to the surrounding rock mass ($\Delta\sigma_2$) can be estimated from the allowable tensile stresses in the rock bolts. For example, for a tunnel at depth of 500 m with hydro-static in situ stress condition, and assuming the density of rock mass of 2500 kg/m$^3$, the in situ stress field can be obtained as:

$$\sigma_{ix} = \sigma_{iy} = \sigma_{iz} = 12.5\,\text{MPa} \tag{8.1}$$

The uniaxial compressive strength of the rock material ($\sigma_c$) is assumed as 15.0 MPa, the uniaxial compressive strength of the rock mass strength ($\sigma_{\text{mass}}$) is assumed as 5.0 MPa ($1/3\sigma_c$) and the internal friction angle ($\phi_{\text{mass}}$) is 35°. The cohesion of rock mass can be derived as:

$$c_{\text{mass}} = \sigma_{\text{mass}}(1 - \sin\phi_{\text{mass}})/2\cos\phi_{\text{mass}} = 1.31\,\text{Mpa} \tag{8.2}$$

If the compressive strength of the properly bolted rock mass ($c'_{\text{mass}}$) is 50% greater than that of unbolted rock mass, then $c'_{\text{mass}} = 1.5\,c_{\text{mass}} = 1.97$ MPa. Assuming that the length of bolt is 60 cm and the allowed tensile load is 100 kN, the reaction stress on the rock surface provided by the bolt can be calculated as $\Delta\sigma_2 = 0.28$ MPa.

For the end-anchored bolts, the reinforcement is applied through a reaction force at the rock wall. The reinforcement effect of a bolt is therefore on the rock wall and the bolt anchor while the other parts of the bolt are not in contact with the rock mass.

For full-length grouted or frictional bolts, the reinforcement bolts increase the shear-bearing capacity of the rock masses, in addition to the reaction force at the rock wall generated by the bolts.

Assuming that the stresses $\sigma_1$ and $\sigma_2$ on the opening wall reach the rock mass shear strength, the stress condition of the rock wall can be expressed by the Mohr circle $\sigma_2 A \sigma_1$ with centre O. As shown in Figure 8.1, this circle is tangential to the strength envelope BCA at A.

The reaction force of rock bolts to the rock wall can be defined as $\Delta\sigma_2$ for the end-anchored bolt. In this case the stress circle change to the small Mohr circle $\sigma'_2 A' \sigma_1$ with centre O′, and the strength envelope is B′C′A′. The new minor principal stress is $\sigma_2' = \sigma_2 + \Delta\sigma_2$. From Figure 8.1, the shift of the strength envelope,

$$\Delta OA = (c'_{\text{mass}} - c_{\text{mass}})\cos\phi \tag{8.3}$$

For the full-length grouted bolts, the shear strength of surrounding rock is enhanced by the bolts, and the shear envelope shifts up to the position of B″C″A″ (assuming no change in friction angle). The increased cohesion equals to c″−c′. The strength

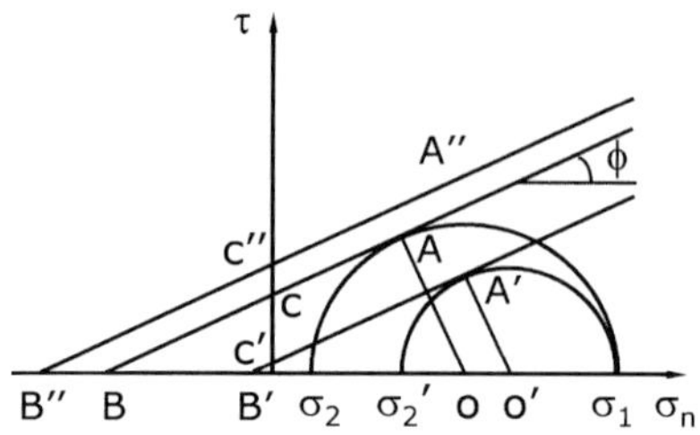


**Figure 8.1.** Strength analysis of rock bolt effects.

envelope of full-length grouted bolt is increased by an offset $\Delta O'A' = (c'' - c')\cos\phi$ from the end-anchored bolt.

The analysis above is under the assumption that the strength offset is based on the shortest relative distance from centre of the Mohr circle to the envelope. The analysis clearly shows that the full-length grouted bolts achieve high strength storage than the end-anchored bolts. The difference of the effects of two types of rock bolts can be significant, due to the effects of grouting.

## 8.2. PHYSICAL MODELLING OF ROCK BOLTS

Rock bolts enhance and reinforce the rock masses. However, the mechanisms of reinforcement, and the deformation and damage of bolts in the surrounding rock mass subjected to change of stress field have not been fully understood [449,463,467,477]. This section presents studies on the rock bolt mechanism using physical modelling.

When the surrounding rock mass is soft and/or subjected to high in situ stresses, displacement of the tunnel wall can often be as high as 50–100 cm. It is of great interest to have a scientific approach in selecting the rock bolt to overcome such large deformation problems. At present, the common way of applying bolts is to install bolts perpendicular to the rock surface. The systematic studies are conducted on the installation angle, density and other parameters, and results are of interest to engineers.

### *8.2.1 Similarity of model materials*

Rock masses close to the surface are treated as separable blocky materials. The model materials should have similar characteristics as the real rock masses, such as stress–strain relation, properties of dilation and softening. After testing different model materials, the selected model material (made of sand and white glue) has elastic modulus of 230 MPa. The bolt is modelled by the bamboo material with

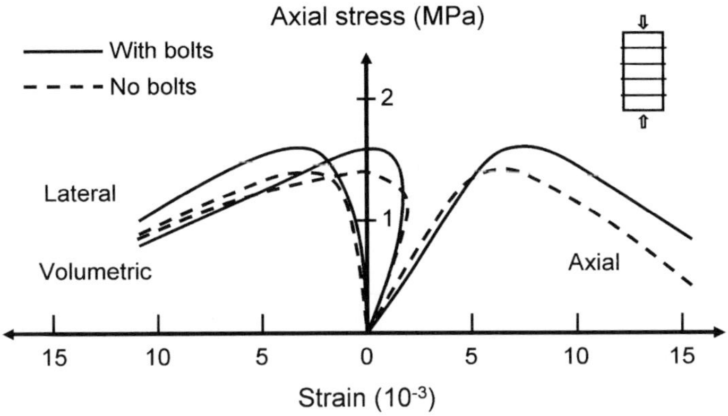


**Figure 8.2.** Results of compression tests.

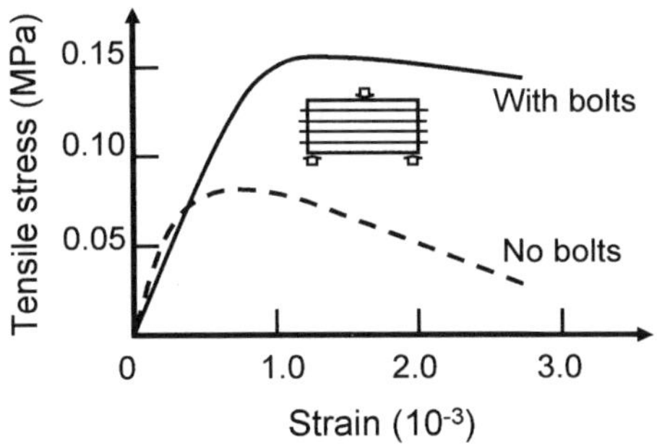


**Figure 8.3.** Results of tensile tests.

elastic modulus of 100–170 MPa. A series of tests are conducted including uniaxial, biaxial and triaxial compression, tensile test, and cyclic loading test for the model specimens with and without bolts, as shown in Figures 8.2–8.4. From the results, it can be seen that the ratio of modulus to strength of the model material is 178 and that of the rock mass is 200. The ratio of modelling material compressive strengths to rock mass strength is 1/15. From the bulk deformation curves, it also can be seen that the modelling material has the similar dilation property as the modelled rock mass.

### *8.2.2 Comparison of different bolting methods*

The effects of different bolt materials, bolting methods, bolt densities and installation angles are systematically compared in the tests. The strength of the material

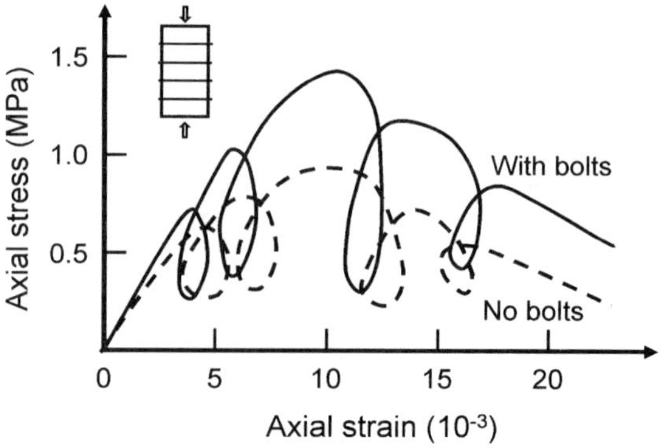


**Figure 8.4.** Results of cyclic compression tests.

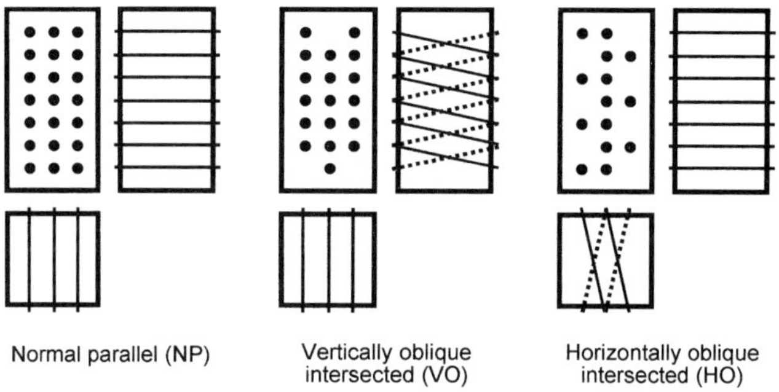


**Figure 8.5.** Different bolt arrangement used in the tests.

modelling the rock mass is 1.0 MPa. Bamboo ($E = 10 \times 10^3$ MPa) and plexiglass ($E = 3 \times 10^3$ MPa) are used as modelling materials of the bolts. Both full-length grouting and end-anchored bolts are modelled. Different densities are used at 8, 10, 12, and 36 bolts per 200 $cm^2$, equivalent to the actual bolt spacing of 1.2 to 0.4 m. The bolting methods in the tests include end-anchored and fully grouted parallel bolts perpendicular to the rock wall, fully grouted perpendicular, vertically and horizontally intersected bolts. The details of various arrangements are summarised in Figure 8.5 and Table 8.1.

Figures 8.6–8.8 indicate the various load-deformation curves. From the figures, it can be seen that high density of bolts can improve the strength of surrounding rock mass. The horizontally obliquely intersected full-length grouted bolts give high strength of the rock mass and small number of bolts. Figure 8.8a illustrates that for the same bolt spacing and length, the rock mass can have the highest

**Table 8.1.** Various types of bolts and their parameters.

| Type | No bolt | Normal parallel | Normal parallel | Normal parallel | Vertically oblique | Vertically oblique | Horizontally oblique |
|---|---|---|---|---|---|---|---|
| Symbols | NB | NP-E10 | NP-F10 | NP-F36 | VO-F8 | VO-F12 | HO-F12 |
| Bolt density (per 200 $cm^2$) | – | 10 | 10 | 36 | 8 | 12 | 12 |
| Inclination to wall | – | 90° | 90° | 90° | 73° | 68° | 75° |
| Anchor and grouting | – | Two ends | Full-length | Full-length | Full-length | Full-length | Full-length |
| Bolt material | – | Bamboo | Bamboo | Bamboo | Bamboo | Bamboo | Bamboo |
| Peak strength (MPa) | 1.0 | 1.5 | 1.5 | 3.0 | 1.54 | 2.2 | 2.06 |
| Softening behaviour | Yes | Yes | No | No | Yes | No | No |

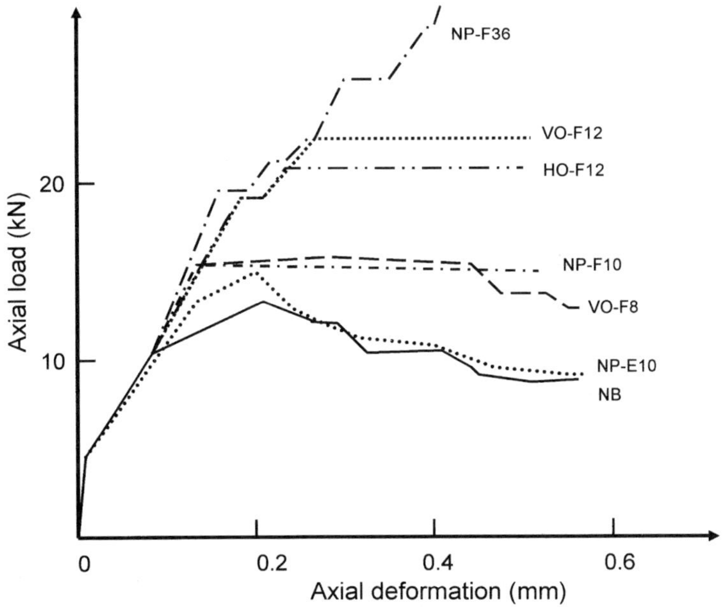


**Figure 8.6.** Axial load – axial deformation of different bolt arrangement.

strength and the lowest dilation when the bolts are installed at 22.5° and 67.5°, to the rock wall surface.

For the same bolting angle and the same total lengths of bolts, the test results shown in Figure 8.8 suggest that mixed length bolts (total 9 bolts of 5.41 and 10.82 cm long and evenly spaced) provide the best reinforcement results, while short bolts with high density (total 12 bolts of 5.41 cm long and evenly spaced) give the

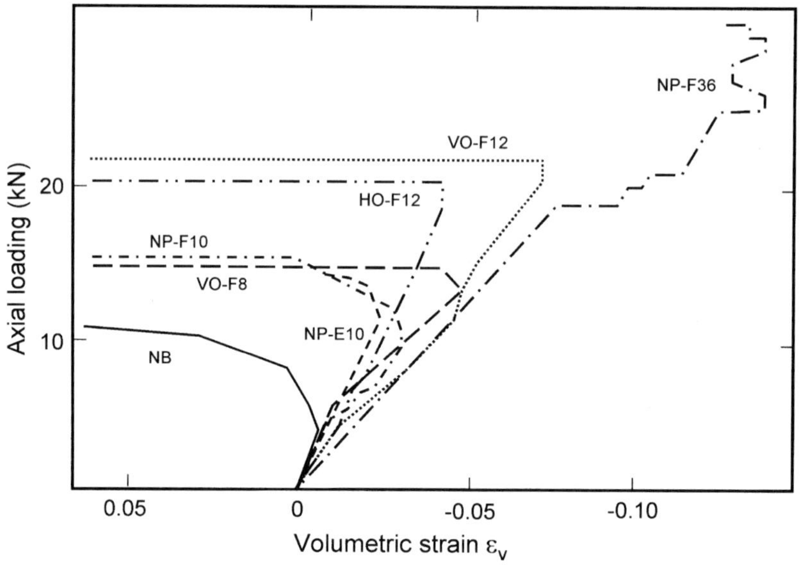


**Figure 8.7.** Axial load – volumetric strain for different bolt arrangement.

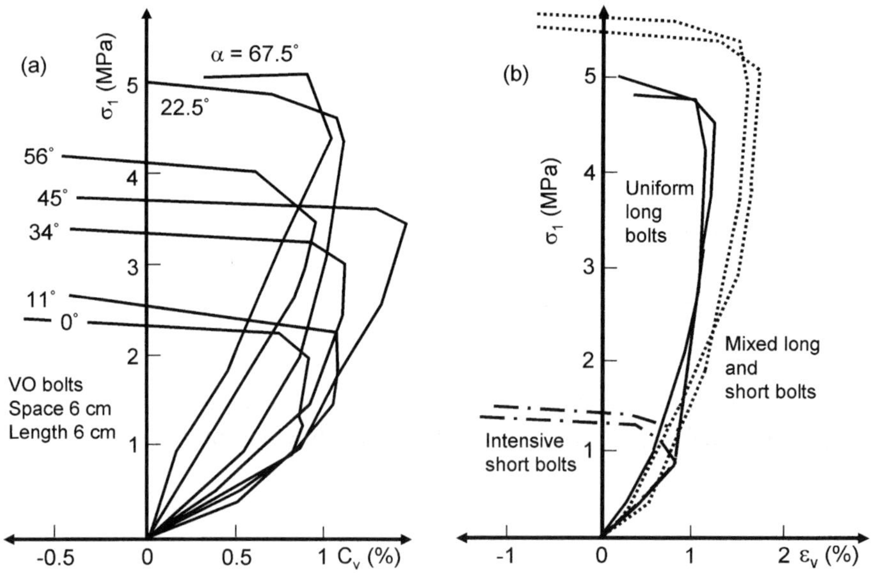


**Figure 8.8.** Axial load – volumetric deformation for different (a) bolting angle, and (b) bolt length.

worst reinforcement results, as compared with uniformly long bolts (total 6 bolts of 10.82 cm long and evenly spaced).

### *8.2.3 Analysis of test results*

From the test results shown in Figures 8.6, 8.7 and 8.8 and Table 8.1, following observations are obtained:

(a) The peak uniaxial compressive strength of the rock mass with bolts increases significantly (up to 20% from the model tests) compared with that without bolts while the residual strength and the tensile strength increase up to 100%. In biaxial case, the peak compressive strength can increase by 50–100% (Figure 8.6) when the density of bolts is high, rock mass strength can increase by 3 times with little volumetric dilation.
(b) Under the plane strain condition, the strength of the models is affected not only by the bolt densities but also by bolt installation angles, types, and shear strength and lateral stiffness of bolt.
(c) The obliquely intersected bolt distribution enhances the rock mass strength and limits the dilation significantly. The best arrangement appears to the bolts at 65° angle to the wall. But the bolts are required to possess a high lateral stiffness in this case.
(d) Compared to the end-anchored bolts, the improvement of full-length grouted bolts on the peak strength of the surrounding rock mass is not obvious. However, the bulk displacement curves are different for the two different bolt types. The dilation of the fully grouted bolts starts at a later stage. The post-peak softening phenomenon is not obvious for the fully grouted bolts while it is significant for the end-anchored bolts. Therefore, end-anchored bolts have low post-peak strength.

## 8.3. NUMERICAL MODELLING OF BOLT

In this section, non-linear finite element numerical modelling is applied to study effects of rock bolts on the stability of tunnels in soft rocks. Parameters obtained from physical modelling are used as input to the numerical modelling.

### *8.3.1 Basic parameters of the numerical model*

A simple circular opening model is used in the parametric study. Two in situ stress conditions are modelled: (a) far field stresses $\sigma_1 = \sigma_2 = 20\,\mathrm{MPa}$, and (b) far field

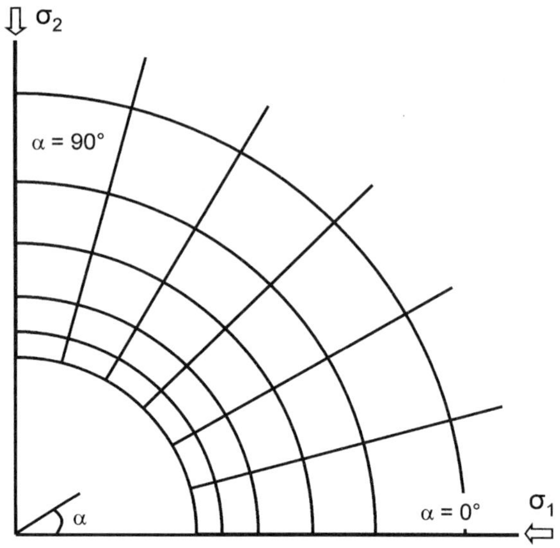


**Figure 8.9.** Element division near the opening.

stresses $\sigma_1 = 2\sigma_2 = 20$ MPa. The surrounding rock mass is regarded as an elasto-plastic material. The problem is treated as a plain stress one. Based on the symmetry, one quarter of problem is analysed by a finite element program. A total of 136 elements and 162 nodes are used in the model, as shown in Figure 8.9. Three different cases are studied: opening without bolts, opening with bolts that are normal to the wall surface, and opening with obliquely intersected bolts. The radius of the opening is 2 m and the thickness of bolted region equals to the opening radius. The mechanical parameters of surrounding rock mass are adopted from the physical model tests described in the previous section, and is summarised in Table 8.2.

### 8.3.2 *Models with far field stresses $\sigma_1^\circ = \sigma_2^\circ = 20$ MPa*

Stress distributions of models I, II and III when $\sigma_1^\circ = \sigma_2^\circ$ are presented in Figure 8.10, and the largest displacements of the wall are summarised in Table 8.3. The obliquely intersected bolting gives the smallest displacement, and it is less than half of that of the opening without bolts. Figure 8.11 and Table 8.3 illustrate the range of damage zones in the surrounding rock mass occurring near the wall. The results show that the surrounding rock mass with bolts is generally less damaged than that without bolts. The surrounding rock mass with obliquely intersected bolt system is the least damaged.

**Table 8.2.** Mechanical parameters of rock mass.

| Parameter | Model I: No bolts | Model II: Normal bolts | Model III: Obliquely intersected bolts |
|---|---|---|---|
| $\sigma_c$ (MPa) | 26 | $1.5\sigma_c=39$ | $2.0\sigma_c=52$ |
| $\sigma_t$ (MPa) | $1/15\sigma_c=1.73$ | $1.2\sigma_t=2.03$ | $1.6\sigma_t=2.77$ |
| $E$ (MPa) | $0.4\times10^4$ | $1.15E=0.58\times10^4$ | $1.5E=0.75\times10^4$ |
| $\nu$ (MPa) | 0.25 | 0.25 | 0.25 |
| $\phi$ (MPa) | 35° | 35° | 35° |
| $\phi_r$ (MPa) | 35° | 35° | 35° |
| $\sigma_r$ (MPa) | 200.8 | 39 | 52 |
| $c$ (MPa) | 3.58 | 5.36 | 7.15 |
| $c_r$ (MPa) | 2.86 | 5.36 | 7.15 |

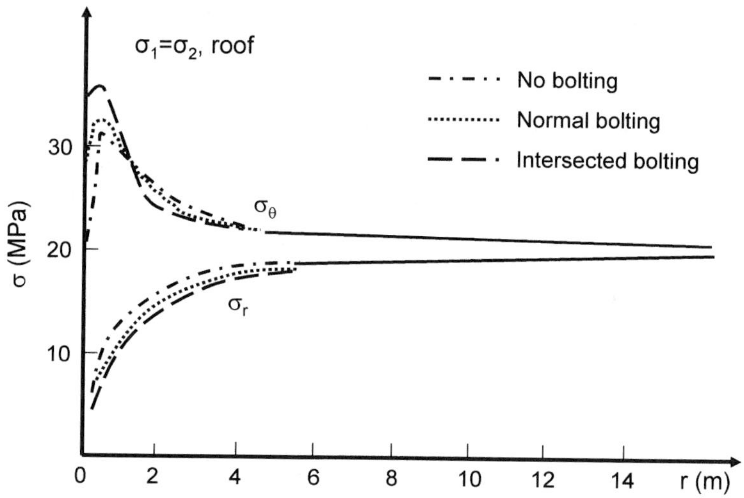


**Figure 8.10.** Stress distribution of models I, II and III, at roof, when $\sigma_1^\circ=\sigma_2^\circ$.

**Table 8.3.** Largest displacement and damage of models I, II and III when far field $\sigma_1=\sigma_2$.

| Model | No bolts | Normal bolts | Oblique intersection bolts |
|---|---|---|---|
| Maximum displacement (cm) | 1.80 | 1.11 | 0.84 |
| Number of damage elements | 24 | 16 | 8 |

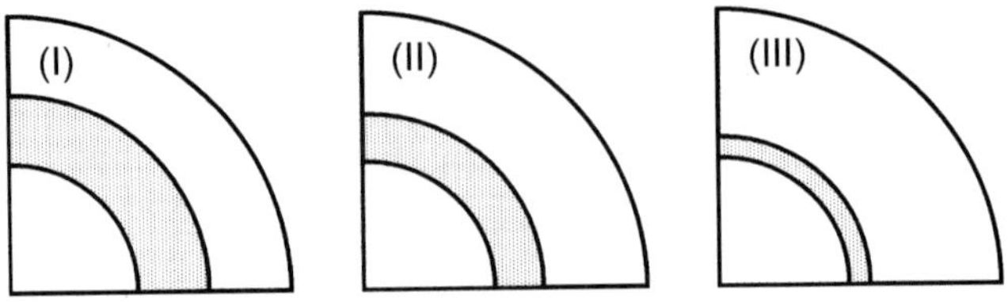


**Figure 8.11.** Damage around the opening of models I, II and III when $\sigma_1^\circ = \sigma_2^\circ$.

### 8.3.3 *Models with far field stresses* $\sigma_1 = 2\sigma_2 = 20$ *MPa*

The modelling results for the models I, II and III are presented in Figures 8.12 and 8.13, showing the stress distribution in the roof and sidewall respectively. Figure 8.14 shows the damaged zones near the wall surface for the three models. Again, it can be concluded that obliquely intersected bolt system gives the best results. Table 8.4 shows the maximum displacement on the wall surfaces and the numbers of damaged elements of the three different models. The obliquely intersected bolted rock mass sustains less than one-third of the damaged area than that of the unbolted rock mass. The bolt quantity of the obliquely intersected is 20% more than that of normal bolt system. However, the effects of the obliquely intersected bolts on reducing the damaged zones and the displacements of the surrounding rock are significant compared with that of normal bolt system. The finite element modelling uses elasto-plastic approach, and post-failure dilation of the rock mass is not considered. By considering the post-failure dilation, the effects of the obliquely intersected bolt system are expected to be more significant.

## 8.4. SCALED ENGINEERING MODEL TEST

In order to study the reinforcement effects of rock bolts on the surrounding rock mass in an underground excavation, a series of biaxial compression tests on engineering physical model were conducted. The effects of various bolt parameters are compared. The modelling materials and mechanical parameters are selected according to modelling scale to model an excavation in a rock mass.

The actual excavation is in a sedimentary rock. The surrounding rock mass is faulted and fractured. The in situ horizontal stresses are 19.5 MPa and 12.8 MPa, respectively. The estimated uniaxial compression strength of the rock mass is about 30 MPa. The elastic modulus and Poisson's ratio are approximately $1.0 \times 10^4$ MPa and 0.25 respectively. Both stress and geometry scale ratios are 40. The model is $50 \times 50$ cm, and the bolts are modelled by bamboo of 2 mm diameter and 40 mm long, to match the stiffness ratio. The opening shape is an arched roof with

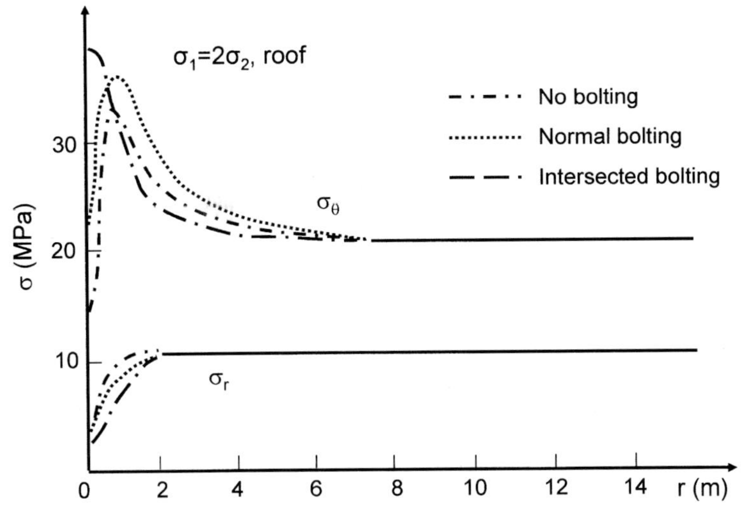


**Figure 8.12.** Stress distribution of models I, II and III, at roof, when $\sigma_1^\circ = 2\sigma_2^\circ$.

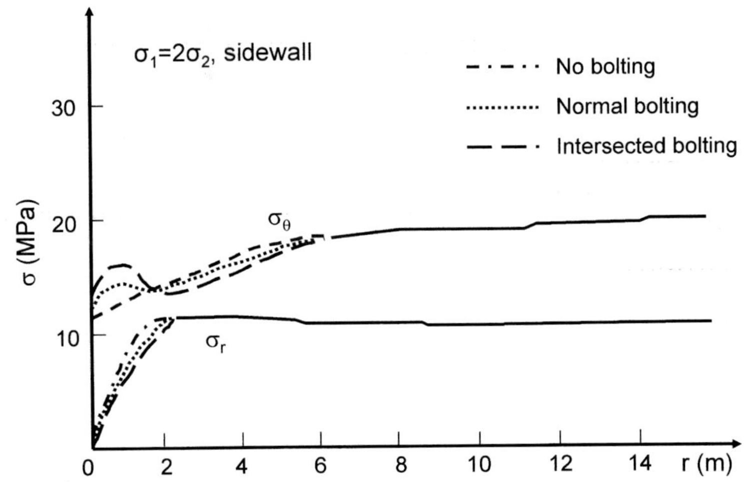


**Figure 8.13.** Stress distribution of models I, II and III, at sidewall, when $\sigma_1^\circ = 2\sigma_2^\circ$.

vertical walls, as shown in Figure 8.15. The tests are performed under the plain strain condition. Three different bolt distributions are simulated: (a) obliquely intersected with bolt at 67.5° to the opening surface, and in the same vertical plane; (b) bolts perpendicular to the opening surface, and (c) no bolts, as shown in

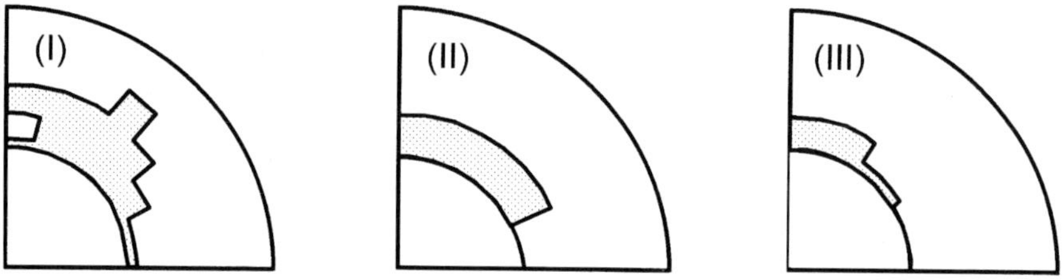


**Figure 8.14.** Damage around opening of models I, II and III when $\sigma_1^\circ = 2\sigma_2^\circ$.

**Table 8.4.** Largest displacement and damage of models I, II and III when $\sigma_1^\circ = 2\sigma_2^\circ$.

| Model | No bolts | Normal bolts | Oblique intersection bolts |
|---|---|---|---|
| Maximum displacement (cm) | 1.45 | 1.23 | 1.03 |
| Number of damage elements | 20 | 11 | 8 |

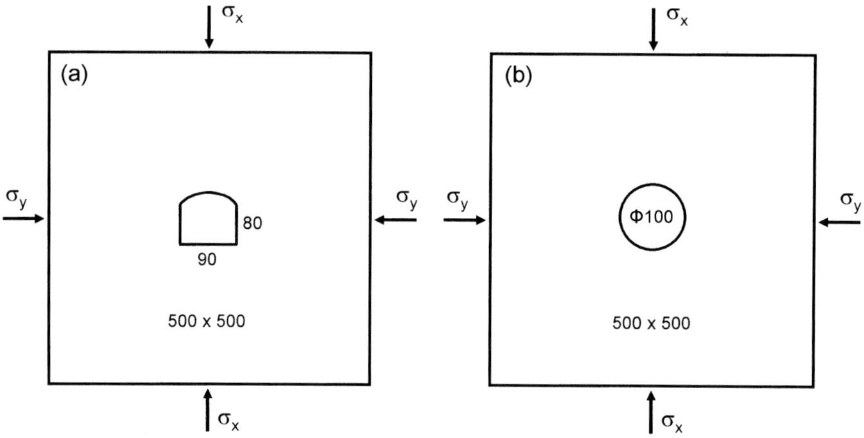


**Figure 8.15.** Scaled engineering model with dimensions.

Figure 8.16. Bolts are fully grouted in the model tests. The test follows the following procedures:

(a) Stresses are applied to the horizontal and vertical directions simultaneously, with vertical stress kept twice that of the horizontal stress. The increment of stress is 0.1 MPa horizontally at each step.

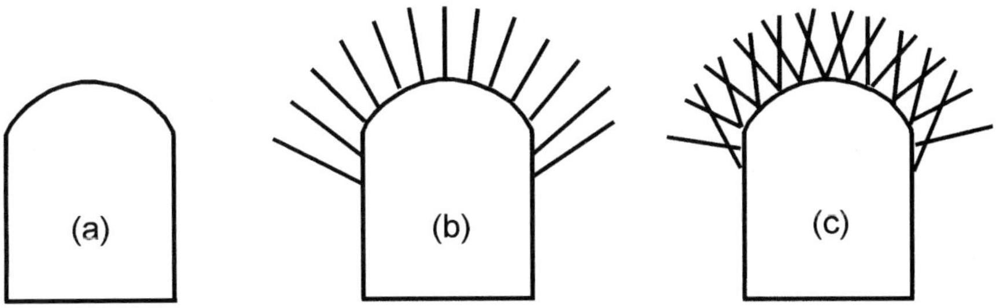


**Figure 8.16.** Arrangement of (a) no bolt, (b) normal bolts and (c) obliquely intersected bolts.

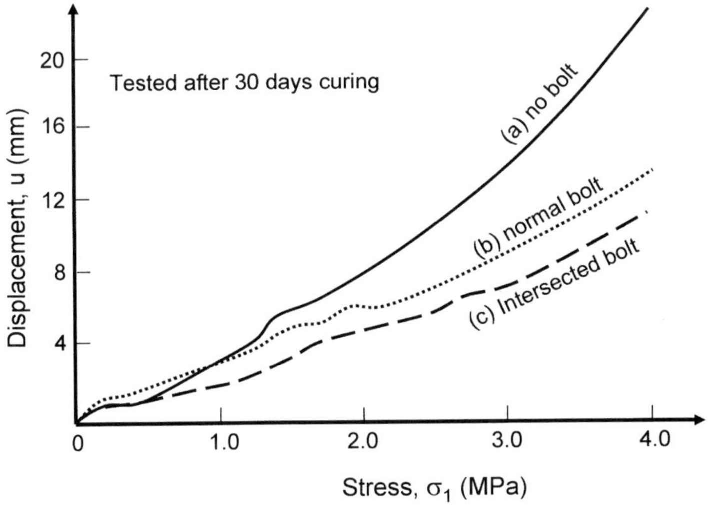


**Figure 8.17.** Convergent displacements of the tunnel monitored during the loading.

(b) Loading is stopped when the horizontal and vertical stresses reach 1.0 MPa and 2.0 MPa, respectively.
(c) The vertical stress is increased gradually to 4.0 MPa while the horizontal stress is kept at 1.0 MPa.

The convergent displacements of the tunnel are monitored during the loading. The monitored displacements are presented in Figure 8.17. It is observed that:

(a) The convergent displacement of the rock mass reinforced by the obliquely intersected bolts is generally smaller than that of the rock mass reinforced by normal bolts and the displacement of the roof is reduced by 16~40%. The plastic damage zone occurring in the sidewalls are also smaller for the normal bolting system.

(b) Bolting in general, as compared to no bolting, improves the surrounding material properties. The convergent displacement of the opening decreases significantly and the plastic damage is greatly reduced by bolting. Normal bolt reduces the displacement of the roof by 14~43%.

(c) Scaled model tests were also performed on circular opening (Figure 8.15) and similar results were obtained.

# References


1. Hudson JA (1992). *Rock Engineering System: Theory and Practice*. Ellis Horwood, Chichester.
2. Hudson JA (1995). *Rock Engineering Systems 2: Implementation*. Ellis Horwood, Chichester.
3. Cai J, Zhao J, Hudson JA (1998). Computerised rock engineering systems with neural networks and expert system. *Rock Mech. and Rock Engng.*, **31**, 135–152.
4. Barton N (1973). Review of a new shear strength criterion for rock joints. *Engng. Geol.*, **8**, 287–332.
5. Barton N (1986). Deformation phenomena in jointed rock. *Geotechnique*, **36**, 147–167.
6. Brown ET, Trollope DH (1970). Strength of a model of jointed rock. *J. Soil Mech. Foundns. Div., ASCE*, **96**, 685–704.
7. Zhao J (1997). Joint matching and shear strength, *Int. J. Rock Mech. Min. Sci.*, **34**, 173–185.
8. Zhu W, He M (1995). *Stability of surrounding rock mass in complex condition and dynamic construction mechanics*. Science Press, Beijing.
9. Zhu W, Li S (2000). Optimizing the construction sequence of a series of underground opening using dynamic construction mechanics and a rock mass fracture damage model. *Int. J. Rock Mech. Min. Sci.*, **37**(2), 517–523.
10. Jing L (2003). A review of techniques, advances and outstanding issues in numerical modeling for rock mechanics and rock engineering. *Int. J. Rock Mech. Min. Sci. Geomech. Abstr.*, **40**, 283–353.
11. Beer G, Watson JO, Swoboda G (1987). Three-dimensional analysis of tunnels using infinite boundary elements. *Comp. & Geotech.*, **3**, 37–58.
12. Brady BHG, Bray JW (1978). The boundary element method for determining stress and displacements around long openings in a triaxial stress field. *Int. J. Rock Mech. Min. Sci. Geomech. Abstr.*, **15**, 21–28.
13. Carter J, Booker JR (1982). Elastic consolidation around a deep circular tunnel. *Int. J. Solids Struct.*, **18**, 1059–1074.
14. Crouch SL, Starfield AM (1983). *Boundary Element Methods in Soil Mechanics*. p. 322, Allen and Unwin, London.
15. Cundall PA (1988). Formulation of a three-dimensional distinct element model – part 1: A scheme to detect and represent contacts in a system composed of many polyhedral blocks. *Int. J. Rock Mech. Min. Sci. & Geomech. Abstr.*, **25**, 107–116.
16. Chen SG, Zhao J (2002). Modelling of tunnel excavation using a hybrid DEM/BEM method. *Computer-Aided Civ. and Infrastr. Engng.*, **17**, 381–386.
17. Amadei B (1988). Strength of a regularly jointed rock mass under biaxial and axisymmetric loading conditions. *Int. J. Rock Mech. Min. Sci. & Geotech. Abstr.*, **25**, 3–13.
18. Bieniawski ZT (1978). Determining rock mass deformability: Experience from case histories. *Int. J. Rock Mech. Min. Sci. & Geomech. Abstr.*, **15**, 237–247.

19. Hoek E (1983). Strength of jointed rock masses – 23rd Rankine Lecture. *Geotechnique*, **33**, 187–223.
20. Shi G (1992). Modeling rock joints and blocks by manifold method. *Proc. of the 33rd US Symp. on Rock Mech.*, Santa Fe, NM, pp. 639–648.
21. Shi G, Goodman RE (1989). The key blocks of unrolled joint traces in developed maps of tunnel walls. *Int. J. Num. Anal. Methods Geomech.*, **13**, 131–158.
22. Kawamoto T, Ichikawa Y, Kyoya T (1988). Deformation and fracturing behaviour of discontinuous rock mass and damage mechanics theory. *Int. J. Num. Anal. Methods Geomech.*, **12**, 1–30.
23. Kim MK, Lade PV (1984). Modelling rock strength in three dimensions. *Int. J. Rock Mech. Min. Sci. & Geomech. Abstr.*, **21**, 21–33.
24. Brown ET (1987). *Analytical and Numerical Methods in Engineering Rock Mechanics*. Allen & Unwin, London.
25. Goodman RE (1973). On modelling techniques for the study of tunnels in jointed rock. *Proc. Int. Conf. Rock Mech.*, New Horizonsoin.
26. Goodman RE (1976). *Methods in Geological Engineering in Discontinuous Rock*, West, St Paul.
27. Drumm EC, Desai CS (1986). Determination of parameters for a model for cyclic behaviour of interfaces. *J. Earthquake Eng. Struct. Diy.*, **14**, 1–18.
28. Ohnishi Y, Mimuro T, Takewaki N, Yoshida J. (1985). Verification of input parameters for distinct elements analysis of jointed rock mass. *Proc. Int. Symp. Fundamentals of Rock Joints*. Bjorkliden, Sweden (ed. Stephansson O), pp. 205–214. Centek, Lulea, Sweden.
29. Ohkami T, Swoboda G (1999). Parameter identification of viscoelastic materials. *Comput Geotech.*, **24**, 279–295.
30. Ledesma A, Gens A, Alonso EE (1996). Parameter and variance estimation in geotechnical back-analysis using prior information. *Int. J. Num. Anal. Methods Geomech.*, **20**(2), pp. 119–141.
31. Ledesma A, Gens A, Alonso EE (1996). Estimation of parameters in geotechnical back analysis I – maximum likelihood approach. *Comput Geotech.*, **18**(1), 1–27.
32. Gens A, Ledesma A, Alonso EE (1996). Estimation of parameters in geotechnical back analysis. II – application to a tunnel problem., *Comput Geotech.*, **18**(1), 29–46.
33. Chatterjee S, Hadi A (1988). *Sensitivity Analysis in Linear Regression*. Wiley, New York.
34. Fiacco A (1984). *Sensitivity, Stability, and Parametric Analysis*. North-Holland, New York.
35. Paterson M (1987). Problems in the extrapolation of laboratory rheological data. *Tectonophysics* **133**, 33–43.
36. Patton TL, Fletcher RC (1998). A rheological model for fractured rock. *J. Struct. Geol.*, **20**(5), 491–502.
37. Cristescu N (1989). *Rock Rheology*. Kluwer, Dordrecht.
38. Cristescu N, Hunsche U (1998). *The time effects in rock mechanics*. Wiley, Chichester.
39. Tan TK (1993). The importance of creep and time-dependent dilatancy, as revealed from case records in China. *Comprehensive Rock Engineering* (ed. Hudson JA, Brown WT, Fairhurst C, Hoek E), Pergamon Press, Oxford. Vol. 3, pp. 709–744.
40. Bieniawski ZT (1970). Time-dependent behaviour of fractured rock. *Rock Mech.*, **2**, 123–137.

41. Hobbs DW (1970). Stress-strain-time behaviour of a number of coal measure rocks. *Int. J. Rock Mech. Min. Sci.*, **7**, 149–170.
42. Kaiser PK, Maloney S, Morgenstern NR (1983). Time-dependent behaviour of tunnels in highly stressed rock. *Proc. 5th Congr. Int. Soc. Rock Mech.*, Melbourne, pp. D329–D336. Balkema, Rotterdam.
43. Kaiser PK, Morgenstern NR (1979). Time-dependent deformation of jointed rock near failure. *Proc. 4th Congr. Int. Soc. Rock Mech.*, Montreux, Balkema, Rotterdam.
44. Peng SS (1973). Time-dependent aspects of rock behaviour as measured by servo-controlled hydraulic testing machine. *Int. J. Rock Mech. Min. Sci. & Geomech. Abstr.*, **10**, 235–246.
45. Wawersik WR (1973). Time-dependent rock behaviour in uniaxial compression. *Proc. 14th US Symp. Rock Mech.*, University Park, PA (ed. Hardy RR, Stefanko R), pp. 85–106. ASCE, New York.
46. Zienkiewicz OC, Pande GN (1977). Time dependent multilaminate model of rocks – A numerical study of deformation and failure of rock masses. *Int. J. Numer. Anal. Meth. Geomech.*, **1**, 219–247.
47. Berest P, Nguyen-Minh D (1983). Time dependent behaviour of lined tunnels in soft rocks. *Proc. Eurotunnel '83 Conf.*, Basle, pp. 57–62.
48. Gilbert MJ, Farmer IW (1981). A time dependent model for lining pressure based on strength concepts. *Proc. Symp. Weak Rock*, Tokyo, Vol. 1, pp. 137–142.
49. Belegundu A, Chandrupatla T (1999). *Optimization Concepts and Applications in Engineering*. Prentice Hall.
50. Ciarlet P (1995). *Introduction to Numerical Linear Algebra and Optimisation*. Cambridge University Press.
51. Horst R, Pardalos P (1995). *Handbook of Global Optimization*. Kluwer Academic Publishers.
52. Zhu W, Wu BL, Lin SS (1986). Optimum non-linear analysis on reasonable excavating order and anchoring treatment of large sized rock chambers. *Int. Symp. On Large Rock Caverns*, Helsinki.
53. Sakurai S, Akutagawa S, Takeuchi K, Shinji M, Shimizu N (2003). Back analysis for tunnel engineering as a modern observational method. *Tunnelling and Underground Space Technology*, **18**(2–3), 185–196.
54. Sakurai S, Takeuchi K (1983). Back analysis of measured displacements of tunnels. *Rock Mech. Rock Eng.*, **16**, 173–180.
55. Gioda G, Sakurai S (1987). Back analysis procedures for the interpretation of field measurements in geomechanics. *Int. J. Num. Anal. Methods Geomech.*, **11**, 555–583.
56. Yang L, Zhang K, Wang Y (1996). Back analysis of initial rock stress and time-dependent parameters. *Int. J. Rock Mech. Min. Sci. Geomech. Abstr.*, **33**(6), 641–645.
57. Shimizu N, Sakurai S (1983). Application of boundary element method for back analysis associated with tunneling problems. *Proc. 5th Int. Conf. Boundary Elements*, Hiroshima, pp. 645–654.
58. Yang L, Sterling RL (1989). Back analysis of rock tunnel using boundary element method. *J. Geotech. Eng. Div. Am. Soc. Civ. Eng.*, **115**, 1163–1169.
59. Hatzor YH, Benary R (1998). The stability of a laminated Vousoir beam: back analysis of a historic roof collapse using DDA. *Int. J. Rock Mech. Min. Sci.*, **35**(2), 165–181.

60. Sakurai S, Ine T, Shinji M (1988). Finite element analysis of discontinuous geological materials in association with field observations. *Proc. 6th Int. Conf. Numerical Methods in Geomech.*, Innsbruck (ed. Swoboda G) Vol. 3, pp. 2029–2034.
61. Sakurai S (1970). Stability of tunnel in viscoelastic-plastic medium. *Proc. 2nd Congr. Int. Soc. Rock Mech.*, Belgrade, Vol. 2, pp. 521–529.
62. Sakurai S, Ine T (1989). Strain analysis of jointed rock masses for monitoring the stability of underground openings. *Proc. Int. Symp. Computer and Physical Modelling in Geotechnical Engineering*, Bangkok, pp. 221–228.
63. Dolezelova M (1979). The influence of construction work sequence on the stability of underground openings. *Proc. 3rd Int. Conf. on Numerical Methods in Geomech.*, Aachen, pp. 561–569.
64. Whittaker BN, Frith RC (1990). *Tunnelling: Design, Stability and Construction*. p. 460. Institution of Mining and Metallurgy, London.
65. Wittke W (1979). Fundamentals for the design and construction of tunnels in swelling rock. *Proc. 4th Int. Congr. Rock Mech.*, Montreux, pp. 719–729. Balkema, Rotterdam.
66. Wittke W (1990). *Rock Mechanics: Theory and Applications with Case Histories*. Springer-Verlag, Berlin.
67. Sharma JS, Zhao J, Hefny AM (2000). NATM – effect of shotcrete setting time and excavation sequence on surface settlement. *Proc. Int. Conf. on Tunnels and Underground Struct.*, Singapore.
68. Yang Q, Li QQ (1985). Effect of excavation sequence on the stability of underground powerhouse. *Conf. on Structures in Rock Masses*, Beijing, PRC.
69. Hudson JA (2001). Rock engineering case histories: key factors, mechanisms and problems. *Rock Mechanics – a Challenge for Society, Proc. of the ISRM Regional Symp. EUROC2001* (ed. Sarkka, Eloranta), Espoo, Finland. Rotterdam: Balkema, pp. 13–20.
70. Zhao J, Bian HY, Cai JG (2003). A computerised Tunnelling Engineering System integrating neural network and expert system. *Tunnelling and Underground Space Technology*, **19**.
71. Beer G, Pousen A (1995). Efficient numerical modelling of faulted rock using the boundary element method. *Int. J .Rock Mech. Min. Sci.*, **32**(3), 117A.
72. Cundall PA, Lemos JV (1990). Numerical simulation of fault instabilities with a continuously-yielding joint model. *Rockbursts and Seismicity in Mines, Proc. 2nd Int. Symp. On Rockbursts and Seismicity in Mines* (ed. Fairhurst S), Balkema, Rotterdam, pp. 147–152.
73. Lei XY, Swoboda G, Zenz G (1995). Application of contact – friction interface element to tunnel excavation in faulted rock. *Comput Geotech.*, **17**, 349–370.
74. Amadei B, Goodman RE (1981). Formulation of complete plane strain problems for regularly jointed rocks. *Rock Mechanics from Theory to Application, Proc. 22nd US Symp. Rock Mech.*, Mass. Inst. Technol., Cambridge, Mass. pp. 245–251.
75. Bandis SC, Barton NR, Christianson M (1985). Application of a new numerical model of joint behavior to rock mechanics problems. *Proc. Int. Symp. Fundamentals of Rock Joints* (ed. Stephansson O), Cantek, Lulea, 345–355.
76. Barton N (1991). Modeling jointed rock behaviour and tunnel performance. *World Tunnelling*. **4**(7), 414–416.
77. Chen SH (1986). Analysis of elasto-plastic and visco-elastic jointed rock mass by finite element method. *J. Wuhan Hydr. And Elect. Inst.*, **6**, 79–86.

78. Desai CS, Fishman KL (1989). Plasticity based constitutive model with associated testing for joints. *Int. J. Rock Mech. Min. Sci. & Geomech. Abstr.*, **28**, 15–26.
79. Desai CS, Ma Y (1992). Modeling of joints and interfaces using the disturbed state concept. *Int. J. Numer. Anal. Methods Geomech.* **16**, 623–653.
80. Einstein HH, Hirschfield RC (1973). Model studies on mechanics of jointed rock. *J. Soil Mech. Found. Div., Am. Soc. Civ. Eng.*, **99**, 229–248.
81. Fossum AF (1985). Effective elastic properties for a randomly jointed rock mass. *Int. J. Rock Mech. Min. Sci. & Geomech. Abstr.*, **22**, 467–470.
82. Gens A, Carol I, Alonso EE (1995). Rock joints: implementation and applications. *Mechanics of geomaterial interfaces* (ed. Selvadurai APS, Boulon M), pp. 395–420. Elsevier, Amsterdam.
83. Gerrard CM (1982). Elastic models of rock masses having one, two and three sets of joints. *Int. J. Rock Mech. Min. Sci. & Geomech. Abstr.*, **19**, 15–23.
84. Goodman RE, Taylor RL, Brekke TL (1968). A model for the mechanics of jointed rock. *J. Soil Mech. Found. Div., Am. Soc. Civ. Eng.*, **94**, 637–659.
85. Heuze FE, Barbour TG (1982). New models for rock joints and interfaces. *J. Geotech. Eng. Div., Am. Soc. Civ. Eng.*, **108**, 757–776.
86. Jaeger JC (1970). Behavior of closely jointed rock. *Proc. 11th U.S. Symp. Rock Mech.*, Berkeley, CA (ed. Somerton WH), pp. 57–68.
87. Jing L (1987). Characterization of jointed rock mass by boundary integral equation method. *Proc. Int. Symp. on Min. Tech. and Sci.*, pp. 794–9. Xuzhou, Trans Tech Publications, China.
88. Lee JS, Pande GN (1999). A newjoint element for the analysis of media having discrete discontinuities. *Mech. Cohes-Frict. Mater.*, **4**, 487–504.
89. Ohnishi Y, Chan T, Jing L (1996). Constitutive models of rock joints. *Coupled thermohydro-mechanical processes of fractured media, Developments in Geotechnical Engineering* (ed Stephansson O, Jing L, Tsang CF), Vol. 79, pp. 57–92. Elsevier, Amsterdam.
90. Pariseau WG (1999). An equivalent plasticity theory for jointed rock masses. *Int. J. Rock Mech. Min. Sci.*, **36**, 907–918.
91. Ramamurthy T, Arora VK (1991). A simple stress-strain model for jointed rocks. *Proc. 7th Congr. Int. Soc. Rock Mech.*, Aachen (ed. Wittke W). Balkema, Rotterdam, Vol. 1, pp. 323–326.
92. Chen SG, Zhao J (1998). A study of UDEC modelling for blast wave propagation in jointed rock masses. *Int. J. Rock Mech. Min. Sci.*, **35**, 93–99.
93. Singh B (1973). Continuum characterization of jointed rock masses: Part I – the constitutive equations. *Int. J. Rock Mech. Min. Sci. & Geomech. Abstr.*, **10**, 311–335.
94. Wei L, Hudson JA (1988). A hybrid discrete-continuum approach to model hydro-mechanical behaviour of jointed rocks. *Engng. Geol.* **49**, 317–325.
95. Wu FQ (1988). A 3D model of a jointed rock mass and its deformation properties. *Int. J. Min. Geol. Eng.*, **6**, 169–176.
96. Wu FQ, Wang SJ (2001). A stress-strain relation for jointed rock masses. *Int. J. Rock Mech. Min. Sci.*, **38**, 591–598.
97. Brown ET (Ed) (1981). *Rock Characterization, Testing and Monitoring: ISRM Suggested Methods.* Pergamon Press, Oxford.
98. Desai CS, Drumm EC, Aman MM (1985). Cyclic testing and modelling of interfaces. *J. Geotech. Eng. Div., Am. Soc. Civ. Eng.*, **111**, 793–815.

99. Sun Z, Gerrard C, Stephansson O (1985). Rock joint compliance tests for compression and shear loads. *Int. J. Rock Mech. Min. Sci. & Geomech. Abstr.*, **22**(4), 197–213.
100. Brady BHG, Cramer ML, Hart RD (1985). Preliminary analysis of a loading test on a large basalt block. *Int. J. Rock Mech. Min. Sci. & Geomech. Abstr.*, **22**, 345–348.
101. Gaziev EG, Erlikmann SA (1971). Stresses and strains in anisotropic rock foundations (model studies). *Rock Fracture, Proc. Int. Symp. Rock Fracture.* Nancy, Paper 2-1.
102. Vutukuri VS, Lama RD, Saluja SS (1974). *Handbook on Mechanical Properties of Rocks*, Vol. 1, Trans. Tech. Publications, Clausthal.
103. Chen EP (1989). A constitutive model for jointed rock mass with orthogonal sets of joints. *J. Appl. Mech. Trans, ASME*, **56**, 25–32.
104. Ghaboussi J, Wilson EL, Isenberg J (1973). Finite element for rock joints and interfaces. *J. Soil Mech. Div., ASCE 99, SM10*, pp. 833–848.
105. Jing L, Nordlund E, Stephansson O (1994). A 3-D constitutive model for rock joints with anisotropic friction and stress dependency in shear stiffness. *Int. J. Rock Mech. Min. Sci. Geomech. Abstr.*, **31**(2), 173–178.
106. Wei ZQ, Hudson JA (1986). The influence of joints on rock modulus. *Proceedings of the International Symposium Rock Engineering in Complex Rock Formations.* Beijing, China: Science Press, pp. 54–62.
107. Lama RD, Vutukuri VS (1978). *Handbook on Properties of Rocks*, Vol. IV.
108. Hoover WG, De Groot AJ, Hoover CG (1992). Massively parallel computer simulation of plane-strain elastic-plastic flow via nonequilibrium molecular dynamics and Lagrangian continuum mechanics. *Computers Phys.*, **6**, 155–167.
109. Wheel MA (1996). A geometrically versatile finite volume formulation for plane elastostatic stress analysis. *J. Strain Anal.*, **31**(2), 111–116.
110. Clough RW (1960). The finite element method in plane stress analysis. *Proc. 2nd ASCE Conf. Electronic Computation*, Pittsburg, PA.
111. Crouch SL (1976). Solution of plane elasticity problems by the displacement discontinuity method. *Int. J. Numer. Methods Eng.*, Vol. 10, pp. 301–43.
112. Griffith AA (1921). The phenomena of rupture and flow in solids. *Phil. Trans. Roy. Soc.*, Vol.A221, 163–197.
113. Griffith AA (1924). Theory of rupture. *Proc. 1st Congr. Mech.*, Delft, pp. 55–63.
114. Atkinson BK (ed) (1987). *Fracture Mechanics of Rock*. Academic Press, London.
115. Bieniawski ZT (1967). Mechanism of brittle fracture of rock. *Int. J. Rock Mech. Min. Sci.*, **4**, 95–430.
116. Brace WF (1960). An extension of Griffith theory of fracture to rocks. *J.Geophys. Res.* Vol. 65, 3477–3480.
117. Brace WF, Byerlee J (1967). Recent experimental studies of brittle fracture of rocks. *Failure and Breakage of Rock*, p. 57. AIME, New York.
118. Walsh JB, Brace WF (1964) A fracture criterion for brittle anisotropic rock. *J. Geophys. Res.* **69**, 3449–3456.
119. Walsh JB (1965). The effect of cracks on the compressibility of rock. *J. Geophys. Res.* **70**, 2225–2236.
120. Brown ET (1974). Fracture of rock under uniform biaxial compression. *Advances in Rock Mechanics, Proc. 3rd Congr., Int. Soc. Rock Mech.*, Denver, Vol. 2A, Nat. Acad. Sciences, Washington, DC, pp. 111–117.
121. Wang S, Ge X (1997). Application of manifold method in simulating crack propagation. *Chin. J. Rock. Mech. Eng.* **16**(5), 405–410.

122. Chong KP, Hoyt PM, Smith JW (1980). Effects of strain rate on oil shale fracturing. *Int. J. Rock Mech. Min. Sci. & Geomech. Abstr.*, **17**, 35–43.
123. Cloos E (1990). Experimental analysis of fracture patterns. *Geol. Soc. Am. Bull.*, **66**, pp. 241–256.
124. Hoek E (1964). Fracture of anisotropic rock. *J.S. Afr. Inst. Min. Metall.*, **64**, 501–518.
125. Hoek E (1965). *Rock Fracture under Static Stress Conditions*. Nat. Mech. Engg. Res. Inst. Report MEG383, CSIR.
126. Hoek E, Bieniawski Z (1965). Brittle fracture propagation in rock under compression. *Int. J. Fract.*, **1**, 137–155.
127. Jaeger JC (1978). *Elasticity, Fracture and Flow*. Chapman & Hall, London.
128. Mogi K (1972). Effect of the triaxial stress system on the fracture and flow of rocks. *Phys. Earth planet. Int.*, **5**, 318–324.
129. Zhao J (2000). Applicability of Mohr-Coulomb and Hoek-Brown strength criteria to dynamic strength of brittle rock materials. *Int. J. Rock Mech. Min. Sci.*, **37**, 1115–1121.
130. Murrell SAF (1971). Micromechanical basis of the deformation and fracture of rocks. *Struct., Solid Mech. Eng. Des., Proc. Southampton Civ. Eng. Mater. Conf.*, Southampton, 1969 (Edited by Te'eni, M), Vol. 1, pp. 239–248. Wiley, Chichester.
131. Wawersik WR, Fairhurst C (1970). A study of brittle rock fracture in laboratory compression experiments. *Int. J. Rock Mech. Min. Sci. & Geomech. Abstr.*, **7**, 561–575.
132. Lockner DA, Byerlee JD, Kuksenko V, Pomonarev A, Sidorin A (1991). Quasi-static fault growth and shear fracture energy in granite. *Nature (London)*, **350**, 39–42.
133. Wang R, Zhao Y, Chen Y, Yan H, Yin Y, Yao C, Zhang H (1987). Experiment and finite element simulation of X-type shear fractures from a crack in marble. *Tectonophysics*, **144**, 141–150.
134. Yang SQ, Zeng YW, Zhao DS (1986). Support of underground powerhouse at Buluge hydropower station. *J. Rock Mech. and Engg* (in Chinese), **5**(4).
135. Gu JC, Liao XB (1989). Model test of support scheme for stress tunnels of Xiaolangdi powerhouse (in Chinese). *Proc. 2nd Conf. Rock Mech. and Engg*, Knowledge publisher.
136. Gu YQ (1981). *Stability Model for Underground Caverns in Sandstones* (in Chinese). Hydraulics Department, Tsing Hua University.
137. Zhu W, Wang K, Peng G (1984). The stability of underground power chambers in brittle rock. *Int. Conf. on Case Histories in Geotech. Engng.*, Rolla, Missouri.
138. Zienkiewicz OC (1977). *The finite element method in engineering sciences*, 3rd edn. McGraw-Hill, New York.
139. Zienkiewicz OC (1977). *The Finite Element Method*, 3rd edn., McGraw-Hill, London.
140. Zienkiewicz OC, Taylor RL (1989). *The Finite Element Methods*. McGraw-Hill, New York.
141. Zienkiewicz OC, Best B, Dullage C, Stagy KG (1968). Analysis of nonlinear problems in rock mechanics with particular reference to jointed rock systems. *Proc. 2nd Congr. Int. Soc. Rock Mech.*, Belgrade, Vol. 3, pp. 501–509.
142. Zienkiewicz OC, Cormeau IC (1974). Viscoplasticity and creep in elastic solids: A unified numerical solution approach. *Int. J. Numer. Meth. Eng.*, **8**, 821–845.
143. Zienkiewicz OC, Emson C, Bettess P (1983). A novel boundary in finite element. *Int. J. Numer. Methods Engng.*, **19**, 393–404.
144. Beer G (1984). BEFE – A combined boundary-finite element computer program. *Advances in Engineering Software*, **6**(2), 103–109.

145. Beer G, Meck JL (1981). Infinite domain elements. *Int. J. Numer. Methods Engng.*, **17**(1), 43–52.
146. Beer G, Watson JO (1992). *Introduction to Finite Boundary Element Method for Engineers.* Wiley, New York.
147. Brebbia CA, Dominguez J (1977). Boundary element method for potential problems. *J. Appl. Math. Modelling.*,**1**, 372–78; Metallurgy Publication, London.
148. Cundall PA (1971). A computer model for simulating progressive large-scale movements in blocky rock systems. *Proc. Symp. Int. Soc. Rock Mech.*, Nancy **1**, Paper II-8.
149. Starfield AM, Cundall PA (1988) Towards a methodology for rock mechanics modeling. *Int. J. Rock Mech. Min. Sci. & Geomech. Abstr.*, **25**, 99–106.
150. Cundall PA (1990). Distinct element models of rock and soil structure. *Analytical and Computational Methods in Engineering Rock Mechanics* (ed Brown ET), 129–162. Allen and Unwin, London.
151. Cundall PA, Board M (1988). A microcomputer program for modeling large-strain plasticity problems. *Proc. 6th Int. Conf. Numer. Meth. Geotech.*, Innsbruck, Balkema, Rotterdam, pp. 2201–2208.
152. Cundall PA, Hart RD (1989). Numerical modelling of discontinua. Keynote address., *1st US Conf. Discrete Element Method.*
153. Cundall PA, Hart RD (1992). Numerical modelling of discontinua. *Engng. Comput.*, **9**, 101–113.
154. Cundall PA, Marti J (1979). Some new developments in discrete numerical methods for dynamic modelling of jointed rock masses. *Proc. Conf. Rapid Excavation Tunnelling*, **2**, 1464–1466.
155. Curran JH, Ofoegbu GI (1993). Modeling discontinuities in numerical analysis. *Comprehensive Rock Engineering* (ed. Hudson JA, Brown WT, Fairhurst C, Hoek E), Vol. 1, pp. 443–68. Pergamon Press, Oxford.
156. Goodman RE, Shi G (1985). *Block Theory and Its Applications in Rock Engineering.* Prentice Hall, New Jersey.
157. Shi G (1988). Discontinuous deformation analysis, a new model for atatics and dynamics of block systems, PhD Thesis. University of California, Bekeley.
158. Shi G (1991). Manifold method of material analysis. *Transaction of the Ninth Army Conference on Applied Mathematics and Computing, Minneapolis*, MN. pp. 57–76.
159. Shi G, Goodman RE (1985). Two dimensional discontinuous deformation analysis. *Int. J. Numer. Anal. Methods Geomech.*, **9**, 541–56.
160. Brady BHG (1987). Boundary element and linked methods for underground excavation design. *Analytical and Computational Methods in Engineering Rock Mechanics* (ed. Brown ET), 164–204. Allen and Unwin, London.
161. Brady BHG, Wassyng A (1981). A coupled finite element-boundary element method of stress analysis. *Int. J. Rock Mech. Min. Sci. & Geomech. Abstr.*, Vol. 18, pp. 475–485.
162. Pan E, Amadei B, Kim YI (1998). 2D BEM analysis of anisotropic half-plane problems—application to rock mechanics. *Int. J. Rock Mech. Min. Sci.*, Vol. 35, 69–74.
163. Gerrard CM (1982). Equivalent elastic moduli of a rock mass consisting of orthorhombic layers. *Int. J. Rock Mech. Min. Sci. & Geomech. Abstr.*, **19**, 9–14.
164. Ingraffea A, Heuze F (1980). Finite element models for rock fracture mechanics. *Int. J. Numer. Anal. Methods Geomech.*, **4**, 24–43.

165. Tan XC, Kou SQ Lindqvist PA (1998). Application of the DDM and fracture mechanics model on the simulation of rock breakage by mechanical tools. *Eng. Geol.* **49**(4), 277–84.
166. Chaboche JL (1988). Continuum damage mechanics: Part I – General Concepts. *J. Appl. Mech.*, **55**, 59–64.
167. Chaboche JL (1988). Continuum damage mechanics: Part II – Damage growth, crack initiation, and crack growth. *J. Appl. Mech.*, **55**, 65–72.
168. Hart R, Cundall P, Lemos J (1988). Formulation of a three-dimensional distinct element model – part II: Mechanical calculations for motion and interaction of a system composed of many polyhedral blocks. *Int. J. Rock Mech. Min. Sci. & Geomech. Abstr.*, **25**, pp. 117–126.
169. Hart RD (1993). An introduction to distinct element modelling for rock engineering. *Comprehensive Rock Engineering* (ed. Hudson JA, Brown WT, Fairhurst C, Hoek E), Vol. 2, pp. 245–261. Pergamon Press, Oxford.
170. Jing L, Stephansson O (1994). Topological identification of block assemblage for jointed rock masses. *Int. J. Rock Mech. Min. Sci. Geomech. Abstr.*, **31**(2), 163–172.
171. Jing L, Stephansson O, Nordlund E (1993). Study of rock joints under cyclic loading conditions. *Rock Mech. Rock Engng.*, **26**(3), 215–232.
172. Brandis S, Lumsden AC, Barton NR (1981). Experimental studies of scale effects on the shear behaviour of rock joints. *Int. J. Rock Mech. & Min. Sci. & Geomech. Abstr.*, **18**, 1–21.
173. Barton N (1971). A relationship between joint roughness and joint shear strength. *Proc. Int. Symp. Rock Fracture*, Nancy, France. Paper I.8.
174. Barton N, Bandis S (1990) Review of predictive capabilities of JRC-JCS model in engineering practice. *Proc. Int. Conf. Rock Joints*, Loen, Norway (ed. Barton N and Stephansson O), Balkema, Rotterdam.
175. Barton N, Bandis S and Bakhtar K (1985). Strength, deformation and conductivity coupling of rock joints. *Int. J. Rock Mech. Min. Sci. & Geomech. Abstr.*, **22**, 121–140.
176. Barton N, Bandis SC (1983). Effects of block size on the shear behavior of jointed rock. *Proc. 23rd U.S. Symp. Rock Mech.*, Berkeley, CA (ed. Goodman RD and Heuze FE). Am. Soc. Min. Eng., New York, pp. 739–760.
177. Barton N, Choubey V (1977). The shear strength of rock joints in theory and practice. *Rock Mech.*, **10**, 1–54.
178. Barton N, Lien R, Lande J (1974). Engineering classification of jointed rock masses for the design of tunnel support. *Rock Mech.*, **6**, 189–236.
179. Bandis SC (1990). Scale effects in the strength and deformability of rocks and rock joints. *Proc. 1st Int. Workshop on Scale Effects in Rock Masses*, Loen, Norway (ed. Pinto da Cunha A), Balkema, Rotterdam, pp. 59–76.
180. Bandis SC, Lumsden AC, Barton NR (1981). Experimental studies of scale effects on the shear behavior to rock joints. *Int. J. Rock Mech. Min. Sci. & Geomech. Abstr.*, **18**, pp. 1–21.
181. Bandis SC, Lumsden AC, Barton NR (1983). Fundamentals of rock joint deformation. *Int. J. Rock Mech. Min. Sci. & Geomech. Abstr.*, **20**, 249–268.
182. Tsang YW, Witherspoon PA (1982). Correlation between fracture roughness characteristics and fracture mechanical and fluid flow properties. *Proc. 23rd US Symp. on Rock Mech.*, pp. 560–567.

183. Snow DT (1969). Rock fracture spacings, openings and porosities. *J. Soil Mech. Found Div., Am. Soc. Civ. Eng.*, **94**, 73–91.
184. Zhao J, Brown ET (1992). Hydro-thermo-mechanical properties of joints in the Carnmenellis granite. *Quart. J. Engng. Geol.*, **25**, 376–388.
185. Blyth FGH, De Freitas MH (1974). *A Geology for Engineers*, 6th edn. Edward Arnold, London.
186. Brady BHG, Brown ET (1985). *Rock Mechanics for Underground Mining*. Allen & Unwin, London.
187. Walsh JB, Grosenbaugh MA (1979). A new model for analysing the effect of fractures on compressibility. *J. Geophys. Res.*, **84**, 3532–3536.
188. Plesha ME, Ni D (2001). Scaling of geological discontinuity normal load-deformation response using fractal geometry. *Int. J. Numer. Anal. Methods Geomech.*, **25**, 741–756.
189. Plesha ME (1987). Constituitive models for rock discontinuities with dilantancy and surface degradation. *Int. J. Numer. Anal. Methods Geomech.*, **11**, 345–362.
190. Mahtab MA, Bolstad DD, Alldredge JR, Shanley RJ (1972). *Analysis of Fracture Orientations for Input to Structural Models of Discontinuous Rock*, US Bur. Mines Re. Invest. 7669.
191. Derahowitz WS, Einstein HH (1988). Characterizing rock joint geometry with joint system models. *Rock Mech. Rock Engng.*, **20**(1).
192. Donath FA (1961). Experimental study of shear failure in anisotropic rocks. *Geol. Soc. Am. Bull.* **72**, 985–990.
193. Goodman RE (1989). *Introduction to Rock Mechanics*, 2nd edn. Wiley, New York.
194. Hoek E, Brown ET (1980). *Underground Excavation in Rock*. p. 527, Inst. of Min. and Metall., London.
195. Hudson JA (1989). *Rock Mechanics Principles in Engineering Practice*. Butterworth, London.
196. Hudson JA, Harrison JP (1994). *Engineering Rock Mechanics*. Elsevier, Oxford.
197. Hudson JA, Priest SD (1979). Discontinuities and rock mass geometry. *Int. J. Rock Mech. Min. Sci. & Geomech. Abstr.*, **16**, 339–362.
198. Jaeger JC (1972). *Rock Mechanics and Engineering*. Cambridge University Press, Cambridge.
199. Jaeger JC, Cook NGW (1979). *Fundamentals of Rock Mechanics*, 3rd edn. Chapman & Hall, London.
200. Jaeger JC, Rosengren KJ (1969). Friction and sliding of joints. *Proc. Aust. Inst. Min. Metall.*, No. 229, pp. 93–104.
201. Koch GS, Link RF (1971). *Statistical Analysis of Geological Data*. Vol. 2. Wiley, New York.
202. Priest SD (1992). *Discontinuity Analysis for Rock Engineering*. Chapman & Hall, London.
203. Yeung MR, Loeng LL (1997). Effects of joint attributes on tunnel stability. *Int. J. Rock Mech. Min. Sci.*, **34**(3–4), Paper No. 348.
204. Cawsey DC (1977). The measurement of fracture patterns in the Chalk of Southern England. *Eng. Geol.* (Amsterdam), **11**, 210–215.
205. Brown ET (1970). Strength of models of rock with intermittent joints. *J. Soil Mech. Foundns. Div., ASCE*, **96**, 1935–1949.
206. Brown ET (1976). Volume changes in models of jointed rock. *J. Geotech. Eng. Div., ASCE*, **102**, 273–276.

207. Sirat M, Talbot CJ (2001). Application of artificial neural networks to fracture analysis at the Aspo HRL, Sweden: fracture sets classification. *Int. J. Rock Mech. Min. Sci.*, **38**(5), 621–639.
208. Drucker DC, Prager W (1952). Soil mechanics and plastic analysis of limit design. *Q. Appl. Math.*, **10**, 157–165.
209. Prager W (1959). *An Introduction to Plasticity*, Addison-Wesley, Reading, Mass.
210. Attewell PB and Farmer IW (1976). *Principles of Engineering Geology*. Chapman & Hall, London.
211. Kovari K (1979). Basic considerations on the design of underground openings. *Proc. 3rd Int. Assoc. Bridge and Struct. Engng.*
212. Hoek E (1981). Geotechnical design of large openings at depth. *Proc. Conf. Rapid Excavation and Tunnelling*, pp. 1167–1180. AIME, New York.
213. Hoek E (1982). Geotechnical considerations in tunnel design and contract preparation. *Trans, Instn. Min. Metall.*, **91**, pp.A101–A109.
214. Franklin JA, Dusseault MB (1989). *Rock Engineering*, p. 601. McGraw-Hill, Toronto.
215. Brown ET, Bray JW, Ladanyi B, Hoek E (1983). Ground response curves for rock tunnels. *J. Geotech. Engng.*, **109**, 15–39.
216. Broch E (1982). Designing and excavating underground power plants. *Water Power and Dam Construction*, **34**, 19–25.
217. Broch E (1982). The development of unlined pressure shafts and tunnels in Norway. In *Rock Mechanics: Caverns and Pressure Shafts* (ed. Wittke W), pp. 545–554. Balkema, Rotterdam.
218. Broch E, Sorheim S (1984). Experiences from the planning, construction and supporting of road tunnel subjected to heavy rockbursting. *Rock Mech. Rock Eng.*, **17**, 15–35.
219. Horii H, Nemat-Nasser S (1986). Brittle failure in compression: splitting, faulting and brittle-ductile transition. *Philos. Trans. R. Soc., London, Ser.* A., **1549**, 337–374.
220. Reyes O, Einstein HH (1991). Failure mechanics of fractured rock – A fracture coalescence model. *Proc. 7th Int. Congr. Rock Mech.* (ed. Wittke E), Balkema, Rotterdam.
221. Einstein HH, Thompson DE, Azzouz AS, O'Reilly KP, Schultz MS, Ordun S (1983). Comparison of five empirical tunnel classification methods – Accuracy, effect of subjectivity and available information. *Proc. 5th Int. Congr. Rock Mech., Melbourne*, pp. C303–C313. Balkema, Rotterdam.
222. Gerrard CM (1977). Background to mathematical modeling in geomechanics. *Finite Elements in Geomechanics* (ed. Guhehus G), pp. 33–120. Wiley, London.
223. Goodman RE, Christopher SJ (1977). Finite element analysis for discontinuous rocks. *Numerical Methods in Geotechnical Egineering* (ed. Desai CS, Christian JT), pp. 148–175. McGraw-Hill, Toronto.
224. Karanagh K, Clough RW (1971). Finite element application in the characterization of elastic solids. *Int. J. Solids Struct.*, **7**, 11–13.
225. Kawai T (1980). Some considerations on the finite element method. *Int. J. Numer. Methods Eng.*, **6**, 81–120.
226. Naylor DJ, Pande GN, Simpson B, Tabb R (1981). *Finite Elements in Geotechnical Engineering*. Pineridge Press, Swansea, UK.
227. Owen DRJ, Hinton E (1980). *Finite Elements in Plasticity: Theory and Applications. Swansea*, Pineridge Press, UK.

228. Reyes SF, Deere DU (1966). Elastoplastic analysis of underground openings by finite element methods. *Proc. 1st Int. Congr. Rock Mech.*, Vol. II, Lisbon.
229. Sharma KG, Varadarajan A, Srivastava RK (1985). Elasto-viscoplastic finite element analysis of tunnels. *Proc. 5th Int. Conf. Numerical Methods in Geomechanics, Nagoya* (ed. Kawamoto T, Ichikawa Y). Balkema, Rotterdam, 1141–1148.
230. Swenson DV, Ingraffea AR (1988). Modelling mixed dynamic crack propagation using finite elements: theory and applications. *Computational Mechanics*, Vol. 3, pp. 187–192.
231. Barton N, Hansteen H (1979). Very large span openings at shallow depth: deformation magnitudes from jointed models and finite element analysis. *Proc. 4th Rapid Excavation and Tunnelling Conf.*, Vol. 2, Atlanta, pp. 1331–1353.
232. Krauland N, Nilsson G, Jonasson P (1981). Comparison of rock mechanics observations and measurements with FEM calculations. *Applications of Rock Mechanics to Cut and Fill Mining* (ed. Stephansson O, Jones MJ), Instn. Min. Metall., London, pp. 250–260.
233. Pande GN, Beer G, Williams JR (1990). *Numerical Methods in Rock Mechanics*, Wiley, London.
234. Brebbia CA, Telles JCF, Wrobel LC (1984). *Boundary Element Techniques: Theory and Applications in Engineering*. Springer, Berlin.
235. Beer G (1988). Application of 3-D boundary element and coupled analysis in geomechanics case studies. *Proc. 6th Int. Conf. Numer. Modelling in Geomech.*, Innsbruck, pp. 2209–2216.
236. Shou KJ (2000). A three-dimensional hybrid boundary element method for non-linear analysis of a weak plane near an underground excavation. *Tunnelling Underground Space Technology*. **16**(2), 215–26.
237. Venturini WS, Brebbia CA (1983). Some applications of the boundary element method in geomechanics. *Int. J. Numer. Methods Engng.*, **7**, 419–43.
238. Wang BL, Ma QC (1986). Boundary element analysis methods for ground stress field of rock masses. *Comput Geotech.*, **2**, 261–74.
239. Watson JO (1979). Advanced implementation of the boundary element method for two- and three-dimensional elastostatics. *Developments in Boundary Element Methods* (ed. Banerjee PK, Butter.eld R), **1**, Applied Science Publishers, London. pp. 31–63.
240. Cerrolaza M, Garcia R (1997). Boundary elements and damage mechanics to analyze excavations in rock mass. *Eng. Anal. Boundary Elements*, **20**, 1–16.
241. Brady BHG, Brown ET (1981). Energy changes and stability in mine structures: design applications of boundary element methods. *Trans. Instn. Min. Metall.*, **90**, pp. A61–68.
242. Jing L (1998). Formulation of discontinuous deformation analysis (DDA) – an implicit discrete element model for block systems. *Eng. Geol.* **49**, 371–81.
243. Lemos JV, Hart RD, Cundall PA (1985). *A Generalized Distinct Element Method for Modelling Jointed Rock Mass. Fundamentals of Rock Joints* (ed. Stephansson O), pp. 335–343. Centak, Lulea.
244. Ghaboussi J (1988). Fully deformable discrete element analysis using a finite element approach. *Int. J. Comput. Geotech.*, **5**, 175–195.
245. Dusseault MB, Rothenburg L, Kornelson PK (1991). Transient response simulation using a discrete element approach. *Proc. KALI '91*, Hamburg.

246. Chen SG, Cai JG, Zhao J, Zhou YX (2000). Discrete element modelling of underground explosions in jointed rock mass. *Geol. and Geotech. Engng.*, **18**, 59–78.
247. Swoboda G, Mertz W, Beer G (1986). Application of coupled FEM-BEM analysis for three-dimensional tunnel analysis. *Proc. Int. Conf. Boundary Elements* (ed. Du QH), Beijing. Pergamon Press, Oxford.
248. Gioda G, Carini A (1985). A combined boundary element–finite element analysis of lined openings. *Rock Mech. Rock Eng.* **18**, 293–302.
249. Swoboda G, Mertz W, Beer G (1987). Rheological analysis of tunnel excavations by means of coupled finite element (FEM)–boundary element (BEM) analysis. *Int. J. Numer. Anal. Methods Geomech.*, **11**, 15–129.
250. Larig, LJ, Brady BHG, Cundall PA (1986). Hybrid distinct element boundary element analysis of jointed rock. *Int. J. Rock Mech. Min. Sci. & Geomech. Abstr.*, **23**, pp. 303–312.
251. Lorig LJ, Brady BHG (1982). A hybrid discrete element-boundary element method of stress analyss. *Issues in Rock Mechanics, Proc. 23rd US Symp. Rock Mech.* (ed. Goodman RE, Heuze FE), AIME, New York, 628–636.
252. Chen, SG (1999). Discrete element modelling of rock cavern under dynamic loading. PhD thesis, Nanyang Technological University, Singapore.
253. Pan XD, Reed MB (1991). A coupled distinct element–finite element method for large deformation analysis of rock masses. *Int. J. Rock Mech. Min. Sci. Geomech. Abstr.*, **28**(1), 93–99.
254. Bedford T, Cooke R (2001). *Probabilistic Risk Analysis: Foundations and Methods.* Cambridge University Press, Cambridge.
255. Kovari K, Hagedorn H, Fritz P (1976). Parametric studies as a design aid in tunnelling. *Proc. 2nd Int. Conf. Numer. Methods in Geomech.*, Blacksburg, VA (ed. Desai CS). ASCE, New York.
256. Sun J, Lee YS (1985). A viscous elasto-plastic numerical analysis of the underground structure interacted with families of multi laminate rock mass using FEM. *Proc. 5th Int. Conf. Numer. Methods in Geomech., Nagoya* (ed. Kawamoto T, Ichikawa Y). Balkema, Rotterdam, pp. 1127–1134.
257. Wu C, Hao H, Zhao J, Zhou YX (2001). Statistical analysis of anisotropic damage of the Bukit Timah granite. *Rock Mech. Rock Engng.*, **34**, 23–28.
258. Wu C, Hao H, Zhou Y (1999). Fuzzy-random probabilistic analysis of rock mass response to explosive loads. *Comput. Geotech.*, **25**, 205–225.
259. Baecher GB (1983). Statistical analysis of rock mass fracturing. *J. Math. Geol.*, **15**, 329–347.
260. Baecher GB, Lanney NA, Einstein HH (1978). Statistical description of rock properties and sampling. *Proc. 18th US Symp., Rock Mech. Keystone* (ed. Wang FD and Clark GB), pp. 5C1–8, University of Nevada.
261. Bardsley WE, Major TJ, Selby MJ (1990). Note on a Weibull property for joint spacing analysis. *Int. J. Rock Mech. Min. Sci. & Geomech. Abstr.*, **27**, 133–134.
262. Evans HE (1984). *Mechanisms of Creep Fracture.* Elsevier, New York.
263. Langer M (1979). Rheological behaviour of rock masses. *Proc. 4th Int. Congr. Rock Mech.*, Vol. 1, pp. 29–62. Montreux, Balkema, Rotterdam
264. Griggs D (1939). Creep of rocks. *J. Geo.*, **47**, 225–251.

265. Cividini A, Gioda G, Barla G (1985). Calibration of a rheological material model on the basis of field measurements. *Proc. 5th Int. Conf. Numer. Methods in Geomech.*, Nagoya (ed. Einsenstein Z) pp. 1621–1628.
266. Ito H (1983). Creep of rock based on long-term experiments. *Proc. 5th Int. Congr. Rock Mech.*, Melbourne, Section A, pp. 117–120. Balkema, Rotterdam.
267. Hofer KH, Knoll P (1971). Investigations into the mechanism of creep deformation in Carnallitite and practical applications. *Int. J. Rock Mech. Min. Sci.*, **8**, 1–73.
268. Nair K, Boresi AP (1970). Stress analysis of time-dependent problems in rock mechanics. *Proc. 2nd Congr. Int. Soc. Rock Mech.*, Belgrade, **2**, 531–536.
269. Akai K, Ohnishi Y (1983). Strength and deformation characteristics of soft sedimentary rock under repeated and creep loading. *Proc. 5th Congr. Int. Soc. Rock Mech.*, Melbourne, Vol. 1, pp. A121–A124. Balkema, Rotterdam.
270. Kranz RL (1979). Crack growth and development during creep of Barre granite. *Int. J. Rock Mech. Min. Sci. & Geomech. Abstr.*, **16**, 23–25.
271. Singh DP (1975). A study of creep of rocks. *Int. J. Rock Mech. Min. Sci. & Geomech. Abstr.*, **12**, 271–275.
272. Schwartz CW, Kolluru S (1981). Laboratory investigation of creep of jointed rock materials. *Proc. 22nd US Symp. Rock Mech.*, Boston, MA (ed. Einstein HH), pp. 283–289.
273. Carter NL, Kirby SH (1978). Transient creep and semi-brittle behaviour of crystalline rocks. *Pure Appl. Geophys.*, Vol. 116, pp. 807–839.
274. Carter NL, Anderson DA, Hansen FD, Kranz RL (1981). Creep and creep rupture of granitic rocks. Mechanical Behavior of Rocks, *Geophys. Monogr., Am. Geophys. Union.* **24**, pp. 61–82.
275. Cruden DM (1971). The form of the creep law for rock under uniaxial compression. *Int. J. Rock Mech. Min. Sci.*, **8**, 105–126.
276. Semple RM, Hendron AJ, Mesri G (1973). *The Effect of Time-dependent Properties of Altered Rock on Tunnel Support Requirements.* Final Report No. FRA-ORDD-74–30, Department of Civil Engineering, University of Illinois at Urbana-Champaign.
277. Chin HP, Rogers JD (1987). Creep parameters of rocks on an engineering scale. *Rock Mech. Rock Engng.*, **20**, 137–146.
278. Shao JF, Duveau G, Hoteit N, Sibat M, Bart M (1997). Time dependent continuous damage model for deformation and failure of brittle rock. *Int. J. Rock. Mech. Min. Sci.*, **34**(3/4), 385.
279. Crawford AM, Curran JH (1983). A displacement discontinuity approach to modelling the creep behavior of rock and its discontinuities. *Int. J. Numer. Anal. Methods Geomech.*, **7**, 245–268.
280. Sulem J, Panet M, Guenot A (1987). An analytical solution for time-dependent displacements in a circular tunnel. *Int. J. Rock Mech. Min. Sci. & Geomech. Abstr.*, **24**, pp. 155–164.
281. Tharp TP (1997). Time-dependent compressive failure around an opening. *Int. J. Rock. Mech. Min. Sci.* **34**(3/4), p. 380.
282. Walton RJ, Worotnicki G (1986). A comparison of three borehole instruments for monitoring the change of stress with time. *Proc. Int. Symp. on Rock Stress and Rock Stress Measurement*, Stockholm (ed. Stephansson O), Cantek, Lulea, pp. 479–488.

283. Gill DE, Ladanyi B (1987). Time-dependent ground response curves for tunnel lining design. *Proc. 6th Congr. Int. Soc. Rock Mech.*, Montreal (ed. Herget G and Vargpaisal S). Vol. 2, pp. 917–921. Balkema, Rotterdam.
284. Howing KD, Kutter HK (1985). Time-dependent shear deformation of filled rock joints – a keynote lecturer. *Proc. Int. Symp. Fundamentals of Rock Joints*. Bjorkliden, Sweden (ed. Stephansson O), pp. 113–122. Centek, Lulea, Sweden.
285. Ladanyi B, Gill DE (1984). Tunnel lining design in a creeping rock. *Design and Performance of Underground Excavations*, Vol. 1, 19–26, Cambridge University Press, Cambridge.
286. Ladanyi B, Gill DE (1988). Design of tunnel linings in a creeping rock. *Int. J. Min. Geol. Eng.*, **6**, 113–126.
287. Cristescu N (1985). Viscoplastic creep of rocks around horizontal tunnels. *Int. J. Rock Mech. Min. Sci. & Geotech. Abstr.*, **22**, 453–459.
288. Cristescu N (1988). Viscoplastic creep of rocks around a lined tunnel. *Int. J. Plasticity*, **4**, 393–412.
289. Gill DE, Ladanyi B (1983). The characteristic line concept of lining design in creeping ground. *Proc. CIM Symp. Underground Support Systems*, Sudbury, Canada.
290. Nawrocki PA, Mroz Z (1999). A constitutive model for rock accounting for viscosity and yield stress degradation. *Comput. Geotech.*, **25**, 247–280.
291. Desai CS, Zhang D (1987). Viscoplastic model for geologic materials with generalized flow rule. *Int. J. Numer. Anal. Methods Geomech.*, **11**, 603–620.
292. Desai CS, Hashmi QSE (1989). Analysis, evaluation, and implementation of a nonassociative model for geological materials. *Int. J. Plasticity*, **5**, 397–420.
293. Sakurai S (1974). Determination of initial stresses and mechanical properties of visco-elastic underground medium. *Proc. 3rd ISRM Cong.*, Denver, pp. 1169–1174.
294. Abousleiman Y, Cheng AHD, Jiang C, Roegiers JC (1993). A micromechanically consistent poroviscoelasticity theory for rock mechanics application. *Int. J. Rock Mech. Min. Sci. Geomech. Abstr.*, **30**(7), 1177–1180.
295. Hill R (1950). *The Mathematical Theory of Plasticity*. Oxford University Press, Oxford.
296. Dragon A, Mroz Z (1979). A continuum model for plastic brittle behaviour of rock and concrete. *Int. J. Rock Mech. Min. Sci. & Geomech. Abstr.*, **16**, 253–259.
297. Nedjar B (2002). Frameworks for finite strain viscoelastic-plasticity based on multiplicative decompositions. Part I: continuum formulations. *Comput. Methods Appl. Mech. Engng.* **19**, 1541–1562.
298. Nedjar B (2002). Frameworks for finite strain viscoelastic-plasticity based on multiplicative decompositions. Part II: computational aspects. *Comput. Methods Appl. Mech. Eng.*, **19**, 1563–1593.
299. Nemat-Nasser S (1983). On plasticity of crystalline solids and geomaterials. *J. Appl. Mech. Trans., ASME*, **50**, 1114–1126.
300. Gioda G, Cividini A (1981). Viscous behaviour around an underground opening in a two-phase medium. Technical note. *Int. J. Rock Mech. Min. Sci. & Geomech. Abstr.*, **18**, 437–773.
301. Giode G, Maiey G (1980). Direct search solution of an inverse problem in elasto-plasticity, identification of cohesion, friction angle and in situ stress by pressure tunnels tests. *Int. J. Numer. Methods Engng.*, **15**, 1822–1848.
302. Perzyna P (1966). Fundamental problems in viscoplasticity. *Adv. Appl. Mech.*, **9**, 243–377.

303. Perzyna P (1963). The constitutive equation for rate sensitive plastic materials. *Q. Appl. Math.*, **20**, 321–332.

304. Petukhov IM, Linkov AM (1979). The theory of post-failure deformations and the problem of stability in rock mechanics. *Int. J. Rock Mech. Min. Sci. & Geotech. Abstr.*, **16**, 57–76.

305. Swoboda G (1985). Interpretation of field measurements under consideration of the three-dimensional state of stress, the viscoelasticity of shotcrete, and the viscoplastic behaviour of rock., *Proc. 5th Int. Conf. Numerical Methods in Geomechanics, Nagoya* (ed. Kawamoto T, Ichikawa Y). Balkema, Rotterdam, pp. 1127–1134.

306. Shao JF (1998). Poroelastic behaviour of brittle rock materials with anisotropic damage. *Mech Mater*. **30**, 41–53.

307. Zhang W, Zhang XH (1987). Elastic model of jointed rock mass (in Chinese). *J. Geotech. Engng.*, **9**(4), 33–44.

308. Bray JW (1986). Some applications of elastic theory. *Analytical and Computational Methods in Engineering Rock Mechanics* (ed. Brown ET), Allen & Unwin, London, pp. 32–94.

309. Chen WF, Saleeb AF (1982). Constitutive Equations for Engineering Materials – Vol. 1 Elasticity and Modelling, p. 580, Wiley, New York.

310. Brown ET, Hoek E (1978). Trends in relationships between measured in situ stresses and depth. *Int. J. Rock Mech. Min. Sci. & Geomech. Abstr.*, **15**, 211–215.

311. Crawford AM, Bray JW (1978). Influence of the in situ stress field and joint stiffness on rock wedge stability in underground openings. *Can. Geotech. J.*, Vol. 20, 276–287.

312. Haimson BC (1978). The hydrofracturing stress measuring method and recent field trials. *Int. J. Rock Mech. Min. Sci. & Geomech. Abstr.*, **15**, 167–178.

313. Enever JR, Chopra PN (1986). Experience with hydraulic fracture stress measurements in granite. *Proc. Int. Symp. On Rock Stress and Stress Measurement*, Stockholm (ed. Stephansson O), Centek Lulea, pp. 411–420.

314. Ewy RT, Kemeny JM, Zheng Z, Cook NGW (1987). Generation and analysis of stable excavation shapes under high rock stresses. *Proc. 6th Congr., Int. Soc. Rock Mech.*, (Montreal), Balkema, Rotterdam, pp. 875–881.

315. Feng ZL, Lewis RW (1987). Optimal estimation of in situ ground stresses from displacement measurement. *Int. J. Numer. Anal. Methods Geomech.*, **11**, 391–408.

316. Ortlepp WD, Gay NC (1984). Performance of an experimental tunnel subjected to stresses ranging from 50 Mpa to 230 Mpa. *Design and Performance of Underground Excavations*, Cambridge, British Geotech. Soc., London, pp. 337–346.

317. Rutter EH, Atkinson BK, Mainprice DH (1978). On the use of the stress relaxation testing method in studies of the mechanical behaviour of geological materials. *Geophys. J.R. Astron. Soc.*, **55**, 155–170.

318. Sauer G, Sharma B (1977). A system for stress measurement in construction in rock. *Field Measurements in Rock Mechanics* (ed. Kovari K), Vol. 1, pp. 317–329. Balkema, Rotterdam.

319. Zhou YX, Zhao J, Cai JG, Zhang XH, (2003). Effects of high horizontal stresses on the deformation of large-span rock caverns. *Tunnelling and Underground Space Technology*, Vol. 19.

320. Cook NGW (1967). The design of underground excavations. *Failure and Breakage of Rock, Proc. 8th US Symp. Rock Mech.* (ed. Fairhurst C.), AIME, New York, pp. 167–193.

321. Deere DU, Hendron Aj, Patton ED, Cording EJ (1967). Design of surface and near-surface construction in rock. *Failure and Breakage of Rock* (ed. Fairhurst C), pp. 237–302. Society of Mining Engineers of AIME, New York.
322. Maury V (1987). Observation, researches and recent results about failure mechanism around single openings. Report of the Rock Failure Mechanism around Underground Openings SRM Comm. *Proc. 6th Int. Soc. Rock Mech. Congr.*, Montreal (ed. Herger G and Vangpaisal S), Vol. 2, pp. 1119–1128. Balkema, Rotterdam.
323. Hammett RD, Hoek E (1981). Design of large underground caverns for hydroelectric projects, with reference to structurally controlled failure mechanisms. *Proc. Am. Soc. Civ. Eng. Int. Conv. Recent Development in Geotechnical Engineering for Hydro Projects*, pp. 192–206. ASCE, New York.
324. Hudson JA, Brown ET, Fairhurst C (1972). Soft, stiff, and servo-controlled testing machines: a review with reference to rock failure. *Engng. Geol.*, **6**, 155–189.
325. Wawersik WR, Brace WF (1971). Post-failure behavior of a granite and diabase. *Rock Mech.* **3**, 61–85.
326. Kovari K, Tisa A (1975). Multiple failure state and strain controlled triaxial tests. *Rock Mech.*, **7**, 17–33.
327. Patton FD (1966). Multiple modes of shear failure in rock. *Proc. 1st Congr., Int. Soc. Rock Mech.*, Lisbon, **1**, 509–513.
328. Mogi K (1971). Effect of the triaxial stress systems on the failure of dolomite and limestone. *Tectonophysics*, **11**, 111–127.
329. Mogi K (1971). Fracture and flow of rocks under high triaxial compression. *J. Geophys. Res.* **76**, 1255–1269.
330. Hoek E (1990). Estimating Mohr-Coulomb friction and cohesion values from the Hoek-Brown failure criterion. *Int. J. Rock Mech. Min. Sci. & Geomech. Abstr.*, **27**, 227–229.
331. Hoek E, Brown ET (1980). Empirical strength criterion for rock masses. *J. Geotech. Eng. Div., ASCE*, **106**, 1030–1035.
332. Hoek E, Brown ET (1988). The Hoek-Brown criterion – a 1988 update. *Proc. 15th Can. Rock Mech. Symp.*, Toronto University Press, Toronto, pp. 31–38.
333. Hoek E, Brown ET (1997). Practical estimates of rock mass strength. *Int. J. Rock Mech. Min. Sci.*, **34**(8), 1165–1186.
334. Sheorey PR (1997). *Empirical Rock Failure Criteria*. Balkema, Rotterdam.
335. Kumar P (1998). Shear failure envelope of Hoek–Brown criterion for rock mass. *Tunnelling Underground Space Technology*, **13**(4), 453–458.
336. Lai YS, Wang CY, Tien YM (1999). Modified Mohr–Coulomb-type micro-mechanical failure criteria for layered rocks. *Int. J. Numer. Anal. Methods Geomech.*, **23**, 451–460.
337. Lee C, Sterling R (1992). Identifying probable failure modes for underground openings using a neural network. *Int. J. Rock Mech. Min. Sci.*, **29**(1), 49–67.
338. Yudhbir ET (1983). An empirical failure criterion for rock masses. *Proc. 5th Int. Congr. Rock Mech.* Vol. 1, Melbourne, pp. B1–B8, Balkema, Rotterdam.
339. Brown ET, Hudson JA (1974). Fatigue failure characteristics of some models of jointed rock. *Earthquake Eng. And Struct. Dyn.*, **2**, 379–386.
340. Daemen JJK, Fairhurst C (1972). Rock failure and tunnel support loading. *Proc. Int. Symp. Underground Openings*, Lucerne.

341. Voegele M, Fairhurst C, Cundall P (1978). Analysis of tunnel support loads using a large displacement, distinct block model. *Storage in Excavated Rock Caverns* (ed. Bergman M), Vol. 2, Pergamon, Oxford, pp. 247–252.
342. Wahlstrom EE (1973). *Tunnelling in Rock*. Elsevier, Amsterdam.
343. Ward WH (1980). 18th Rankine Lecture: Ground Supports for tunnels in weak rocks. *Geotechnique*, **28**, 133–170.
344. Wickham GE, Tiedemann HR, Shinner EH (1974). Ground support prediction model-RSR concept. *Proc. 2nd Rapid Excavation and Tunnelling Conf.*, San Francisco, Vol. 1, pp. 691–707.
345. Wilson AH (1980). A method of estimating the closure and strength of lining required in derivages surrounded by a yield zone. *Int. J. Rock Mech. Min. Sci. & Geotech. Abstr.*, **17**, 349–355.
346. Hoek E, Wood D (1986). Rock Support. *World Tunnelling and Subsurface Excavation.* **2**, 131–134.
347. Windsor CR, Thompson AG, Choi SK, (1988). Rock reinforcement research for hard rock mining. *Proc. WASM Conference '88 – R&D for the Minerals Industry*. Kalgoolie, pp. 113–121.
348. Barton N (1976). Recent experiences with the Q-system of tunnel support design. *Proc. Symp. Exploration for Rock Engineering*, Johannesburg (ed. Bieniawshi ZT). ASCE, New York, Vol. 1, 163–177; discussion 234–241.
349. Barton N (1983). Application of Q-system and index tests to estimate shear strength and deformability of rock masses. *Proc. Int. Symp. On Engng. Geol. and Underground Constr.*, Vol. 2, Lisbon, pp. II.51–70.
350. Barton N (1988). Rock mass classification and tunnel reinforcement selection using the Q-system. *Rock Classification Systems for Engineering Purposes* (ed. Kirkaldie L). ASTM, Philadelphia, pp. 59–88.
351. Daemen JJK (1975). Tunnel support loading caused by rock failure. *Tech. Rep. MRD-3–75*, Missouri River Div., US Corps Engrs, Omaha.
352. Daemen JJK (1977). Problems in tunnel support mechanics. *Underground Space*. Vol. 1, pp. 163–172.
353. Franklin JA (1976). An observation approach to the selection and control of rock tunnel linings. *Proc. Conf. Shotcrete for Ground Support, ASCE*, Easton, MD, pp. 556–596.
354. Gen HS, Goodman RE (1981). A new concept for support of underground and surface excavations in discontinuous rocks based on a keystone principle. *Rock Mechanics from Research to Application, Proc. 22nd US Symp. Rock Mech.*, Mass. Inst. Techol., Cambridge, Mass., pp. 290–296.
355. Grimstad E, Barton N (1988). Design and methods of rock support. *Norw. Tunnelling Today*, **5**, 59–64.
356. Heuer A (1973). Use of shotcrete for underground structural support. *Proc. Conf. Engineering Foundation*, South Berwick, ME, American Concrete Institute Publication SP-45.
357. Jethwa JL, Dube AK, Singh B, Mithal RS (1982). Evaluation of methods for tunnel support design in squeezing rock conditions. *Proc. 4th Int. Congr. International Association of Engng. Geol.*, Vol. 5, pp. 125–134, Balkema, Rotterdam.
358. Kaiser PK, McCreath DR (1992). *Rock Support*. Balkema, Rotterdam.

359. Douglas TH, Arthur LJ (1983). *A Guide to the Use of Rock Reinforcement in Underground Excavation*. CIRIA, London.
360. Muir Wood AM (1979). Ground behaviour and support for mining and tunneling. *Tunnelling '79*, pp. xi–xxiii. Institution of Mining and Metallurgy, London.
361. Farmer IW, Shelton PD (1980). Review of underground rock reinforcement systems. *Trans. Inst. Min. Metall.*, **89**, A68–A83.
362. Rabciewics L (1969). Stability of tunnels under rock load. *Water Power.*, **21**, 266–273.
363. Kimura F, Okabayashi N, Ono K, Kawamoto T (1985) Rock-mechanical discussion for the mechanism of supporting systems in severe swelling rock tunnel. *Proc. 5th Int. Conf. Numer. Methods in Geomech.* (ed. Kawamoto T, Ichikawa Y), Nagoya, Balkema, Rotterdam, pp. 1265–1271.
364. Kovari K, Madsen FT, Amstad C (1981). Tunnelling with yielding supports in swelling rocks. *Proc. Int. Symp. Weak Rock*, Tokyo, pp. 1019–1026.
365. Laubscher DH (1984). Design aspects and effectiveness of support systems in different mining conditions. *Trans. Instn. Min. Metall.*, **93**, pp. A70–A81.
366. Matthews SM, Worotnicki G, Tillmann VH (1983). A modified cable bolt system for the support of underground openings. *Proc. Aust. Inst. Min. Metall. Conf.*, Broken Hill, pp. 243–255.
367. Ortlepp WD (1970). An empirical determination of the effectiveness of rock bolt support under impulse loading. *Proc. Int. Symp. Large Permanent Underground Openings*. Oslo (ed. Brekke TL and Jorstad FA), pp. 197–205, Scandanavian University Books, Oslo, Norway.
368. Ortlepp WD (1983). Considerations in the design of support for deep hard-rock tunnels. *Proc. 5th Congr., Int. Soc. Rock Mech.*, Melbourne, Vol. 2, pp. D179–187. Balkema, Rotterdam.
369. Ortlepp WD, Hagan TO (1986). Rockburst: understanding and control – past, present and future. *Proc. 13th CMMI Congr. Aus. Inst. Min. Met.*, Victoria, **3**, 77–84.
370. Ortlepp WD, More O'Ferrall RC, Wilson JW (1972). Support methods in tunnels. *Assn. Mine Managers S. Afr., Papers and Discns*, pp. 167–194.
371. Pender EB, Hosking AED, Mattner RH (1963). Grouted rock bolts for permanent support of major underground works. *J. Inst. Eng. Aust.*, **35**, 129–150.
372. Rose D (1989). Shotcrete for support of underground openings. *Underground Structures* (ed. Sinha RS), pp. 295–318. Elsevier, Amsterdam.
373. Skinner EH (1988). A ground support prediction concept: The Rock Structure Rating (RSR) model. *Rock Classification Systems for Engineering Purposes, Proc. ASTM STP*, Vol. 984, pp. 35–49.
374. Sofianos AI, Kapenis AP (1998). Numerical evaluation of the response in bending of an underground hard rock voussoir beam roof. *J Rock Mech. Min. Sci.*, **35**(8), pp. 1071–86.
375. Lang TA (1961). Theory and practice of rock bolting. *Trans. Soc.Min. Engrs*, AIME, Vol 220, pp. 333–348.Terzaghi K (1946). Rock defects and loads on tunnel support. *Rock Tunnelling with Steel Supports* (ed Proctor RV and White T), 15–99. Commercial Shearing Co., Youngstown, OH.
376. Lang TA (1972). Rock reinforcement. *Bull. Assoc. Eng. Geol.* **9**, 215–239.
377. Lang TA, Bischoff JA (1982). Stabilization of rock excavations using rock reinforcement. *Issues in Rock Mechanics, Proc. 23rd US Symp. Rock Mech.*, (ed. Goodman RE, Heuze FE), AIME, New York, pp. 935–943.

378. Lorig LJ (1985). A simple numerical representation of fully bonded passive reinforcement for hard rocks. *Computers and Geotechnics*, **1**, 79–97.
379. Lorig LJ (1988). Rock reinforcement: mechanical representation and use in finite difference schemes. *Proc. Int. Symp. Underground Engineering*, New Delhi.
380. Mathews KE, Meek JL (1975). Modelling rock reinforcement systems in cut-and-fill mining. *Proc. 2nd Aust.-NZ Conf. Geomech.*, Brisbane, pp. 42–47.
381. Brady B, Lorig L (1988). Analysis of rock reinforcement using finite difference methods. *Comp. & Geotech.*, **5**, 123–149.
382. Krajcinovic D (2000). Damage mechanics: accomplishments, trends and needs. *Int. J. Solids Struct.*, **37**(1–2), 267–77.
383. Li N, Chen W, Zhang P, Swoboda S (2001). The mechanical properties and a fatigue-damage model for jointed rock masses subjected to dynamic cyclical loading. *Int. J. Rock Mech. Min. Sci.*, **38**, 1071–1079.
384. Swoboda G, Shen XP, Rosas L (1998). Damage model for jointed rock mass and its application to tunnelling. *Comput. Geosci.*, **22**(3–4), 183–203.
385. Shao JF, Rudnicki JW (2000). A microcrack-based continuous damage model for brittle geomechanics. *Mech. Mater.*, **32**, 607–619.
386. Simo JC, Ju JW (1987). Strain- and stress-based continuum damage models – I. Formulation. *Int. J. Solids Struct.*, **23**, 821–840.
387. Zhao Y (1998). Crack pattern evolution and a fractal damage constitutive model for rock. *Int. J. Rock Mech. Min. Sci.*, **35**(3), 349–366.
388. Aubertin M, Simon R (1997). A damage initiation criterion for low porosity rocks. *Int. J. Rock Mech. Min. Sci.*, **34**(3/4), p. 554.
389. de Borst R (2002). Fracture in quasi-brittle materials: a review of continuum damage-based approaches. *Engng. Fract. Mech.*, **26**, 95–112.
390. Peck RB (1969). Advantages and limitations of the observational method in applied soil mechanics. Rankine lecture, *Geotechnique*, **19**, 171–187.
391. Muir Wood AM (1974). The observational method revisited. *Proc. 10th SEA Asian Geotechnical Conference*, Taipei, **2**, 37–42.
392. Sakurai S, Shimizu N (1986). Initial stress back analysed from displacements due to underground excavation. *Proc. Int. Symp. Rock Stress and Rock Stress Measurements*, Stockholm, pp. 679–686.
393. Sakurai S, Shinji M (1984). A monitoring system for the excavation of underground openings based on microcomputers. *Proc. ISRM Symp. Design and Performance of Underground Excavation*, pp. 471–476. Cambridge.
394. Noam H, Nagano S, Sakurai S (1987). The monitoring of a tunnel excavated in shallow depth. *Proc. 2nd Int. Symp. Field Measurements in Geomech.*, pp. 851–859. Kobe.
395. Yang Z, Lee CF, Wang S (2000). Three-dimensional back-analysis of displacements in exploration adits-principles and application. *Int. J. Rock Mech. Min. Sci.*, **37**, pp. 525–533.
396. Yang Z, Wang Z, Zhang L, Zhou R, Xing N (2001). Back-analysis of viscoelastic displacements in a soft rock road tunnel. *Int. J. Rock Mech. Min. Sci.*, **38**, 331–341.
397. Chen SG, Zhao J (1998). A data processing and management method for cavern deformation monitoring. *Rock Mech. Rock Engng.*, **31**, 125–131.
398. Franklin JA (1977). The monitoring of structures in rock. *Int. J. Rock Mech. Min. Sci. & Geomech. Abstr.*, **14**, 163–192.

399. Kirstem HA (1991). Determination of rock mass moduli by back analysis of deformation measurements. *Exploration for Rock Engineering* (ed. Bieniawski ZT), Balkema, Vol. 1, pp. 165–172.
400. Chi SY, Chen JC, Lin CC (2001). Optimized back-analysis for tunnelling-induced ground movement using equivalent ground loss model. *Tunnelling Underground Space Technology*, **16**, 159–65.
401. Detournay C, Hart R (1999). FLAC and numerical modelling in geomechanics. *Proc. Int. FLAC Symp. on Numer. Modelling in Geomech.*, Minneapolis, Balkema, Rotterdam.
402. Eberhardt E (2001). Numerical modelling of three-dimensional stress rotation ahead of an advancing tunnel face. *Int. J. Rock Mech. Min. Sci.*, **38**, 499–518.
403. Liu LQ (1984). *Optimisation of Engineering Structures* (in Chinese). Huazhong Univ. Press, Wuhan.
404. Yu YT, Li QQ (1983). Numerical methods for underground excavation in rheologic rock masses (in Chinese). *Chinese J. Geotech. Engng.*, **6**(2).
405. Amadei B (1984). Underground excavation in anisotropic rock masses. *Proc. Inst. Soc. Rock Mech. Symp. Design and Performance of Underground Excavations.* Cambridge (ed. Brown ET and Hudson JA), pp. 27–37. British Geotechnical Society, London.
406. Bieniawski ZT (1992). *Design Methodology in Rock Engineering.* Balkema, Rotterdam.
407. Einstein HH, Baecher GB (1983). Probabilistic and statistical methods in engineering geology. *Rock Mech. Rock Engng.*, **16**, 39–72.
408. Krumbein WC, Graybill FA, (1965). *An Introduction to Statistical Methods in Geology.* McGraw-Hill, New York.
409. Long JCS, Billaux D, Hestir K, Chiles JP (1987). Some geostatistical tools for incorporating spatial structure in fracture network modelling. *Proc. 6th Int. Congr. Rock Mech.*, Montreal (ed. Herget G and Vangpaisal S), pp. 171–176. Balkema, Rotterdam.
410. Till R (1974). *Statistical Methods for the Earth Scientist: An Introduction.* MacMillan, London.
411. Villaescusa E, Brown ET (1990). Characterizing joint spatial correlation using geostatistical methods. *Rock Joints* (ed. Barton N, Stephansson O). Balkema, Rotterdam, pp. 115–122.
412. Cheeney RF (1983). *Statistical Methods in Geology.* Allen & Unwin, London.
413. Priest SD, Samaniego JA (1988). The Statistical analysis of rigid block stability in jointed rock masses. *Proc. 5th Australia-New Zealand Conf. Geomechanics*, Sydney, pp. 398–403. Ann Arbor, MI.
414. Deng J (1985). *Grey System.* National Defence Press, Beijing.
415. Deng J (1987). *Basic Method of Grey System.* Huazhong Univ. Press, Wuhan.
416. Wu K (1988). Grey predicting back analysis for displacement measured of geotechnical tunnel. *Proc. 3rd Nat. Symp. Numer. and Anal. Method in Geotechnique.* Beijing, China.
417. Liu K, Zhang G (1991). Application of grey prediction system to Pan Dao Ling Tunnel. *Proc. 4th Nat. Symp. Numer. and Anal. Method in Geotechnique.* Beijing, China.
418. Kaiser PK (1981). A new concept to evaluate tunnel performance – influence of excavation procedure. *Proc. 22nd US Symp. Rock Mech.*, Cambridge, MA (ed. Einstein HH), pp. 264–271. MIT Press, Cambridge, MA.

419. Ladanyi B (1974). Use of the long-term strength concept in the determination of ground pressure on tunnel linings. *Advances in Rock Mechanics, Proc. 3rd Congr., Int. Soc. Rock Mech.*, Denver, Vol. 2A, Nat. Acad. Sci., Washington, DC, pp. 1150–1156.
420. Lorig LJ, Brady BHG (1983). An improved procedure for excavation design in stratified rock. *Rock Mechanics-Theory-Experiment-Practice, Proc. 24th US Symp. Rock Mech.*, College Station, Texas, AEG, New York, pp. 577–585.
421. Lorig LJ, Brady BHG (1984). A hybrid computational scheme for excavation and support design in jointed rock media. *Design and Performance of Underground Excavation* (ed. Brown ET and Hudson JA), pp. 105–112, British Geotechnical Society, London.
422. Chen SG, Ong HL, Tan KH, Zhao J (2001). A study on working face effect in tunnel excavation. *Proc. AITES-ITA 2001 World Tunnel Congress*, Milan, Italy, pp. 199–206.
423. Chen SG, Ong HL, Tan KH, Zhao J (2001). UDEC modeling of rock tunnel excavation and supports. *Frontiers of Rock Mechanics and Sustainable Development in the 21th Century, Proc. 2nd Asian Rock Mech. Symp.*, Beijing, China, pp. 341–345.
424. Megaw TM, Bartlett JV (1981). *Tunnels: Planning, Design, Construction*. p. 284. Ellis Horwood, Chichester.
425. Sakurai S (1997). Lessons learned from field measurements in tunnelling. *Tunnelling Underground Space Technology*, **12**(4), 453–460.
426. Sakurai S (1983). Displacement measurements associated with the design of underground openings. *Proc. Int. Symp. Field Measurements in Geomechanics*, Zurich, Vol. 2, pp. 1163–1178.
427. Sakurai S (1978). Approximate time-dependent analysis of tunnel support structure considering progress of tunnel face. *Int. J. Numer. & Anal. Methods Geomech.*, **1**, 159–175.
428. Peng SS, Podnieks ER (1972). Relaxation and behaviour of failed rock. *Int. J. Rock Mech. Min. Sci.*, **9**, 699–712.
429. Lo KY, Yuen CMK (1981). Design of tunnel lining in rock for long term time effects. *Can. Geotech. J.*, **18**, 24–39.
430. Salamon MDG (1974). Rock mechanics of underground excavations. *Advances in Rock Mechanics, Proc. 3rd Congr., Int. Soc. Rock Mech.*, Denver, **1B**, Nat. Acad. Sci., Washington, DC, pp. 951–1099.
431. Yoichi H, Yamashita R (1985). Study on the stability of caverns. *5th Int. Conf. on Numerical Methods in Geomech.*, Nagoya, 1201–1206.
432. Yang SQ (1986). Stability analysis of underground caverns during excavation. *J. Wuhan Hydr. And Elect. Inst.*, pp. 73–82.
433. Yu XF, Yu J (1993). *Concept of Rock Memory during Excavation.* Metallurgy Industry Publisher.
434. Zhu W, Wang P (1991). Predicting deformation and stability of a large power plant under construction by monitoring in openings. *Proc. 3rd Int. Symp. on Field Measurements in Geomech.*, Oslo.
435. Wang P, Yuan C, Zhu W (1992). A displacement prediction model based on measured results of underground engineering. *Rock and Soil Mech.*, **13**(4).
436. Chen Z (1991). Time-series analysis – a possible way for predicting mechanical behaviour of rock mass. *Chinese J. Geotech. Eng.*, **13**(4).

437. Li S (1991). The time series analysis model for displacement of gallery support. *Proc. 4th Nat. Symp. Numer. and Anal. Method in Geotechnique*, Beijing, China.
438. Muller L (1978). Remove misconceptions on the New Austrian Tunnelling Method. *Tunnels and Tunnelling*, October 1978, pp. 29–32.
439. Bellman R (1957). *Dynamic Programming*. Princeton Univ. Press, Princeton, N.J.
440. Milar DL, Hudson JA (1993). Rock engineering system performance monitoring using neural networks. *Artificial Intelligence in the Mineral Sector*, Proceedings of an Institution of Mining and Metallurgy Conference, April.
441. Singh VK, Singh D, Singh TN (2001). Prediction of strength properties of some schistose rocks from petrographic properties using artificial neural networks. *Int. J. Rock Mech. Min. Sci.*, **38**(2), 269–284.
442. Cai J, Zhao J, Hudson JA, Wu X (1996). Using neural networks in rock engineering systems for cavern performance auditing. *Eurock'96 – Int. Symp. on Prediction and Performance in Rock Mech. and Rock Engng.*, Torino, Italy, pp. 965–972.
443. Kacewicz M (1994). Model-free estimation of fracture apertures with neural networks. *Math Geol.*, **26**(8), 985–94.
444. Kim CY, Bae GJ, Hong SW, Park CH, Moon HK, Shin HS (2001). Neural network based rediction of ground surface settlements due to tunnelling. *Comput. Geotech.*, **28**(6–7), 517–547.
445. Kovari K (2003). History of the sprayed concrete lining method. *Tunnelling and Underground Space Technology*, **18**, 57–83.
446. Schach R, Garachol K, Heltzen AM (1979). *Rock Bolting: A Practical Handbook*. Pergamon, Oxford.
447. Stacey TR, Page CH (1986). *Practical Handbook for Underground Rock Mechanics*. Trans. Tec. Clausthal-Zellerfeld.
448. Stillborg B (1986). *Professional Users Handbook for Rock Bolting*. Trans Tech Publication, Series on Rocks and Soil Mechanics, Vol. 15.
449. Huang Z, Broch E, Lu M (2002). Cavern roof stability – mechanism of arching and stabilization by rockbolting. *Tunnelling and Underground Space Technology*, **17**(3), pp. 249–261.
450. Beer G, Meek JL (1982). Design curves for roofs and hangingwalls in bedded rock based on voussoir beam and plate solutions. *Trans. Instn. Min. Metall.*, **91**, pp. A18–A22.
451. Lee IM, Park JK (2000). Stability analysis of tunnel keyblock: a case study. *Tunnelling Underground Space Technology*, **15**(4), 453–462.
452. Lin D, Fairhurst C (1988). Static analysis of the stability of three-dimensional blocking systems around excavations in rock. *Int. J. Rock Mech. Min. Sci. & Geomech. Abstr.*, **25**, 139–148.
453. Lin D, Fairhurst C, Starfield AM (1987). Geometrical identification of three-dimensional rock block systems using topological techniques. *Int. J. Rock Mech. Min. Sci. & Geomech. Abstr.*, **24**(6), 331–338.
454. Gerrard CM, Pande GN (1985). Numerical modelling of reinforced jointed rock masses. *Comput. Geotech.*, **1**, 293–318.
455. Bjurstrom S (1979). Shear strength on hard rock joints reinforced by grouted untensioned bolts. *Proc. 4th Congr., Int. Soc. Rock Mech.*, Vol. 2, Balkema, Rotterdam, pp. 1194–1199.
456. Fairhurst C, Singh B (1974). Roof bolting in horizontally laminated rock. *Eng. Min. J.*, February, pp. 80–90.

457. Bieniawski ZT (1973). Engineering classification of jointed rock masses. *Trans. S. Afr. Inst. Civ. Engrs.*, **15**, 335–344.
458. Bieniawski ZT (1976). Rock mass classifications in rock engineering. *Exploration for Rock Engineering*. Vol. 1, Balkema, Cape Town, pp. 97–106.
459. Bieniawski ZT (1979). The Geomechanics Classification in rock engineering applications. *Proc. 4th Int. Congr. Rock Mech.*, Montreux, Vol. 2, pp. 41–48. Balkema, Rotterdam.
460. Bieniawski ZT (1984). *Rock Mechanics Design in Mining and Tunnelling*. Balkema, Rotterdam.
461. Bieniawski ZT (1989). *Engineering rock mass classifications*, p. 251. Wiley, New York.
462. Obert LE, Duvall WI (1967). *Rock Mechanics and the Design of Structures in Rock*. Wiley, New York.
463. St John CM, Van Dillen DE (1983). Rockbolts: a new numerical representation and its application in tunnel design. *Rock Mechanics, Theory, Experiments, Practice, Proc. 24th US Symp. Rock Mech.*, AEG, New York, pp. 13–26.
464. Haas CJ (1981). Analysis of rock bolting to prevent shear movement in fractured ground. *Min. Engng.* (June), pp. 698–704.
465. Indraratna B, Kaiser PK (1990). Design for grouted rock bolts based on the convergence control method. *Int. J. Rock Mech. Min. Sci. & Geomech. Abstr.*, **27**, 269–281.
466. Littlejohn GS, Rodger AA, Mothersille DKV, Holland DC (1989). Dynamic response of rock bolt systems. *Proc. Int. Conf. on Foundation and Tunnels*. University of London, Vol. 2, pp. 57–64.
467. Moy D (1975). *The Design and Monitoring of Rock Bolt and Dowel Systems in Underground Excavations*, PhD Thesis, University of London.
468. Pells PJ, Snyder VW, Viegelahn GL, Parker J (1974). The behaviour of fully bonded rock bolts. In *Proc. 3rd Int. Congr. Rock Mech.*, Denver, Vol. 2B, pp. 1212–1217.
469. Shi G, Goodman RE (1983). Key block bolting. *Proc. Int. Symp. Rock Bolting*, Abisko, Sweden (ed. Stephansson O), pp. 143–167.
470. Thomas E (1962). Stabilisation of rock by bolting. *Reviews in Engineering Geology I & II*, New York.
471. Crawford AM, Ng L, Lau KC (1985). The spacing and length of rock bolts for underground openings in jointed rock. *Proc. 5th Int. Conf. Numer. Methods in Geomech.*, Nagoya, Japan (ed. Einsenstein Z) pp. 1293–1300.
472. Gerdeen JC, Snyder VW, Viegelahn GL, Parker J (1977). Design criteria for rock bolting plans using fully resin-grouted non-tensioned bolts to reinforce bedded mine roof. *USBM OFR 46(4)-80*.
473. Franklin JA, Woodfield PF (1971). Comparison of a polyster resin and a mechanical rockbolt anchor. *Trans. Instn. Min. Metall.*, **80**, pp. A91–A100.
474. Ortlepp WD (1968). A yielding rock bolt. *Research Organisation Bulletin of the Chamber of Mines South Africa*, **14**, August, 6–8.
475. Connell JP (1982). State of the art of shotcrete. *Underground mining Methods Handbook* (ed Hustralid WA), pp. 1561–1566. AIME, New York.
476. Brekke TL (1972). Shotcrete in hard-rock tunneling. *Bull. Assoc. Eng. Geol.*, **9**, pp. 241–264.
477. Dowding CH, Rozen A (1978). Damage to rock tunnels from earthquake loading. *J. Geotech. Engng. Div., ASCE*, **104**(GT2), 175–191.

# Index